Real Estate, Construction and Economic Development in Emerging Market Economies

Real Estate, Construction and Economic Development in Emerging Market Economies examines the relationships between the real estate and construction sectors and explores how each sector, and the relationships between them, affects economic development in emerging market economies (EMEs).

Throughout the book, the international team of contributors discuss topics as diverse as real estate finance and investment, housing, property development, construction project management, valuation, sustainability and corporate real estate. In doing so, the book demonstrates how the relationship between construction and real estate impacts on economic development in countries such as Argentina, Brazil, Colombia, China, Ghana, Nigeria, Turkey, Lithuania, Hungary and Slovenia.

Topics include:

- the role of real estate brokerage in improving the living standards of citizens;
- the effect of a mineral boom on construction cycles, real estate values and the socio-economic conditions of people in boom towns and cities;
- corporate real estate management practices and how they affect economic growth; and
- the synergies between construction and real estate and how they, in turn, affect economic development.

This book will be of interest to those studying and researching real estate, construction, development studies, urban economics and emerging market economies.

Raymond T. Abdulai is a Senior Lecturer in Real Estate at the Department of Built Environment, Liverpool John Moores University (LJMU), UK.

Franklin Obeng-Odoom is a Senior Lecturer in Property Economics at the School of the Built Environment, University of Technology, Sydney, Australia.

Edward Ochieng is a Senior Lecturer in Project Management at the Department of Built Environment, Liverpool John Moores University (LJMU), UK.

Vida Maliene is a Reader in Property and Planning Studies at the Department of Built Environment, Liverpool John Moores University (LJMU), UK.

Routledge Studies in International Real Estate

The Routledge Studies in International Real Estate series presents a forum for the presentation of academic research into international real estate issues. Books in the series are broad in their conceptual scope and reflect an inter-disciplinary approach to Real Estate as an academic discipline.

Oiling the Urban Economy
Land, labour, capital, and the state in Sekondi-Takoradi, Ghana
Franklin Obeng-Odoom

Real Estate, Construction and Economic Development in Emerging Market Economies
Edited by Raymond T. Abdulai, Franklin Obeng-Odoom, Edward Ochieng and Vida Maliene

Real Estate, Construction and Economic Development in Emerging Market Economies

Edited by Raymond T. Abdulai, Franklin Obeng-Odoom, Edward Ochieng and Vida Maliene

LONDON AND NEW YORK

First published 2016
by Routledge
2 Park Square, Milton Park, Abingdon, Oxon OX14 4RN

and by Routledge
711 Third Avenue, New York, NY 10017

First issued in paperback 2018

Routledge is an imprint of the Taylor & Francis Group, an informa business

British Library Cataloguing-in-Publication Data
A catalogue record for this book is available from the British Library

Library of Congress Cataloging in Publication Data
Real estate, construction and economic development in emerging market
economies / edited by Raymond T. Abdulai, Franklin Obeng-Odoom,
Edward Ochieng and Vida Maliene. — First Edition.
pages cm. — (Routledge studies in international real estate)
Includes bibliographical references and index.
1. Real property—Developing countries. 2. Construction industry—
Developing countries. 3. Economic development—Developing countries.
I. Abdulai, Raymond T., editor.
HD1375.R433 2016
338.9009172'4—dc23
2015017532

ISBN 13: 978-1-138-62603-4 (pbk)
ISBN 13: 978-0-415-74789-9 (hbk)

Typeset in Times New Roman
by Swales & Willis Ltd, Exeter, Devon, UK

Contents

Figures

Tables

Contributors

Raymond T. Abdulai, PhD, MPhil (Cantab), PGCTLHE, BSc (Hons) [First Class], MRICS, MGhIS, FHEA, is a Senior Lecturer in Real Estate at the Department of Built Environment, Liverpool John Moores University (LJMU) in the United Kingdom. Prior to his appointment at LJMU, he worked at the School of Technology, Wolverhampton University, as a Postdoctoral Research Fellow and Lecturer. He lectured at Kwame Nkrumah University of Science and Technology in Ghana where he also practised as a chartered general practice surveyor before commencing his academic career in the United Kingdom. Raymond's research interests span various facets of real estate including real estate economics, management, finance, investment, valuation and development; land dispute resolution, especially in the developing world; emerging real estate markets; and landed property rights. He has published extensively in reputed international journals, edited textbooks and conference proceedings. He has three textbooks to his credit and has edited four textbooks. Raymond is currently the Editor-in-Chief of the *Journal of International Real Estate and Construction Studies*. He is a reviewer to various international journals as well as serving on the editorial boards of various international journals.

Divine K. Ahadzie, PhD, MSc, BSc (Hons), MGIOC, is a Senior Research Fellow at the Centre for Settlements Studies, Kwame Nkrumah University of Science and Technology (KNUST), Kumasi, Ghana. Ahadzie has a strong knowledge of the construction industry, including project management practices, and especially in emerging economies. He has published widely in both local and international journals. He is regular reviewer for the *International Journal of Project Management* (IJPM), *Construction Management and Economics* (CME) and *Engineering, Construction and Architectural Management* (ECAM). In 2010, Ahadzie was a consultant to the World Bank on the Ghana Skills and Technology Development Project (GSTDP) as a construction industry analyst. In 2011, he was a local representative for GHK Consult (a UK-based international development consultancy) in holding the Africa Regional Workshop for the preparation of *The Global Handbook on Flooding*, held in Accra, Ghana. Ahadzie also assisted the College of Architecture and Planning of the KNUST as conference manager, hosting the

1st International Conference on Infrastructure Development in Africa, held in Kumasi, Ghana, from 22–24 March 2012.

Nii Ankrah, PhD, MSc [Distinction], PGCHE, BSc (Hons) [First Class], ICIOB, FHEA, is a Senior Lecturer in Quantity Surveying, Commercial Management and Construction Project Management at the University of Wolverhampton, UK. Nii is a quantity surveyor by profession and spent the early years of his career working for a building and civil engineering contractor in Ghana. His current research interests focus on organisational behaviour within the construction industry as well as issues relating to tendering and procurement of construction projects, and contract administration.

Isabel Atkinson, MSc, BA (Hons), AFHEA, is a postgraduate research student studying towards a PhD at the School of the Built Environment, Liverpool John Moores University. Between 2008 and 2012, she studied Urban Planning [BA (Hons)] and Environmental Planning [MSc] at Liverpool John Moores University. Her current area of study surrounds the topic of British high street decline and sustainable communities. Isabel has published materials relating to the English planning system with regard to regeneration, sustainable communities and real estate management.

Meine Pieter van Dijk, PhD, is an Economist and Professor of Water Services Management at UNESCO-IHE Institute for Water Education in Delft and Professor of Urban Management in Emerging Economies at the Economic Faculty of the Erasmus University in Rotterdam (EUR), both in the Netherlands. His educational activities include teaching topics related to economics, regulation and public–private partnerships. Meine Pieter also supervises several MSc and PhD research students on a wide range of topics. Meine Pieter has extensive experience in curriculum development for water services management. He is a member of the research school CERES and the European Institute for Comparative Urban Research (Euricur). He has worked in developing countries since 1973, in particular on the role of small enterprises in urban development, water and sanitation, and urban finance issues. Meine Pieter has also worked as a consultant for NGOs, the Asian Development Bank, the Inter-American Development Bank and the World Bank as well as different bilateral donors and UN agencies. He has published extensively in his areas of expertise.

Robert Dixon-Gough retired in 2009 as a Lecturer and Researcher at the University of East London. He is currently a Visiting Researcher and an Affiliate of the University of Agriculture in Kraków. His principal research activity since 1980 has been in the general field of land management, which encompasses a holistic view of the land and all the actions and activities, past and present, with a particular interest in cultural landscapes and land drainage. Externally-funded research includes a comparative analysis of European land administration systems; post-industrial landscapes; evaluation of the historical and current role of the Austro-Hungarian cadastre from 1918 to 2009; and land ownership patterns and land management changes in English National Parks. He was the UK representative on the Management Committee of the EU-COST/EFS Project "Modelling Real Estate Transaction" and has been an active member of the European Faculty of Land Use and Development (Strasbourg), now the European Academy of Land Use Development (Zürich) since 1992, and involved in the publication of proceedings since 1998. Between 1978 and 2009 he was the author of over 100 scientific publications on cartography, remote sensing and land management and has written, co-written, edited and co-edited 17 books, the majority of which are in the field of sustainable land management. Since 2009, his publications have largely involved collaboration with colleagues from the University of Agriculture in Kraków.

Işıl Erol, PhD (Cantab), MPhil (Cantab), MSc, BSc (Hons), is a Lecturer at the Business School, Ozyegin University, Istanbul, Turkey. She obtained her BSc from Middle East Technical University (METU) in Ankara, Turkey, in 1998, her MSc in Economics from METU in 2000, her MPhil in Land Economy from University of Cambridge in 2001, and her PhD in Real Estate Finance from the University of Cambridge in December 2004. After completing her PhD, Erol joined the Department of Economics at Middle East Technical University (METU) as a Lecturer. She has also taught at Adelaide University in Australia and Bogazici University in Turkey. From June 2009 to June 2010 Erol worked as a consultant for İş Bank in the area of mortgage pricing. Her areas of speciality are real estate finance, mortgage markets, pricing mortgage contracts with option pricing models and real estate investment trusts. Her publications have appeared in journals such as the *Journal of Housing Economics*, the *Journal of Real Estate Finance and Economics*, *Urban Geography* and the *Review of Urban and Regional Development Studies*.

Rimvydas Gaudėšius, MSc in Environmental Engineering and Land Management, is a postgraduate research student studying towards a PhD at the Institute of Land Management and Geomatics at Aleksandras Stulginskis University in Lithuania. Also, Rimvydas is a Chief Specialist with the National Land Service, under the Ministry of Agriculture in Lithuania. Between 2011 and 2012 he worked at the headquarters where he was responsible for co-ordinating the activities of the territorial offices of the National Land Services, preparing scientific papers about real estate management, land use and sustainable urbanisation. Rimvydas' research interest is in urbanisation, real estate and geomatics.

Virginija Gurskienė, PhD (Technology Science, Environmental Engineering and Landscape Management), is an Associate Professor and Director of Land Management and Geomatics Institute at Aleksandras Stulginskis University in Lithuania. Her field of research includes land use planning, real estate cadastre and valuation, and the planning and management of protected areas. She has published over twenty scientific papers within Lithuanian and international academic journals and conference proceedings and four books. She is a member of the Association of Lithuanian Engineers of Land and Water Management, and sits on the Editorial Board of the scientific journal *Agricultural Sciences*.

Józef Hernik, Dr. hab. inż., is employed at the Department of Land Management and Landscape Architecture at the Faculty of Environmental Engineering and Land Surveying, University of Agriculture, Krakow, Poland. He was the Project Co-ordinator of four international research projects. Furthermore, he is the author of more than 100 scientific publications on land management, cultural landscapes and environment development as well as the editor and co-editor of several research monographs on cultural landscapes protection and a partner in several international research projects in this discipline. Currently he is actively involved in scientific co-operation in the field of development and protection of cultural landscapes and land use with universities and research institutes in Europe, the USA and Asia.

Maruška Šubic Kovač, PhD, MSc, BSc, is an Associate Professor and Head of the Municipal Economics Institute at the Faculty of Civil and Geodetic Engineering of Ljubljana University, Republic of Slovenia. Her research is focused on property and planning and other related subjects, for instance spatial and urban planning, urban regeneration, land management and economics, property valuation and taxation. She has published around 70 scientific papers in Slovenian and international peer-reviewed academic journals and conference proceedings, books/monographs, and several other scientific works. She is a fellow of several associations in Slovenia and member of the Executive Board of the European Academy of Land Use and Development (EALD) of Zürich.

Vida Maliene, PhD, MSc, BSc (Hons), FHEA, is a Reader in Property and Planning Studies at the Department of the Built Environment and an active member of the Built Environment and Sustainable Technologies (BEST) Research Institute, both at Liverpool John Moores University in the UK. Vida is a Visiting Professor at the Institute of Land Management and Geomatics at Aleksandras Stulginskis University, Lithuania. Her research focuses on property and planning, including such related subjects as spatial and urban planning, urban regeneration and sustainable communities, land management and economics, property valuation and taxation. Vida has published over fifty scientific papers in international peer-reviewed academic journals and conference proceedings, a book/monograph, eleven book chapters and several journal editorials. She is a member of the Executive Board of the European Academy of Land Use and Development (EALD) in Zürich, and a member of the European Group of Operational Research, the International Council of Research and Innovation in Building and Construction (CIB) and the Lincoln Institute of Land Use Policy in the US.

Joseph Mante, PhD, LLM, BL, LLB is a Lawyer with over 10 years' experience in Property, Contract and Commercial law practice and a lecturer at the Law School, Robert Gordon University, Scotland. He is currently the Deputy Course leader for the LLM/MSc Construction Law and Arbitration programme and teaches commercial law, construction law and arbitration at the postgraduate level. He recently undertook research at the University of Wolverhampton in the UK into procurement strategies and the resolution of disputes arising from major infrastructure projects in developing countries with a focus on Ghana. His research interests are in the areas of procurement, contract formation and administration, construction law generally and dispute resolution.

Judit Nyiri Mizseiné, PhD, "dr univ", MSc, is an Associate Professor at the University of Óbuda Geoinformatics Institute of "Alba Regia" Technical Faculty, Hungary. She has 43 years of teaching experience. Her educational activity includes teaching land management and real estate valuation. Her special field of study is land consolidation and land valuation. She worked as a technician at a mapping and surveying company from 1968 until 1972. Following this, she taught at the University of West Hungary's Faculty of Geoinformatics until 2014. She regularly consults on the diploma and scientific work of students. Her several publications appear in international conference proceedings and as book chapters. Judit's main field of research surrounds land consolidation and real estate valuation processes using GIS analysis. She is a fellow of the European Academy of Land Use and Development (EALD) in Zürich.

Claudia B. Murray, PhD, MA, MSc Arch, is a Research Fellow at the School of Real Estate and Planning, Henley Business School, University of Reading. Prior to this, Claudia worked for several architectural firms specialising in international architectural competitions in Buenos Aires. She was also a part-time Lecturer at the Faculty of Architecture, Design and Urbanism, Universidad de Buenos Aires, until 1997. Her research interests focus on the socio-cultural and economic implications of architectural and urban design. Claudia has received several research grants from councils including the Arts and Humanities Research Council and the Visiting Arts Board. She is an active member of the Academy of Urbanism, a Fellow of the Royal Society for the Encouragement of Arts, Manufactures and Commerce (RSA), an Associate of the Walker Institute and a member of the Architectural Association. She is also principal investigator in urban laboratories in Latin America that involve international aid organisations such as Architecture for Humanity and Engineers without Borders.

Issaka Ndekugri, PhD, MSc, PGCSN, LLB (Hons), BSc (Hons), MRICS, MCIOB, is Professor of Construction and Engineering Law and Director of MSc Construction Law and Dispute Resolution at Wolverhampton University in the United Kingdom. His book (co-authored with Mike Rycroft) on UK standard building contracts in 2001 won the Gold Award of the CIOB's International Literary Scheme and has been compared favourably with *Keating on Construction Contracts* in a review by a leading London QC. Towards development of best practice, he is heavily engaged in applied research in collaboration with major companies and leading construction lawyers, arbitrators and other dispute resolution practitioners across the globe. He is a prolific contributor on construction contracts and dispute resolution to many journals of the highest international standing and has authored over 100 publications. He also consults and runs training courses for industry in the areas of construction law, contract administration, dispute resolution, and project management. He has been appointed external examiner by many UK universities including the Univeristy of Cambridge, Salford University and Manchester University. Issaka is a CEDR-accredited Mediator and member of the College of Peers of the UK's Engineering and Physical Sciences Research Council.

Franklin Obeng-Odoom, PhD, MSc, BSc (Hons) [First Class], is a Senior Lecturer in Property Economics at the School of the Built Environment, University of Technology, Sydney. His books include *Oiling the Urban Economy: Land Labour, Capital and the State in Sekondi-Takoradi, Ghana* (Routledge, London). He is the substantive Editor of *African Review of Economics and Finance* and the Book Review Editor of the *Journal of International Real Estate and Construction Studies.* He serves on the editorial board of Urbani Izziv and The Extractive Industries and Society. Franklin is an elected Fellow of the Ghana Academy of Arts and Sciences.

Edward Ochieng, PhD, PGCHE, MSc, BSc (Hons), FHEA, FAPM, is a Senior Lecturer in Project Management at LJMU in the United Kingdom. Edward's research interests include multicultural project team performance, value creation, portfolio management, return on investment in projects, stakeholder management, project governance, project culture, project integration, modelling technology and sustainability in construction. He has presented at both national and international conferences, such as those of the Association of Researchers in Construction Management (ARCOM), Australian Universities Building Educators Association (AUBEA), CIB World Congress, American Society for Engineering Education (ASEE) and International World of Construction Project Management, during which he shared his knowledge of "Global Project Teams, Project Complexity, Project Performance and Project Team Integration". Edward has authored, co-authored and contributed to one book and over 40 refereed papers. His research into project management is mainly concerned with people experiences and organisational challenges of managing projects in developing and developed nations. He has extensive experience in all aspects of the heavy engineering development project lifecycle.

Timothy Tunde Oladokun, MSc, MPhil, BSc (Hons), is a Lecturer in the Department of Estate Management, Obafemi Awolowo University, Ile Ife,

Nigeria. Timothy is currently pursuing a doctoral degree in the area of corporate real estate and facilities management at Obafemi Awolowo University. Prior to joining academia, Timothy had over twelve years' experience in active professional practice. He is a registered estate surveyor and valuer and an associate member of the Nigerian Institution of Estate Surveyors and Valuers. He has published widely and presented papers in conferences at home and abroad and has also won the Emerald Group Publishing (UK) Property Management Best Paper Award (2010).

Anthony Owusu-Ansah, PhD, MSc, BSc (Hons) [First Class], AHEA, is a Senior Lecturer in Finance at GIMPA Business School in Accra, Ghana. He has previously taught at Royal Institute of Technology (KTH) in Stockholm, Sweden, Aberdeen University in the UK and Kwame Nkrumah University of Science and Technology, Kumasi, Ghana. Anthony has published in a number of reputed international journals and conference proceedings. His doctoral studies at Aberdeen University focused on measuring and understanding the house price dynamics of the Aberdeen housing market, which has led to the first ever published quality-adjusted local house price indices in Scotland.

Maria Pazdan, MSc in Engineering, MSc in Agricultural Land Surveying and Property Appraisal [distinction], is a Scientific and Teaching Assistant in the Department of Land Management and Landscape Architecture at the University of Agriculture in Krakow in Poland. Maria is currently pursuing a doctoral degree in predicting the land use changes in areas with valuable landscapes. Maria scientific interests include land use changes, transformation of rural landscapes and the legal and administrative aspects of spatial planning.

Andrea Pödör, PhD, MSc, is a cartographer, GIS expert and an Associate Professor at the University of Óbuda Institute of Geoinformatics, Hungary. One of her main research fields is visualisation possibilities in land consolidation processes as a possible tool in disputes with citizens. She has worked on research projects regarding spatial data and IT infrastructure to enhance the efficiency of land consolidation processes and as a team leader of geo-database development and visualisation for urban ecological projects. She is an active member of the European Academy of Land Use and Development (EALD) and a member of the Executive Board of the Hungarian Society of Surveying, Mapping and Remote Sensing.

Andrew Price, DSc, PhD, BSc (Hons), FCIOB, FICE, CEng, is Professor of Project Management with over 30 years of design, construction and industry-focused research experience, and has considerable experience in major collaborative funded research projects. He has acted as a Visiting Professor at four overseas Universities. Andrew has published six books and over 300 papers in refereed journals and conference proceedings. He has been a principal or co-investigator on 24 completed research projects. He is Co-Investigator with the SUMoT

Consortia (£1.6 million) and Co-Investigator with Loughborough's IMCRC, with access to research funds of £13 million over the past 10 years. Andrew is Co-Director of the EPSRC-funded (£13 million over seven years) Health and Care Infrastructure Research and Innovation Centre (HaCIRIC). He has previously been a member of eleven DETR/DoE/DTI/CIRIA steering groups and nine European Construction Institute steering groups and task forces.

Eric Yeboah, PhD, BSc (Hons) [First Class] is a Lecturer in the Department of Land Economy, Kwame Nkrumah University of Science and Technology, Kumasi, Ghana. His PhD studies at Liverpool University in the United Kingdom were in Land Management and Urban Planning. Eric's research interests include land use planning, land policy, urbanisation and property rights.

1 Real estate, construction and economic development

Context, concepts and (inter) connections

*Raymond T. Abdulai, Franklin Obeng-Odoom,
Edward Ochieng and Vida Maliene*

1. Background

Existing studies in the fields of real estate (RE) and construction have typically had more pragmatic concerns [see, for example, volume 29(3) of *Property Management* (2011)]. Such studies have also tended to approach RE and construction issues as though they were in watertight compartments. Part of the reason for this emphasis is historical. Since its beginning in the twentieth century, RE economics has traced its roots to mainstream economics and hence has tended to regard construction as necessary but not integral to its epistemology (Weimer, 1966). More recently, disciplinary subdivisions and specialisms (Malpezzi, 2009), the teaching of RE and construction courses (Boydell, 2007) and the focus of academic journals (*Emerald Built Environment Newsletter*, 2012) have perpetuated this schism.

The few respectable exceptions that tend to examine housing and economic development (ED), include Tibaijuka's (2009) well-known book, *Building Prosperity: Housing and Economic Development* and the work of Canadian-based housing scholar Godwin Arku and his team (e.g., Harris and Arku, 2006; Arku, 2006) and a few others (Anaman and Osei-Amponsah, 2007). Within this cohort of studies, there is a strand that considers housing remittances and urban economic development (e.g., Obeng-Odoom, 2010) or housing, migration, and economic development (e.g., Firang, 2011). Another strand (see for instance, Dujardin and Goffette-Nagot, 2009) tends to focus exclusively on housing provision and its social and economic impacts. Admittedly, the *Journal of International Real Estate and Construction Studies* has tried to fuse together the two sub-disciplines, but it will take more time to widen the breadth of the scholarship it publishes.

These exceptions offer a substantial advance on mainstream areas in RE and construction studies, but not even these studies sufficiently address the interrelated issues of (a) the role of RE brokerage in improving the living standards of people; (b) the effect of mineral boom on construction cycles, RE and the socioeconomic conditions of people in boom towns and cities; (c) corporate RE and facilities management practices and how they affect economic growth; and (d) the synergies between RE and construction and how they, in turn, affect ED.

This edited volume, therefore, attempts to fill the noted lacunae by focusing on "non-traditional" themes in emerging market economies (EMEs). The dual aim of the book is to first examine the relationships between RE and construction sectors, and secondly to explore how each sector and the relationship between the sectors affects ED in EMEs. ED in this collection refers to the sustained, concerted actions of communities and policymakers that improve the general standard of living and economic health of a specific geographical location; it is, therefore, about the quantitative and qualitative changes in an existing economy, which involves development of human capital; increasing the literacy ratio; and improvement in important infrastructure, health and safety and others areas that generally aim at increasing the general welfare of the citizenry (Todaro, 2014). The book's emphasis is on the whole RE and construction services processes, the relationships between the sectors and how all these contribute to ED in EMEs.

The importance of concentrating on EMEs cannot be overemphasised since such nations normally receive aid and guidance from international donor community institutions such as Bretton Woods and also provide an outlet for expansion for foreign investors from the developed world.

The rest of the chapter is organised as follows. The next section gives an overview of EMEs stressing why it is a useful focus for this book. Following that section is a section that looks at the role of institutional arrangements in RE and construction markets. Section four describes the demographics and economic indicators of the case study countries whilst the last section distills and evaluates what the authors have covered in their respective chapters.

2. An overview of EMEs

EME is a term that was first used in 1981 by Antoine, an economist at the International Finance Corporation (IFC) of the World Bank. As explained by Antoine, EMEs are countries that are progressing towards becoming more advanced, usually by means of rapid growth and industrialisation, and such countries experience an expanding role both in the world economy and in global politics – they tend to have lower per-capita incomes, above-average socio-political instability, higher unemployment and lower levels of business or industrial activity in comparison with the well-established economies with mature markets such as the USA, Canada, Japan and the United Kingdom. EMEs constitute approximately 80% of the global population, represent about 20% of the world's economies and are considered to be fast-growing economies (Heakal, 2009).

Heakal (2009) has identified the following to be the main features of EMEs. First, they are characterised as transitional, which means they are in the process of moving from a closed economy to an open market economy while building accountability within the system. Thus, as an EME, the country is embarking on an economic reform program that will lead it to stronger and more responsible economic performance levels, as well as transparency and efficiency in the capital market. An EME reforms its exchange rate system because a stable local currency builds confidence in an economy, thereby reducing the desire for local investors to send their capital abroad (capital flight). EMEs normally receive aid and guidance from large donor countries and/or world organisations such as the World Bank and International Monetary Fund.

Secondly, a characteristic of an EME is an increase in both local and foreign investment. A growth in investment in a country often indicates that the country has been able to build confidence in the local economy. Also, foreign investment is a signal that the world has begun to take notice of the EME and when international capital flows are directed towards an EME, the injection of foreign currency into the local economy adds volume to the country's stock market and long-term investment to the infrastructure. An EME provides an outlet for foreign investors or developed-economy businesses to expand by serving, for example, as a new place for a new factory or for new sources of revenue. Regarding the recipient country, employment levels rise, labour and managerial skills become more refined and a sharing and transfer of technology occurs. Thus, in the long run, the EME's overall production levels should rise, increasing its gross domestic product and eventually lessening the gap between the emerged and emerging worlds.

Thirdly, in terms of portfolio investment and risks, because EMEs are in transition and hence not stable, they offer an opportunity to investors who are looking to add some risk to their portfolios. The possibility for some economies to fall back into a not-completely-resolved civil war or a revolution sparking a change in government could result in a return to nationalisation, expropriation and the collapse of the capital market. Since the risk of an EME investment is normally higher than an investment in a developed market, panic, speculation, and knee-jerk reactions are more common, as typified by, for example, the 1997 Asian crisis, during which international portfolio flows into these countries actually began to reverse. However, there is a direct relationship between risks and returns: the higher the risk, the higher the reward. Consequently, emerging market investments have become a standard practice among investors who wish to diversify for higher returns while adding risk. BRICs (Brazil, Russia, India and China) are often considered the largest EMEs in the world.

Allied to EMEs is the concept of frontier market economies (FMEs). FME as an economic term was first used in 1992 by the IFC to refer to smaller EMEs experiencing or poised for strong growth. They are considered to have lower market capitalisation, less liquidity and to be less established in comparison to the larger and more mature EMEs, but still to show signs of stability and openness to investors; they can therefore best be described as up-and-coming EMEs for early-stage investors (Kuepper, n.d.; Financial Times, 2014). FMEs are thus a type of EME and, for that matter, a sub-set of EMEs. Generally, FMEs appeal to investors because they offer potential high returns with low correlation to other markets and it is expected that over time they will become more liquid and take on the characteristics of the majority of EMEs (Financial Times, 2014). Table 1.1 shows country classifications of EMEs and FMEs.

Table 1.1 Country classifications of EMEs and FMEs

Region	EMEs	FMEs
Americas	Brazil, Chile, Mexico, Peru & Colombia	Argentina
Asia/Pacific	China, India, Malaysia, Indonesia, Philippines, South Korea, Taiwan & Thailand	Bangladesh, Sri Lanka, Pakistan, Vietnam & Kazakhstan
Europe	Czech Republic, Russia, Turkey, Hungary & Poland	Bulgaria, Croatia, Cyprus, Estonia, Latvia, Lithuania, Macedonia, Malta, Romania, Serbia, Slovakia, Slovenia & Ukraine
Africa	Egypt, Nigeria, Morocco & South Africa	Kenya, Ghana, Mauritius, Botswana & Tunisia
Middle East	–	Bahrain, Jordan, Kuwait, Lebanon, Oman, Qatar & United Arab Emirates

Source: Dow Jones Indexes (2011); MSCI FactSet (2012), cited in Schroders (2012); Economist Intelligence (2013).

As these EMEs and FMEs currently generate the most global attention and debate, they help enhance an understanding of real estate, construction and the economy.

3. RE, construction and the economy

The construction and RE industries play a critical role in the economies of nations. Raftery (1991) aptly notes that the construction industry exists because people and firms need shelter or RE to carry out various activities. The construction industry supplies RE as space to three categories of clients as identified by Raftery (1991). First, individuals, public institutions and private companies who need building space as part of their production processes – production in this case covers both manufacturing and services and other non-manufacturing processes. Second, investors who demand RE space as part of their investment portfolio and therefore do not wish to use the space for the production of any particular goods and services other than supplying the building space itself. Third, individuals and families who demand housing as largely, but not entirely, a consumer commodity. Therefore in an economy the two industries are inextricably linked. This relationship is diagrammatically represented in Figure 1.1.

It is clear from the preceding discourse that construction and RE are derived demand – the demand for space for various uses leads to the demand for construction of buildings or units of RE.

According to DiPasquale and Wheaton (1992), the analysis of RE markets presents challenges because space and asset markets are interrelated. As noted by Boshoff (2013), the earliest recording of work that distinguishes between use decisions and investment decisions with respect to RE was probably Weimer (1966), but it is Hendershott and Ling (1984) who were the first to develop a

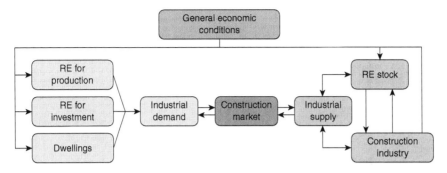

Figure 1.1 Relationship between RE and construction

Source: Adapted from Raftery (1991).

model that integrates space and capital markets into RE. Corcoran (1987) and Fisher (1992) refined the model; Fisher, for example, shows the equilibrium that exists between the short-run and long-run situations of space and capital markets. DiPasquale and Wheaton (1992) and Fisher et al. (1993) refined the model further into what commentators such as Viezer (1999) call a diagrammatic model but which according to Boshoff (2013) has been officialised in a textbook on RE economics by DiPasquale and Wheaton (1992) as the FDW-model.

The FDW-model conceptualises the interrelationships between the market for space, asset valuation, the construction sector and stock adjustment. It is a static quadrant model that traces the relationships between RE market and asset market variables as well as the adjustments that take place to establish equilibrium in the supply of and demand for RE space. Commencing with a given stock of RE space, the authors explain that changes in the macro economy (for example, increases in employment, production or the number of households; that is, market demand and supply forces) increase the demand for RE space and so, given a particular level of RE space, rents rise. This then gets translated into RE prices by the asset market. These asset prices, in turn, generate a new and higher level of construction, which eventually leads to a new and greater level of a stock of RE space.

Archer and Ling (1997) explain that generally the demand for RE space is determined by location, accessibility, level of rent and other economic factors. Regarding capital markets, Fisher et al. (1993) note that RE competes with other assets such as shares for inclusion in an investor's portfolio. However, one benefit of RE in the case of direct RE investment is its relative low correlation with equities. This low correlation means greater diversification potential, which makes it an attractive asset class for investors who wish to optimise their multi-asset portfolio. The required rate of return (RRR) on a RE investment consists of two principal components, the risk-free rate and a risk premium that reflects the risk profile of the RE's cash flow (Archer and Ling, 1997). The required risk premium for investments relative to their risk is determined simultaneously in the RE (specific systematic risk) and capital markets (risk-free rate) and thus the capital market (risk-free interest rate and the RE risk premium) affects the space market by altering equilibrium rents (Hendershott, 1995).

The determination of RE specific discount rates, RE values, capitalisation rates and construction feasibility occur in the RE market where in the initial stages the RE specific discount rate is determined by the interaction of the risk-free rate, market risk premium and the risk profile of the specific RE; beyond this, the market value and capitalisation rate of the RE can be assessed by discounting the expected cash flow of the specific RE, taking into consideration any government income tax effects (Archer and Ling, 1997). According to Archer and Ling, developers can assess the current RE market condition using the information on RE values and construction costs to determine the construction feasibility of a specific development. Thus, Archour-Fischer (1999) describes the FDW-model as an elegant metaphor that integrates the different markets in the built environment, with specific reference to the RE market, the capital market and construction activity.

4. Parts, chapters, and structure

This is a three-part book with 16 chapters. Fourteen of the chapters cover substantive topics, while the remaining two constitute a distillation and further analysis of the 14 chapters. Out of the 14 main chapters, 10 chapters relate to specific EMEs whilst the remaining chapters are non-country specific. Table 1.2 summarises the basic statistics of the EMEs considered in the book.

After this chapter, which provides the relevant context, the book is divided into three main parts. Part I has four chapters (Chapters 2, 3, 4 and 5) that concentrate on the links between RE and construction studies and economic development.

Chapter 2 (by *Franklin Obeng-Odoom*) considers oil, construction and economic development. The cleavage between construction and economic growth takes a

Table 1.2 Basic statistics of considered EMEs

Country	Land area (km²)	Population (million)	Population density (per km²)	GDP (US$ billion)	GDP per capita (US$)
Turkey	769,630	76.67	97	820.21	8,716.68
Slovenia	20,140	2.06	102	45.47	18,629.96
Ghana	227,540	25.9	114	47.93	766.05
Argentina	2,736,690	41.45	15	611.76	11,601.63
Brazil	8,358,140	201	24	2,245.67	5,823.04
Chile	743,532	17.56	24	277.20	9,728.48
Colombia	1,109,500	47.1	44	378.15	4,376.4
China	9,388,211	1361	145	9,240.27	3,583.38
Nigeria	910,770	166	191	522.64	1,097.97
Lithuania	62,674	2.94	47	42.25	10,549.18
Poland	306,220	38.49	126	517.54	10,572.82
Hungary	90,530	9.91	109	126	11,000.2

Sources: World Bank (2014a, b, c).

distinctive form under oil-induced economic development. New construction and RE developments, old RE refurbishment and the brisk construction to complete incomplete RE units generate related RE management services that, in turn, contribute to further expand oil-propelled development of the construction sector. *A priori*, it would be expected that these articulated processes would create excitement among market actors, which, in turn, would generate enhanced RE values, especially in residential and commercial RE sub-sectors. This chapter investigates these intuitive associations. It analyses in what ways and to what extent the surplus generated in the process is shared among the players in the connected sectors of construction and RE. Thus, through the lens of resource extraction, the chapter extends an understanding of the symbiosis between RE and construction and how they are moulded by oil in the process of economic development. The chapter draws out three possible policy implications distinguished by their emphases: market, state or community. Neither has all the answers but a combination of all three, seeking to promote linkages within and across society, economy and environment, can help to foster equitable, efficient and ecologically sustainable outcomes.

Chapter 3 (by *Işıl Erol*) investigates the lead–lag relationship between construction industry, RE mortgage market development and economic growth in Turkey. The chapter uses quarterly data between 1998 Q1 and 2013 Q4 and employs the Granger causality analysis in order to test whether or not the construction sector growth and RE mortgage market development separately Granger-causes economic growth, and vice versa. The main hypothesis that the economy drives the construction industry in Turkey is supported by the Granger causality analysis that revealed that the Turkish economy leads the construction industry in nominal terms. Over the period 1998 to 2013, economic growth in Turkey Granger-caused the construction activities with up to a two-year lag but not vice versa. Hence, unlike the widespread belief that construction plays a crucial role in Turkey's economic development, the study concludes that the construction industry is not a driver of GDP but a follower of fluctuations in the macroeconomic environment in Turkey. It has also been established that it is the expansion of the residential RE mortgage market that affects economic growth interactively, as when the mortgage market grows, there is more construction, a larger volume of banking sector credits, more insurance products, and more appraisal works take place due to higher demand for housing units: the economy expands with more residential RE investment. The RE mortgage market thus includes not only construction activities in the RE sector of the economy, but also banking, insurance and appraisal activities in the financial sector of the economy. The chapter concludes that although the government aims to use the construction industry as the major vehicle for achieving high growth rates in the aggregate economy, the residential RE mortgage market and related industries need to be considered as major drivers of economic growth in addition to construction activities. Making affordable RE mortgage loans available to a large cross-section of the Turkish population and also developing the secondary market for RE mortgage loans will deepen the financial sector and enhance economic growth in Turkey.

In Chapter 4, *Maruška Šubic Kovač* carries out statistical analysis in order to establish the strength of correlation and the causative–resultant interconnection between particular factors in the economy and construction and RE activities in the Republic of Slovenia in the period 2000–2013. The factors considered in the analysis include GDP, unemployment, the value of construction put in place, construction cost, the value added by RE activities and the house prices of existing dwellings. The results show that demand and supply in the Slovenian RE market in the period 2000–2013 are closely connected to the change in the socioeconomic system in 1991, the crediting conditions for households and investors, privatisation of residential RE, the accession to the EU in 2004 and to the Eurozone in 2007, as well as the economic and financial crises. The results of the Granger causality test in the VAR framework show the causality between the time series of the above factors and thus clarify the situation in these sectors in Slovenia. The caveat, however, is that the results are based on given assumptions and these assumptions and the consequent usability of the study results need to be taken into account in planning relevant future actions. It is also necessary to ask whether the GDP, which is a fundamental assumption in the study, is an appropriate criterion for assessing the progress and wellbeing of citizens.

Chapter 5 (by *Meine Pieter van Dijk*) is on the ups and downs of the RE market and its relation to the rest of the economy in China. The Chinese economy has grown very fast since 1978, in particular in the big cities in the eastern part of the country. During the last decade the RE sector expanded more than tenfold. In the past, the western provinces have grown more slowly. However, China now wants to change this regional imbalance by stimulating the economic development of the western part of the country, to which end it has implemented policies promoting investments in those provinces, stressing the availability of cheap labour and space and facilitating loans. The construction sector has played a major role in the development process and RE prices have been booming. Although a bubble has been identified, experts are still watching what kind of landing the RE market will make and what the consequences for the national economy will be. The chapter therefore analyses how China is managing its RE market and the factors that explain current trends. In particular, different rates of urbanisation and economic growth have been shown to influence demand while government policies promote the supply of houses. It has been established that currently the housing market is changing from supply-driven to demand-driven. However, given incomes are not growing as fast and people have to spend more money on repaying their debts, demand for housing is lower and more directed to low cost housing, which is often not available. As a result, a slower growth of the national economy is expected, and this will have repercussions even for the global economy.

The second part of the book (Part II) deals with construction and economic development. It disaggregates the relationship between RE and construction issues and drills into the field of construction and its impact on economic development. This section has four chapters: Chapters 6, 7, 8 and 9.

Chapter 6 (by *Edward Ochieng*) looks at the issue of managing cultural challenges in construction projects and its implications for economic development in emerging economies. The rapid integration of the world economy has had significant impact on the way construction project managers work, frequently bringing them clients, suppliers and peers that they have never worked with before. In an era of globalisation, projects in emerging economies face unique challenges in coordinating among clients, financiers, developers, designers and contractors from different countries. In addition, construction project teams need to cope with the complexities of both local institutions and physical environments. While offering opportunities, globalisation also poses significant challenges for construction organisations in emerging economies, especially when different cultures are involved as a team. Since construction companies are able to move resources to almost any location worldwide and have the capacity to work on a global scale, for many organisations future opportunities to work entail thinking more clearly about cross-cultural issues and more overtly and systematically about multicultural team working. What does this mean for project leaders and international construction organisations in emerging economies? This chapter, therefore, examines the key concepts in cross-cultural management as well as key functions in construction projects and their impacts on economic development. Findings from this chapter suggest that if countries in emerging economies want to accomplish better practice and economic development, then thought must be given to what constitutes best practice. In spite of the current difficulties the industry faces, there is an increasing need to get multicultural construction project teams of different nationalities to work together effectively, thus managing construction projects in emerging economies will require construction organisations to ascertain how to work within an inter-continental environment with multicultural project teams. Project team leaders must actively promote multicultural team working as the means of addressing poor performance on construction projects. In particular, if organisational change is to be effectively introduced in emerging economies, multinational construction organisations will have to ensure that their key decisions are being informed by the knowledge and experience of local or indigenous managers. This will require construction project leaders to have a better understanding of cultural change processes and procedures in developing countries. The proposed strategies in this chapter present a better way of optimising the performance of project-based operations, thus enabling construction organisations to reform their poor performance on projects and empower them to better manage emerging culture challenges in their future projects.

Chapter 7 (by *Andrew Price* and *Edward Ochieng*) examines the application of sustainability principles into construction projects being delivered in emerging economies. There is an increasing demand in the construction industry to understand sustainable construction process. This demand is driven by a realisation that sustainable practices make sense to both contractors and clients. The output of the construction industry (be it in public and commercial buildings or infrastructure) has a major impact on a nation's ability to maintain a sustainable economy and also has a major impact on the environment. Moreover, it is clear that declared environment targets cannot be met without dramatically reducing the environmental impact of buildings and infrastructure construction. The industry in emerging economies needs practitioners with an understanding of the key issues that will enable them to make a broader, sustainability-aware approach to maximise the quality of construction projects while improving competitiveness and sustainability performance. Construction clients and governments in emerging economies recognise the significant impact the design, construction and occupation of buildings have on the environment and society. Governments in emerging economies have a central role in driving the sustainable development agenda. Good sustainable design will deliver buildings with low running costs – an attribute that is highly attractive to both the society and businesses. The chapter has provided key evidence of the link between sustainability and better project performance through the integration of sustainability principles. The principles are: embed sustainability objectives throughout the team and supply chain; set specific and clear sustainability targets from the outset; and create a project structure that supports a collaborative approach. In addition, this chapter has proposed a sustainable framework for better construction project delivery in emerging economies, based on the philosophies of sustainable construction.

Chapter 8 (by *Divine K. Ahadzie*) considers the economic, technological and structural demands of the construction industry in emerging markets south of the Sahara, using Ghana as a case study. Following constitutional democracy since 1992, Ghana has grown into one of the fastest emerging markets in sub-Saharan Africa with a sustained real growth rate of over 4% for a decade or more and reaching a level of 8% in recent times. Within this context, one sector of the economy which has fared tremendously is the construction industry, which has achieved a consistent GDP growth rate in excess of 8% over the last decade. In 2006, it emerged as the fastest growing sector in the economy with a rate of 8.2% as against a national average of 6.2%, and in 2007 it attained a growth rate of 11%, posting 8.9% to GDP. Current data indicates that it continues to be the most consistent and important industrial sector in the wider economy, posting 10.5% to GDP in 2012. Thus, by all standards the Ghanaian construction industry is expanding as the engine of growth of the economy. However, as an emerging economy, this expansive growth comes with huge economic, technological and structural demands (ETSDs) if the full potential of the industry is to be tapped towards achieving the accelerated development of infrastructure needed to stimulate growth and job creation in emerging economies. It is these ETSDs that are critically examined in this chapter in order to provide a construction industry accelerated development framework for engendering job creation and growth in the wider economy. Opportunities and constraints affecting supply and demand within the industry are also addressed. The findings show that the economic indicators are favourable; however, the capacity of domestic contractors plus the skills of artisans are severe constraints on the quick pace of expansion. The consequence is that local firms are losing out in many high-tech projects that make them uncompetitive in an emerging economy. There is a lack of a decisive and radical construction industry development agenda in Ghana and this could be inimical to the future growth and competitiveness of the domestic sector both locally and internationally. The development of the construction industry is a long-term effort, which requires a deliberate and systematic agenda and Ghana as a matter of urgency needs to act quickly in putting in place a strategic and comprehensive action plan in this respect.

Chapter 9 (by *Nii Ankrah*, *Joseph Mante* and *Issaka Ndekugri*) is on the challenges of infrastructure procurement in emerging economies and their implications for economic development. It is well established that there is a strong correlation between investment in infrastructure/built environment and economic development in any country. It is therefore not surprising that nations and international institutions with a development remit have placed a lot of premium on infrastructure development across the globe. In many developing countries, it remains the responsibility of government to invest in infrastructure development. Public procurement thus constitutes a substantial portion of government spending and can be considered an important policy tool. However it is also a business and management activity requiring efficiency in its execution to optimise economic development outcomes. Using Ghana as a case study, the framework for public procurement particularly relating to infrastructure acquisition in emerging economies is explored. It is argued in the chapter that there is currently much inefficiency in the procurement process that undermines consultant and contractor selection, contract execution, and project delivery and completion according to specifications. The chapter concludes by highlighting the urgent need for the implementation of procurement policy reforms, strengthening and streamlining legal and institutional frameworks, ensuring compliance with laid down procurement procedures and fighting corruption to ensure improvement of procurement systems so that economic development benefits can be optimised.

Part III of the book explores how topics in RE can contribute to wealth creation, poverty reduction and economic development. It has six chapters (Chapters 10, 11, 12, 13, 14 and 15).

Chapter 10 (by *Raymond T. Abdulai* and *Anthony Owusu-Ansah*) assesses the management of RE market information and economic development in developing countries. The longstanding argument that it is RE market management information via land registration (LR) that guarantees RE ownership security in developing countries is common knowledge. Based on the "equation of LR to ownership security argument" it is further asserted that LR makes RE suitable collateral and thus guarantees access to formal capital for investment, which leads to wealth creation, poverty reduction and economic development. This has precipitated the conduct of a lot of empirical research that has examined the extent to which the "equation of LR to ownership security argument" is true. Such studies have not been able to establish any discernible link between LR and ownership security. This chapter critically examines the underpinning theoretical principles of LR systems in order to explain why LR per se cannot guarantee security of RE ownership and for that matter cannot "unlock" capital for investment. The chapter establishes that albeit LR plays a critical role in the economies of nations and therefore can contribute to economic development, it plays that role via a completely different route and not through the commonly held view about the purpose that LR serves. LR is a record-keeping system that creates a database of RE owners that can be used for two main purposes: (a) it facilitates the collection of RE related taxes that can be used for developmental projects; and (b) it facilitates RE transactions thereby reducing transaction costs. The chapter has also defined other determinants of economic development to include good governance and prudent management of the resources of nations.

In Chapter 11, *Franklin Obeng-Odoom* treats informal RE brokerage as a socially-embedded market for economic development in Africa. The touchstone in most RE studies in Africa is that informal RE brokerage is the cause of many of the problems bedevilling the operation of RE markets in the sub-continent. A key exponent of this view is the growing cluster of RE professional bodies that posit formal licensure as a way to cure the pathology of informal RE brokerage. The intellectual force deployed to support this line of analysis is the vast body of work in new institutional economics – including the work of Douglas North, Armen Alchian and Harod Demsetz – that emphasises the need to reduce transaction costs through simplifying and formalising institutions. This chapter examines this prevailing argument by recasting the debate in a wider framework of "social embeddedness", drawing selectively on the work of Karl Polanyi. Using examples from cities in West Africa such as Lagos, Accra and Sekondi-Takoradi, this chapter: (a) traces the origins of and assesses the contribution of informal RE brokerage to economic development; (b) investigates the existing RE brokerage practices both in the formal and informal systems; and (c) examines local client experiences with such brokers compared to experiences with more formal brokers. The chapter shows that: (a) RE brokers and brokerage have a long history in Africa, typically playing the role of supporting the RE system by diffusing information among the various actors, who are often related to them as family members or neighbours; however, changing relationships to community, markets and the state have created tiers of brokers in a formal–informal continuum, making the specific contribution of brokers to the process of economic development contingent on *where* they are and *how* they relate to others in this continuum and to society; (b) the charge about widespread problems with informal brokerage services is exaggerated; (c) the preferred formal brokerage system offers an inferior alternative to informality; and (d) social embeddedness of brokerage services in the informal system is better attuned to social and human needs in the RE sector and wider society in Africa.

Chapter 12 (by *Eric Yeboah*) concentrates on student housing investment and economic development. The housing sector remains one key aspects of the RE industry with direct and immense implications for economic development. The concept of housing is complex, with several cleavages such as student housing. Despite this, conventional efforts to map out the nature of linkage between housing and economic development have often treated the former as a homogenous industry. Such an approach is convenient, although it does not offer the needed opportunity to explore how the different variants of housing contribute to economic development in their peculiar way. The chapter seeks to contribute towards addressing this by examining how student housing impacts on both local and national economies. The chapter draws evidence from the rapid student housing development around the Kwame Nkrumah University of Science and Technology in Kumasi, Ghana. The study establishes that student housing development boosts local economies by creating several jobs along both the construction and RE management value chains. It further reinvigorates local economies by creating a dynamic retail market and ensuring the supply of the needed infrastructure to support convenient student living. Such investment becomes a major source of revenue through various taxes and rates. Student housing investment further contributes to a dynamic and functional property insurance industry which also helps in ensuring efficient allocation of resources. As a result of these and other contributions, this chapter argues that there is a need to create an enabling environment to stimulate investment in student housing, and some recommendations to help achieve this are accordingly offered.

Chapter 13 (by *Claudia B. Murray*) examines RE and social inequality in Latin America (Argentina, Brazil, Chile and Colombia). Latin America is known as the most unequal region in the world where extreme displays of wealth and exposure to scarcity lay bare in the urban landscape. Inequality is not just a social issue; it has considerable impact on economic development. This is because social inequality generates instability and conflict, which can create unsettling conditions for investment. At the macro level, social inequality can also present barriers to economic development as most government policies and resources tend to be directed towards solving social conflict rather than promoting and generating growth. This is one of the reasons usually cited in explaining the development gap between Latin America and other emerging economies; for example, East Asia has similar policies to those applied recently in Latin America but East Asia is achieving better growth. The other reason cited is institutional; this includes governance as well as property rights and enforcement of contracts. The latter is the focus of this chapter. Using a range of literature including official documents issued by local governments, peer reviewed academic journals and the latest media articles from the national press of each country, the chapter argues that a defective planning system, scarcity of distributive mechanisms for land taxation and weak municipal powers are incentivising an unscrupulous RE residential market that is profiting from this confusing scenario. This is not only restricting the options for housing the poor but is also fuelling urban inequality and hindering economic development in Latin America.

Using Nigeria as a case study, Chapter 14 (by *Timothy Tunde Oladokun*) considers corporate RE in Africa. The traditional perception of the RE owned by companies not primarily in the RE business as a "factor of production" limits its importance to the provision of space for manufacturing and delivery of goods and services. This belief regards corporate RE (CRE) management (CREM) executives as deal makers whose responsibilities are RE negotiations and transactions, yet many large non-RE firms are, in addition to investing significantly in properties for business purposes, controlling RE portfolios that are comparable in value terms with those owned by mainstream RE companies. The global competitive and information-driven business trend has necessitated efficient use of organisations' RE as a strategic resource for profitability and productivity. This development has triggered various empirical studies that have examined the roles and the contribution of CRE to business performance in the developed economies of the UK, USA and Germany. Unfortunately, such studies have not been known in African emerging economies. The aim of this chapter is to examine the status of business organisations in Nigeria with a view to establishing the contribution of effective CREM to economic development in African countries, using Nigeria as a case study.

The chapter utilises a quantitative research methodology. Primary data collected via questionnaires administered to 50 CRE managers from 21 banks, 5 GSM communication companies and 24 insurance companies that are listed with the Stock and Exchange Commission in Nigeria were analysed with the aid of descriptive and inferential statistics of percentages and cluster analysis. The study revealed three sets of organisations: "active" companies (30.8%) that are conscious of the various key factors of CREM at a very high level, "selective" companies (61.5%) and "passive" companies (7.7%) that have not been using CRE to improve the performance of their organisations and which are hence losing the value addition of CRE to the financial performance of organisations. It is concluded that the contribution of CREM to economic development has not been largely achieved due to the fact that CREM is still being selectively adopted by business organisations in Nigeria. The implication is that the practice requires proper refocusing and nurturing to enable it to make a value contribution to economic development in Nigeria and the African region as a whole.

Chapter 15 (by *Vida Maliene, Isabel Atkinson, Maruška* Šubic *Kovač, Andrea Pödör, Judit Nyiri Mizseiné, Robert Dixon-Gough, Józef Hernik, Maria Pazdan, Rimvydas Gaudėšius and Virginija Gurskienė*) provides an overview of Slovenian, Hungarian, Polish and Lithuanian RE markets including an analysis of the main players/actors, the legal framework and RE valuation practice. The countries in central and eastern Europe (CEE) began a remarkable transition from centrally planned economies towards market economies in 1989 when the Berlin Wall fell and the Iron Curtain lifted. Slovenia, Hungary, Poland and Lithuania joined the European Union at the same time, on 1st May 2004, but their economies including RE markets have developed in some distinctive ways.

In the 1990s, in parallel with political changes, the transformation of economies began in Slovenia, Hungary, Poland and Lithuania. After the collapse of command economy structures and state socialism, a return to private RE ownership was pursued as a potential remedy to the coming challenges. In the process of economic transition of all four countries, two stages of development can be identified. The first stage covers the early years of economic transition until 1995–1996 and it is during this stage that the biggest differences in approach and priorities have been observed. The second stage started in approximately 1997–1998, when the reforms became clearly EU-accession oriented and were thus driven and guided by the set of intermediate and final objectives bilaterally agreed with the EU institutions, which permanently monitored the process and provided very important financial and technical assistance for the timely accomplishment of the needed reform goals. The fundamental economic reforms related to liberalisation of prices, trade and foreign exchange, macroeconomic stabilisation through the control of inflation, the restoration of private RE, reforming of the banking and financial sectors and the attracting of foreign capital and investments.

Solid foundations for the free market economy, including the RE market sector, a RE legal framework and valuation practice based on open market analysis have been built in each case study country. The newly established legal systems governing each country's RE law have similarities to one another and most other European countries in their nature and structure, as the roots derive from the common heritage of the Napoleonic civil law system. RE valuation practices vary from country to country, but worldwide-recognised common valuation techniques based on open market valuation have been successfully adopted in each country. Building property administration systems, each country has an established land register and cadastre system which successfully contributes towards RE market transparency. Over the last 10 years, RE market performance by sector has been very much dependent on unique government policies on taxation and investment. In the first stage of economic development, the residential RE market sector was established quite quickly and effectively in the biggest cities of each country, which heavily contributed to economic growth. The commercial RE market demonstrated acceleration at the same time. Unfortunately, quite well performing RE markets were hit by global RE recession in 2008, mostly affecting the residential RE market sector. The agricultural market sector has been controlled by policies protecting land ownership from private international investors in each country and has not been developed to the same level as residential and commercial RE sectors.

References

Anaman, K. A and Osei-Amponsah, C. (2007) Analysis of the causality links between the growth of the construction industry and the growth of the macro-economy in Ghana, *Construction Management and Economics*, 25(9), pp. 951–961.

Archer, W. R. and Ling, D. C. (1997) The three dimensions of real estate markets: Linking space, capital, and property markets, *Journal of Real Estate Finance*, Fall, pp. 7–14.

Archour-Fischer, D. (1999) An integrated property market model: A pedagogical tool, *Journal of Real Estate Practice and Education*, 2(1), pp. 20–31.

Arku, G. (2006) Housing and development strategies in Ghana, 1945–2000, *International Development Planning Review*, 28(3), pp. 333–358.

Boshoff, D. G. B. (2013) Empirical analysis of space and capital markets in South Africa: A review of the REEFM and FDW models, *South African Journal of Economic and Management Sciences*, 16(4), pp. 383–394.

Boydell, S. (2007) Disillusion, dilemma and direction: The role of the university in property research, *Pacific Rim Property Research Journal*, 13(2), pp. 146–161.

Corcoran, P. J. (1987) Explaining the commercial real estate market, *Journal of Portfolio Management*, 13, pp. 15–21.

DiPasquale, D. and Wheaton, W. C. (1992) The markets for real estate assets and space: A conceptual framework, *Journal of American Real Estate and Urban Economics Association*, 20(1), pp. 181–197.

Dow Jones Indexes (2011) Dow Jones Indexes: Country Classification System, CME Group Index Services LLC.

Dujardin, C. and Goffette-Nagot, F. (2009) Does Public Housing Occupancy Increase Unemployment? *Journal of Economic Geography*, 9, pp. 823–851.

Economist Intelligence (2013) Ghana is among emerging economies [online]. Available at: http://www.ghananewsagency.org/economics/ghana-is-among-emerging-economies-economist-66592 [Accessed 30 June 2014].

Financial Times (2014) Definition of frontier markets [online]. Available at: http://lexicon. ft.com/term?term=frontier-markets [Accessed 1 July 2014].

Firang, D. Y. (2011) Transnational activities and their impact on achieving a successful housing career in Canada: The case of Ghanaian immigrants in Toronto, PhD thesis submitted to the School of Graduate Studies in partial fulfilment of the requirements for the degree of Doctor in Philosophy, Faculty of Social Work, University of Toronto.

Fisher, J. D. (1992) Integrating research on markets for space and capital, *Journal of Real Estate and Urban Economics Association*, 20(1), pp. 161–180.

Fisher, J. D., Wilson, S. H. and Wurtzebach, C. H. (1993) Equilibrium in commercial real estate market: Linking space and capital markets, *Journal of Portfolio Management*, Summer, pp. 101–107.

Harris, R. and Arku, G. (2006) Housing and economic development: The evolution of an idea since 1945, *Habitat International*, 30(4), pp. 1007–1017.

Heakal, R. (2009) What is an emerging market economy? [online]. Available at: http://www.investopedia.com/articles/03/073003.asp [Accessed 30 June 2014].

Hendershott, P. (1995) Real effective rent determination: Evidence from the Sydney office market, *Journal of Property Research*, 12, pp. 127–135.

Hendershott, P. H. and Ling, D. C. (1984) Prospective changes in tax law and the value of depreciable real estate, *Journal of the American Real Estate and Urban Economics Association*, 12, pp. 297–317.

Kuepper, J. (n.d.) Frontier markets: New opportunities around the world [online]. Available at:http://internationalinvest.about.com/od/gettingstarted/a/Frontier-Markets-New-Opportunities-Around-The-World.htm [Accessed 30 June 2014].

Malpezzi, S. (2009) The Wisconsin program in real estate and urban land economics: A century of tradition and innovation, Fall 2009 edition. Wisconsin: The Department of Real Estate and Urban Land Economics and the James A. Graaskamp Center for Real Estate, University of Wisconsin-Madison.

Obeng-Odoom, F. (2010) Urban real estate in Ghana: A study of housing-related remittances from Australia, *Housing Studies*, 25(3), pp. 357–373.

Raftery, J. (1991) *Principles of building economics*, Oxford: DSP Professional Books.

Schroders (2012) *Talking point, frontier markets: The new emerging markets*, London: Schroders.

Tibaijuka, A. (2009) *Building prosperity: Housing and economic development*, London: Earthscan.

Weimer, A. M. (1966) Real estate decisions are different, *Harvard Business Review*, 44, pp. 105–112.

World Bank (2014a) Trading Economics [online]. Available at: www.tradingeconomics.com [Accessed 26 December 2014].

World Bank (2014b) Land area (sq. km.) [online]. Available at: http://data.worldbank.org/indicator/AG.LND.TOTL.K2/countries [Accessed 26 December 2014].

World Bank (2014c) Population density (people per sq. km of land area) [online]. Available at: http://data.worldbank.org/indicator/EN.POP.DNST [Accessed 26 December 2014].

Viezer, T. W. (1999) Econometric integration of real estate's space and capital markets, *Journal of Real Estate Research*, 18(3), pp. 503–519.

Part I

Links between real estate and construction studies and economic development

The contributors to Part I address the relationships between real estate and construction studies and their contribution to economic development.

2   Oil, construction, and economic development

Franklin Obeng-Odoom

1. Introduction

There is an extensive literature on the effect of oil on economic development (see, for example, Frankel, 2007; Carmody, 2011; Sarbu, 2014) and, as shown by Alagidede (2012) and in the other chapters in this book, a huge literature on construction and economic growth, but very little work on oil, construction, and economic development. The bald literature on this tripartite relationship can be classified into two groups. The first concentrates on how oil revenues are invested in construction projects in oil rich economies and how such investments are correlated with economic growth. Writers on this genre, for example Ramsaran and Hossein (2006), Saka and Lowe (2010) and Olatunji (2010) point to the growth-inducing effects of "oil construction" but hardly connect their analysis to broader processes of economic development, apart from making observations about the employment generating effects of such construction projects. So, there is limited systematic attempt to establish a tripartite oil, construction, and economic development relationship, although this literature suggests that the relationship might be positive. The second strand in the literature explicitly considers oil, construction, and economic development. This existing literature tends to focus on Chinese oil exploration and production activities in Africa. The preponderance of the evidence suggests that "oil construction" exerts no economic development effects (Corkin, 2011a, 2011b, 2011c, 2012) and hence dovetails into the broader body of work that considers effects of the oil industry to be "enclaved" (Ackah-Baidoo, 2013; Hilson, 2014), that is, beneficial but only to a small group of people, usually power groups and elites.

The aim of this chapter is to unify this dichotomous literature and problematize it. The chapter is original in this sense but also in two other ways. Unlike existing work, which remains at the building construction level, the analysis in this chapter cascades up from the building to the city, nation, regional, and global dimensions of the tripartite oil–construction–economic development relationship. Additionally, the approach is unlike previous studies that have used either econometrics (Olatunji, 2010; Saka and Lowe, 2010), interviews (Laryea, 2010), or input–output relationships (Bon et al., 1999; Kofoworola and Gheewala, 2008). This chapter adopts an *original* institutional political economy framework that draws on concepts of "staples", "linkage", and "rent", as developed respectively by Innis (1930), Hirschman (1954, 1984), and George ([1879] 2006).

While agreeing with the first strand of the literature that, overall, there is strong oil–construction–economic growth nexus, this chapter points to important deleterious experiences, which tend to metastasise into social inequality and economic instability. Oil production can drive construction which, in turn, can stimulate a socially inclusive economic development, but the process will require the "invisible hand" of the market to give way to the "visible hand" of a developmental state and empower communities holding land *res communis*. The challenge is how to ensure a smooth baton change in this crucial relay.

The rest of the chapter is divided into three sections. The first explains the oil-construction-economic development nexus using the staples thesis and the concept of linkages. The second uses the concept of rent to analyse social inequality and economic instability in the tripartite relationship, while the third evaluates policy options.

2. The oil–construction–economic development nexus explained

Innis' (1930) staples thesis [widely discussed by Watkins (1963)] is one effective way to explain the oil–construction–economic development nexus. According to this view, the oil industry plays a transformative role in an oil economy. Staples analysts point to *spread effects* of the oil sector, which stimulate the growth of interlinked industries to supply inputs (backward linkages) and make use of outputs from the oil industry (forward linkages). From this perspective, oil production unleashes economic effects that link together booming and lagging sectors to improve the overall social conditions of the oil economy (Watkins, 1963). The oil industry tends to stimulate a wide range of construction activities: from roads to buildings, offices to ports, storages and pipes to support industries and many other construction activities tend to be carried out to process the outputs of the oil industry too. In practice, "the most important example of backward linkage is the building of transport systems for collection of the staple, for that can have further and powerful spread effects" (Watkins, 1963, p. 145). Nevertheless, all these activities, in turn, employ a large workforce whose purchases and taxes stimulate further economic activities and generate revenues for the state. Further, most of such activities require energy to run and hence they boost the demand for oil and other energy sources. All these lead to a process which tends to drive economic growth (Alagidede, 2012). However, beyond growth, the process is economic development-inducing because it generates outcomes that can be diffused in the economy for inclusive progress. But even if the activities do not touch substantial aspects of the economy, staples analysts would argue that revenues from backward and forward linkages can be invested in social and economic support services to spread the fruits of economic growth.

Hirschman (1985) substantially developed this idea of interdependent economic processes into a stronger concept of "linkages". Hirschman (1984) considered Innis' (1930) effort too "descriptive" because, for Hirschman (1984), industrial linkages form a bridge from underdevelopment to economic development through unbalanced but sequential industrialisation. That is, to Hirschman

(1958; 1984), the essence of the economic development process is to set in place, first, industries that will trigger a series of interdependent activities either from the rear (backward linkages) or forward linkages. So, he favoured the establishment of industries that had the potential of unleashing further industrialisation. That is, industries with strong backward and forward linkages. Construction then, was a favourite because of its capital formation-enhancing aspects, as well as employment and revenue generation tendencies. Hirschman (1984) later thought of and encouraged other linkages analysis, but particularly fiscal linkage to enable the state to obtain revenues for re-investment in other linkage generation industries. Figure 2.1 synopsises the linkages framework as it applies to the oil–construction–economic development relationship.

Figure 2.1 shows the *iterative* linkages between and across oil, construction, and economic development. Backward linkages are those inputs and support facilities that are constructed to serve the oil industry, so Hirschman (1958, p. 100) called them "input-provision effects". In this sense, economic processes such the construction of roads, housing, hospitals, and hospitality facilities are all considered backward linkages. Forward linkages, on the other hand, are those facilities for which the oil sector provides inputs. In Hirschman's (1958, p. 100) words, they are "output-utilisation effects". Such facilities are accompanied by construction-related activities of their own. For instance, refineries that are set up come along with housing for the refineries' staff. Other small industries may be built to support the operation of refineries. Some of these construction activities are, of course, related to the backward linkages but others are distinct. Industries that support oil drilling may differ from those set up to utilise oil product, but they are both related to the ramifications that oil has in the economy.

The oil industry generates backward and forward linkages in the form of infrastructural development, be it in terms of construction activities such as road transportation, hospitals, schools, hospitals, and offices (backward linkages). These support the oil industry to produce forward linkages in construction activities such as the building of refineries and petrochemical industries. Employment, higher wages, and tax revenues are among the *combined linkages* which, in turn, empower others to go into the construction industry. New real estate (RE) developments, old RE refurbishment, and the brisk construction to complete incomplete RE units generate related RE management services which, in turn, contribute to further expand oil-propelled development of the construction sector. *A priori*, it would be expected that these articulated processes would create excitement in market actors which, in turn, would generate enhanced RE prices, especially in residential and commercial RE sub-sectors. The articulated linkages and drive are also driven by economic growth. In an *iterative* way, the benefits from these linkages include jobs, revenue, spending, and further economic activity, industrialisation, and economic growth.

There is much evidence of tremors of backward, forward, and combined linkages in the oil–construction–economic growth nexus. It has been suggested in the case of Trinidad and Tobago (Lewis, 2004) and concretely demonstrated in the case of Nigeria (Saka and Lowe, 2010). In these cases, growth has been measured

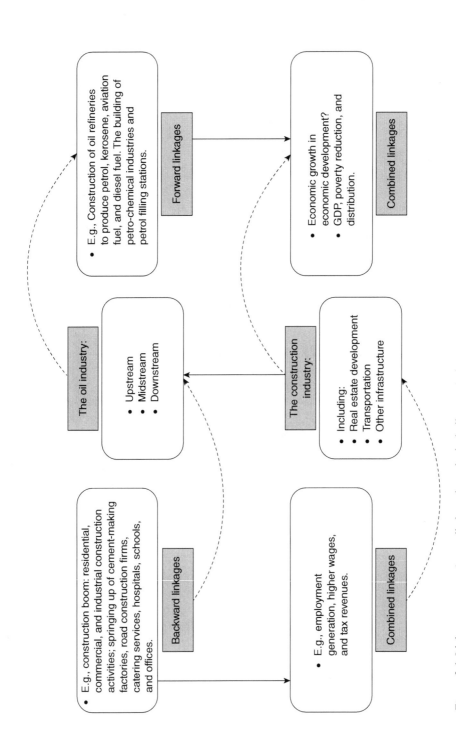

Figure 2.1 Linkages concept (as applied to a hypothetical oil economy)

Source: Adapted from the work of Hirschman (1958: 1984).

in GDP terms. Qualitatively, too, there is much evidence of linkages between oil, construction, and economic growth. Kuwait literally built its economy out of oil and construction. While pre-1957 construction was un-coordinated, since 1957 construction in this oil economy has been massive. Fahad al- Salem Street is one of the most discussed. As one scholar put it recently, "When completed in 1962, the street was flanked by a total of seventy buildings of five stories, comprising 960 ground floor shops along colonnaded pathways with 2200 apartments above – all done in three years" (Al-Nakib, 2013, p. 15). The Kuwait Building Programme from 1947 to 1951 was aimed at providing the requisite infrastructure to support the oil industry. Among the massive building and construction programme was the building of a new city to be called "Ahmadi" and the construction of staff housing for staff of Kuwait Oil Company. Residential RE was to be 1,450 residential units. Office buildings were also constructed in Ahmadi together with hospital, a fire station, a post office, schools, and social amenities (for example, staff clubs), commercial units such as shops, and luxury RE such as pleasure gardens (Alissa, 2013, p. 43).

Other evidence of the linkages between oil and the economy via construction and RE has been recorded in Africa. The Chad–Cameroon Petroleum and Pipeline Development Project is a case in point. Among others, the project consisted of the construction of 315 oil wells, the construction of two collecting stations and one pumping station, a central treatment centre to produce up to 225,000 barrels per day all in Chad and in Cameroon, the construction of off-shore floating storage offloading that could be accessed by 12 km of underwater pipeline, several workers' camps and an office in Douala. Across the two countries, other projects are 1,070 km of pipeline under the earth, 613 km of newly constructed or renovated roads, one new bridge on the border and the construction of a major telecommunications system (Guyer, 2002, p. 109). In turn, there were other spin-off projects whose value is yet to be ascertained. The more direct employment effect has been put at around 7,000 jobs (Guyer, 2002, p. 113). In the frontier oil economy of Ghana, too, Laryea's (2010) interview with contractors showed that they were optimistic of the opportunities that oil and gas present for construction, with one declaring that "There is construction work associated with the emergence of the new industry . . . There is real estate associated with the oil and gas industry."

More recently, the oil boom in Ghana has led to substantial expansion in key aspects of the RE industry. Major growth has been recorded in the residential, commercial, and office market with completed and ongoing construction projects in hospitality, transport, and RE services. These activities have generated expansion in the education sector and employment at various points. Collectively, these forward, backward, and combined linkages have contributed substantially to GDP growth and the expansion of the economy, and in turn have opened up avenues to enhance the economy (Darkwah, 2013; Obeng-Odoom, 2014a; Panford, 2014).

For these same reasons too, construction activity is expected to fall with a decline in oil revenues and prices or a general downturn in the economy. Econometric evidence about this two-way relationship between the economy and construction abounds [see, for example, Anaman and Osei-Amponsah (2007) and

Kovac in this volume]. What some countries do, therefore, is to try to propel the national economy through oil-backed construction projects (Saka and Lowe, 2010; Obeng-Odoom, 2014b). Oil, construction, and growth, then are strongly linked and interdependent.

However, economic development is more than economic growth. In his famous article "The Meaning of Development", Seers (1970) gave three conditions to transform economic growth to economic development, namely, job creation, egalitarian distribution of economic growth, and poverty reduction. While the meaning and objects of economic development has evolved (Arndt, 1989) and become much broader to include indices such as happiness, making it ever more demanding to measure it, any study about economic development must go beyond growth to consider jobs, their size and nature, and the distribution and sustainability of economic growth (Fioramonti, 2013; Obeng-Odoom, 2014c). It is to these broader considerations that the chapter now turns.

3. Rent, social inequality, and economic instability

Rent analysis is central to the political economy of natural resources, especially in the work of George ([1879] 2006), but peripheral among economists specialising in construction. Much writing in the built environment recognises rent but only in terms of payment for the use of housing services. From this perspective, the relationship between rent and social inequality is housing affordability and that between rent and economic instability is reduced to subprime mortgages.

George ([1879] 2006) offered a more nuanced explanation that starts with the generation of rent and goes through the process of how that rent metastasizes into *crisis* in the form of social inequality and economic instability. According to George's ([1879] 2006) theory of crisis, as land is monopolised or privatised, population growth, government and market investment, the application of technology, and knowledge increase rent. Speculation tends to follow which, in turn, causes further increases in rent. Consequently, more and more land is withdrawn from production making, rent goes up, and wages fall – as a consequence. In turn, production falls dramatically as labour becomes dis-incentivized, setting in motion a chain of activities including labour scapegoating and dismissal, which in itself creates another current of deeper and deeper crises and recession.

At the root of this recession-prone system is private property in land. To George ([1879] 2006, p. 192):

> Private property in land – the subjecting of land to that exclusive ownership, which rightfully attaches to the products of labour – is a denial of the true right of property, which gives to each the equal right to exert his labour and the exclusive right to its results. It differs from slavery only in its form, which is that of making property of the indispensable natural factor of production, while slavery makes property of the human factor; and it has the same purpose and effect, that of compelling some men to work for others. Its abolition, therefore, does not mean the destruction of any right but the cessation of a

wrong – that for the future the municipal law shall conform to the moral law, and that each shall have his own.

In short, crises are inherent in capitalism, which from a Georgist perspective is a system of production in which private property is created in land and natural resources. Figure 2.2 is a diagrammatic representation of George's ([1879] 2006) theory of crisis; it shows systemic booms and busts in the land market.

Figure 2.2 shows that George's ideas stress *fundamental* problems within the explanation and discount the role of individuals and their lending practices, important as they are to understanding the system as a whole. The specific application of this crisis theory to contemporary political economic analysis can be found in the work of Mason (2009): *After the Crash: Designing a Depression-Free Economy*. According to him, irresponsible financial lending is widely acclaimed as the cause of the crisis, but it is the structural tendency for speculation in RE markets that he prefers to emphasise. As the cycle recurs, according to Mason (2009), every 18 years, it is the underlying process that ought to be the source of investigation not individual greed of bankers, which has dominated the literature on the current crisis. As a Georgist, it is the cycle of boom and bust engendered by the land market that is emphasised by Mason. Yet, neither George ([1879] 2006) nor Mason (2009) directly addresses the question of oil, construction, and economic development.

It is a gap in Georgist thinking that El-Gamal and Jaffe (2010) – incidentally, non Georgists – fill, in their recent work, *Oil, Dollars, Debt, and Crises: The Global Curse of Black Gold*. According to them, when oil-induced RE prices boom, they fuel speculation by those who invest within the Middle Eastern region, especially those who invest in the shares floated by the RE industry. The realisation that house prices are inflated causes panic and investors withdraw their money, taking it instead to Europe and elsewhere to invest in hedge funds whose managers, paradoxically, invest the money in oil futures. Such bets in turn create conditions for the bubble to trigger further increases in oil prices, resulting in excitement and even more speculation in the Middle East and elsewhere – until there was a sudden burst in 2008. Here, what it means is that there is a bubble and burst story with origins in the Middle East but intensified by systemic problems

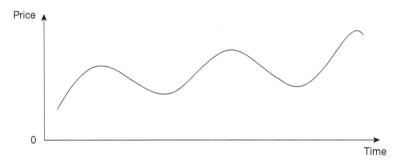

Figure 2.2 The Georgist theory of crisis

in the rest of the world. In this sense the centre of action is not in the USA but in the Middle East.

From these perspectives, speculators also bet on oil-driven RE prices. Through such gambling activities, RE prices soar, but only for a while. When investors realise that the assets are too risky, they panic and in turn move their investment elsewhere. Such panic behaviour also leads to a bust in RE prices and with that comes a ripple effect as unemployment; related problems follow. Notable cases include how RE for the rich lost ground by 19% per month by November 2008 in Dubai, while moderately priced houses lost 5% of their price in October 2008. Emaar Properties PSJC, a leading RE developer, lost 85% of its stock price in 2008 (El-Gamal and Jaffe, 2010, p. 145).

It is easy to forget that RE relations are at the centre of this analysis of crisis and oil. It is such RE relations that bear the outward expression of the systemic processes that begin with rent creation. At the local level, peasant farmers tend to lose their prime land to construction-related activities with oil as the proximate cause. Others who hold a weak position in the housing market miss out too. For instance, poorer tenants are forced to pay higher rent or, failing that, they are forcefully evicted. Some make it to lower tier housing where they pay the same amount or even slightly more because of the overall boom in the RE market (Obeng-Odoom, 2014). Also, while much employment arises with oil-induced construction, these jobs are gendered, insecure, and transient. Globally, women constitute barely 10% of the workforce in the oil industry, as shown in Table 2.1.

As Table 2.1 shows, the region with the most oil, the Middle East, is also the region with the fewest women participants in the industry. Islamic traditions about women and work might be regarded as a possible reason but there are other more plausible reasons. Women are not motivated to participate in the labour force because they have "unearned income" from their husbands, the government (cash transfers), or both (Ross, 2008). For the few that work in the industry, most are concentrated in business development/commercial, health, safety and environment (HSE), and project controls sectors of the oil industry where salaries are

Table 2.1 Regional participation levels of women in the oil and gas industry in 2013

Rank	Region	Percentage of workforce female
1	South America	10.3
2	North America	10.2
3	Australasia	9.1
4	Europe	8.3
4	Commonwealth of Independent States	8.3
6	Asia	6.5
7	Africa	5.6
8	Middle East	3.1

Source: Compiled from *Oil and Gas Global Salary Guide* (Hays Recruiting Experts in Oil and Gas, 2013, p. 18).

Table 2.2 Characteristics of employment in the oil and gas sector in 2013

Type of oil company	Permanent	Permanent/ Part-time	Contracted/ Direct	Contracted through agency
Global super major	52.6	1.5	14.2	31.7
Percentage change 2012–2013	−6.9	−0.6	2.4	5.1
Operators	59.5	1.4	14.9	24.2
Percentage change 2012–2013	−4.7	−1.3	0.1	5.9
EPCM	53.1	1.6	24.6	20.7
Percentage change 2012–2013	−3.1	0.0	1.4	1.6
Equipment manufacturer and supplier	80.7	2.0	10.3	7.0
Percentage change 2012–2013	1.8	−1.6	−0.5	0.3
Oil field services	60.9	3.5	20.2	15.4
Percentage change 2012–2013	0.5	−0.9	0.4	0.0
Consultancy	42.9	3.3	26.4	26.4
Percentage change 2012–2013	−5.3	−0.8	1.3	4.8
Contractors	47.1	2.5	26.4	24.0
Percentage change 2012–2013	−5.3	−0.8	1.3	4.8

Source: Compiled from *Oil and Gas Global Salary Guide* (Hays Recruiting Experts in Oil and Gas, 2013).

some of the lowest (Hays Recruiting Experts in Oil and Gas, 2013). The overall effect of low female participation is that women's empowerment, socially and politically, remains stunted. In turn, the livelihoods of women in oil communities seem to be worsening over time (Adusah-Karikari, 2014), which then can create the conditions for further marginalisation of women at the household level and in the wider society.

Across the board, employment in the oil and gas industry is becoming more and more casualised, as shown in Table 2.2.

Such a global picture is useful, but it hides important local experiences of oil, construction, and economic development. Detailed fieldwork in Ghana by Owusuaa (2012), Darkwah (2013), and Obeng-Odoom (2014) shows that the construction sector has generated a vibrant workforce but most are temporary, so while in the short to medium term, the construction industry generates construction related jobs, in the long run much of the construction-related employment vanishes or declines as the industry enters a phase during which construction related activities are evident but not prominent. But even in the short to medium term, the jobs created are much fewer in practice than are expected.

Additionally, most workers in oil and gas construction complain of overwork. In one study, 320 workers and stakeholders in oil and gas construction projects were interviewed about fatigue in the job. All of them ranked fatigue as the most common cause of accident at the workplace, although there were many other problems too, such as inadequate communication and defective equipment (Chan,

2011). Such findings only complement broader historically documented work-related stress in the oil and gas sector. As shown by Sutherland and Flin (1989, p. 271), oil and gas workers, especially those at sea, typically worry about:

> lack of paid holidays, rate of pay, pay differentials, lack of job security, last minute changes in crew relief arrangements, working with inadequately trained people, unpleasant working conditions due to noise, lack of promotional opportunity, not getting co-operation at work, working 28 days on/off, not being used to my full potential, delay in crew change due to severe weather conditions.

Better off workers who establish their own construction firms face distinct challenges. One of these challenges, especially in the global south, is that indigenous contractors experience payment delays for work done (Laryea, 2010). In such regions, the authorities tend to support foreign-owned companies with the justification that such expatriate-dominated companies are more efficient and provide better services. Often, however, these claims are cover up for a neocolonial mind-set (Obeng-Odoom, 2014b). This preference for "foreign-made or constituted goods" has led to many white elephants. In Ghana, the most recent is the "STX deal" in which the 90% Korean-owned STX Constructions Company Ltd promised but failed to deliver on the construction of 200,000 housing units (Ford, 2011).

This analysis shows that the oil, construction, economic development nexus is contingent not axiomatic. Oil production may trigger construction and economic growth, but not economic development. It is not enough to simply argue that more oil revenues should be invested in construction as construction economists do. "Business as usual" does not work for all RE relations. Rather, it tends to benefit the very rich and wealthy often absorbed in leisure activities. Even worse, this leisure is attained at the peril of the majority "99%", including women – particularly vulnerable women. So, alternative approaches to "oil construction" must be considered.

4. What can be done?

Alternatives to "oil construction" are of three types, namely those that emphasise the market, those that emphasize the state, and those that emphasize the community, as shown in Figure 2.3.

In mainstream economic analysis, policies that emphasize the market (Figure 2.3a) are the most popular. A prominent advocate is Collier (2008), who has consistently argued for the simultaneous extension of market forces and diminution of state and community institutions. From this perspective, the oil sector must be run along market principles that enhance open competition, give clear rules to trade, and provide stronger property rights for oil transnational corporations. State involvement must systematically decline over time, particularly because of the inefficiencies and corruption that tend to confront the state, especially in Africa (Collier, 2008). In this approach the community must be engaged but not put in charge. Indeed, the engagement must be in such a way that the

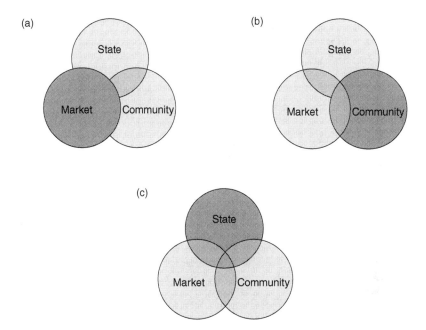

Figure 2.3 Policy frameworks for managing oil resource

Source: Adapted from Stilwell (2000).

community will be sucked into the market economy in a process of proletariani-sation. Such a perspective is rooted in a long-standing view that common land which is left in the hands of tribal, ethnic, traditional, or indigenous society is prone to overuse and abuse – a process better known as "tragedy of the com-mons", the shorthand for how Hardin (1968) understood common property – itself a reflection of the pre-nineteenth-century natural rights school of property rights (for a discussion, see Schlatter, 1951). Contemporary mainstream economists, especially new institutional economists such as Douglas North, Ahmen Alchian, and Harold Demsetz have tended to support this view in their various writings and advocacies (Anderson and Libecap, 2014).

For these economists, empowered by the state – for example, through the development of pro-market policies – the market institution can deal with prob-lems of crises, pollution, distribution, and efficiency. In particular, they claim that such public–private partnerships (PPPs) will benefit the public. While such claims have been embraced by most governments around the world, the preponderance of the evidence reviewed by Stevenson (2014) shows that such empowered markets have worked against the public good and has often enabled state power in cahoots with corporate interest to subordinate the public good to corporate profit. Perhaps, this outcome should not be surprising given that transnational oil corporations go into PPPs to pull down barriers to profit-making, not primarily to enhance socio-economic wellbeing and ecological sustainability (Elbra, 2014).

The antithesis of this view is to emphasise community-based solutions (Figure 2.3b). This view is often held by activist economists and ecologists who suggest that it is oil itself that is the problem. Such activists advocate that oil must be left buried in the ground as its drilling inevitably causes environmental crisis and continues to fuel capitalism's onward march to environmental destruction. In West Africa, the Nigerian ecological activist Nnimmo Bassey (2012) is a strong advocate. According to him, "Africans need soil, not oil. The environment is the cradle in which Africans are nurtured. Crude oil extraction has effectively uprooted the people from the soil. It has polluted their waters and poisoned their air" (p.121). Friends of the Earth Nigeria, the organisation which Bassey leads, recognises that stopping the production of oil could throw oil dependent countries into a crisis and hence offers a carefully developed alternative.

The model is based on a three-step logic. First, calculate how much oil revenue can be obtained per capita. Next, ask the citizens to pay this amount in taxes, so that the state will get the same amount of revenue anyway. Then, as not everyone can pay these taxes, share the remaining "unpaid taxes" among those who can shoulder more. Rich civil society groups can also support the initiative by buying oil under the soil – without actually receiving them, a quasi-Ecuadorian strategy. Leaving oil under the soil has many benefits. According to Bassey (2012, pp. 127–129), it is a sure bet against flaring, pollution, and extraction-related climate change. It will put an end to the displacement of local communities, nip corruption in the bud, put an end to violent conflicts, and restore a clean environment. This view has not had much traction beyond activist circles, although its insights tap into tried and tested indigenous knowledge.

The view that oil can and should be used for development is illustrated in Figure 2.3c in the three Venn diagrams. The African Union takes that view, as does the African Development Bank (African Development Bank et al., 2012), but it is the African Progress Panel that makes the boldest statement. According to the Panel:

> Effectively harnessed and well managed, Africa's resource wealth could lift millions of people out of poverty over the next decade. It could build the health, education and social protection systems that empower people to change their lives and reduce vulnerability. It could generate jobs for Africa's youth and markets for smallholder farmers. And it could put the region on a pathway towards dynamic and inclusive growth.
>
> (African Progress Panel, 2012, p. 11)

Contrary to popular perceptions that regard this standpoint as naïve, it has strong theoretical and even empirical support. The staples thesis developed by Canadian radical political economists posits a strong theoretical connection between resource abundance and development. The successful transformation of the Canadian economy is attributable to mineral dependence (Mills and Sweeney, 2013). Recent advances in geography (Arias et al., 2014), anthropology (Richardson and Weszkalnys, 2014), sociology (Davidson and Dunlap, 2012), and applied local economics (Robbins, 2012) all support the view that it is not naïve to expect that Africa can use its oil resource for prosperity without disease, poverty, and ecological pillage. Geographers use the notion of clustering to show how oil extraction can unleash localisation and urbanisation economies for entire cities and regions. Anthropologists have concretely documented many cases of mining transforming settlements positively, and so have modern sociologists, who take a more critical view of predictions of social disruption made by their forebears, while specialists in local economies stress positive forward and backward linkages between the extractive industries and local economies.

On the other hand, a strategy of investing oil revenue in construction is plausible because investing excess oil rent in other sectors of the economy that are more durable can ensure continued consumption for much longer. This strategy is also called "the Hartwick rule" (Bazilian et al., 2013) and follows Hartwick's (1977, p. 974) seminal work that showed that "the investment of current exhaustible resource returns in reproducible capital implies per capita consumption constant". It shows that resource rents invested in energy generation and other activities can help drive industrialisation, labour-intensive manufacturing, and employment. In all these, construction is key because it meets the "linkage criterion", which Hirschman (1958, p. 96) noted is "priority to investment in industries with strong linkage effects". The whole idea of linkages, in Hirschman's (1958, p. 96) words, is "the idea of favouring industries with strong backward and forward linkages".

Indeed, even if the alternative strategy is investing the rents in the development of refineries so that not all the oil will be sent overseas but some can be turned into gas that will, in turn, power local industries (Bazilian et al., 2013), construction is key. So, a construction–oil nexus can be one route through which resource abundance can be put to socially inclusive uses.

The challenge, then, is to be able to get the state to make three important changes to how it approaches the oil–construction–economic development nexus. The first is to re-orient the state from the so-called "image-driven approach to oil urbanisation" in which oil resources are splashed on fancy RE projects to prove or give the image of modernity or progress (Al-Nakib, 2013, p. 25). Such images serve to excite markets in more than one way, as they are transmitted across the wire or through what Damluji (2013) calls "cinematic spaces". Or, as El-Gamal and Jaffe (2010, p. 144) note with reference to the Gulf region generally, "Construction booms are the mainstay of rent seeking for contractors, cement and steel merchants and various municipal and national administrators."

The second challenge is for the state to change from "business as usual": simply investing in construction because it has been established that there are backward and forward linkages that connect oil-induced construction to the broader society, and assuming that this will naturally work for the good of society. Such an approach can be counter-productive. Indeed, in the Gulf region, it has been noted that "residential and office-building construction grossly exceeded any reasonable expectations of population and economic growth, thus, building up a real estate bubble that burst early in some countries . . . and later in others" (El-Gamal and Jaffe, 2010, pp. 114–115). So, not only does such an approach not undera-chieve, it overshoots and hence becomes the subject of social opprobrium.

The third and real challenge is how to shift state thinking in such a way that it can utilise oil resources to promote efficient, equitable, and ecologically sustain-able outcomes. One useful strategy is to invest socially and ecologically. Socially, public housing that avoids the rent problem can be constructed with the additional advantage of employing more people. To stimulate indigenous input industries, the state can utilise mostly locally made construction materials. This will have the additional advantage of strengthening the backward linkages of the oil industries. To complement this strategy, public schools can be built to support the empower-ment of labour in the workforce and prepare the local workforce to reclaim the commanding heights of the oil sector, while taxes can be levied on heating RE markets to dissuade speculation. Fiscal policies and direct government regulations can be utilised to encourage the recruitment and promotion of women. Revenue from oil can be used in ways that support the empowerment of women and the deconstruction of discourses about strict gender roles. Public education can help to sensitise and to pressure employers to recruit more women into secure and better paid positions. An elaborate analysis of these possibilities has been offered elsewhere (Obeng-Odoom, 2014a). There is also some evidence that in certain oil economies people take to education to learn more about oil and to empower themselves to take positions in the sector (Panford, 2014). So, the state can use this opportunity to train, educate, and empower locals to also go into industries that can use oil as the raw material for the production of soap, for example, as a strategy to enhance the forward linkages of the oil sector.

5. Conclusion

This chapter has offered a systematic analysis of the oil–construction–economic nexus using the ideas of staples, linkages, and rent. It has shown how exploration and production of oil upstream, midstream, and downstream generate forward and backward linkages to construction and then to the rest of the economy. More than only demonstrating linkages, the chapter has problematized the triumphalist view about oil, construction, and economic growth not only because such a view is too simple but also because such a stance deprives one of understanding how diverse RE relations experience oil. Indeed, it romanticizes the idea of construction for construction's sake, which, as the chapter shows, has been beneficial to economy and society but injurious to economy, ecology, and society.

By establishing the tripartite relationship between oil, construction, and eco-nomic development (as opposed to growth), this chapter has extended the litera-ture beyond the usual theme of determining whether there is a strong relationship between construction and economic growth. In turn, this chapter has put forward an alternative approach to studying the ramifications of oil for the construction sector.

This emphasis has important policy implications. It stresses the need for policy makers to take into account linkages and be prepared to use the visible arm of the state to support linkages in a way that will lead to socially, economically, and ecologically sustainable and inclusive societies through the oil, construction, and economic development corridor. In this sense, the tripartite nexus demonstrated in this chapter should not only be used for descriptive but also for prescriptive analysis, such as finding ways of directing the state which focus on oil, the staple, rent, and the price of land, while also directing the state to create linkages or a chain reaction for one key economic vision about starting and sustaining the iterative process of economic development.

Acknowledgements

Many thanks to Dr Augustina Adusah-Karikari of Birmingham University, UK, and GIMPA, Ghana, for helpful comments on an earlier draft of the paper, and to Dr Raymond Talinbe Abdulai for editorial support.

References

Ackah-Baidoo, A. (2013) Enclave development and "offshore corporate social responsibility": Implications for oil-rich sub-Saharan Africa, *Resources Policy*, 37(2), pp. 152–159.

Adusah-Karikari, A. (2014) Black gold in Ghana: Changing livelihoods for women in communities affected by oil production, *Extractive Industries and Society* [online]. Available at: http://dx.doi.org/10.1016/j.exis [Accessed 6 October 2014].

Africa Progress Panel (2013) *African Progress Report 2013: Equity in Extractives: Stewarding Africa's Natural Resources for All*, African Progress Panel, Geneva.

African Development Bank, Development Centre of the Organisation for Economic Co-Operation and Development, United Nations Development Programme, United Nations Economic Commission for Africa (2012) *African Economic Outlook 2012: Promoting Youth Employment*, Paris: OECD Publishing.

Alagidede, P. (2012) "The economy and the construction sector in West Africa", in *Construction in West Africa*, Accra: EPP Book Services Ltd.

Alissa, R. (2013) The oil town of Ahmadi since 1946: From colonial town to nostalgic city, *Comparative Studies of South Asia, Africa and the Middle East*, 33(1), pp. 41–58.

Al-Nakib, F. (2013) Kuwait's modern spectacle: Oil wealth and the making of a new capital city, 1950–90, *Comparative Studies of South Asia, Africa and the Middle East*, 33(1), pp. 7–25.

Anaman, K. A. and Osei-Amponsah, C. (2007) Analysis of the causality links between the growth of the construction industry and the growth of the macro-economy in Ghana, *Construction Management and Economics*, 25(9), pp. 951–961.

Anderson, T. L. and Libecap, G. D. (2014) *Environmental Markets: A Property Rights Approach*, Cambridge: Cambridge University Press.

Arias, M., Atienza, M. and Cademartori, J. (2014) Large mining enterprises and regional development in Chile: Between the enclave and cluster, *Journal of Economic Geography*, 14(1), pp. 73–95.

Arndt, W. (1989) *Economic Development: The History of an Idea*, London and Chicago: The University of Chicago Press.

Bassey, N. (2012) *To Cook a Continent: Destructive Extraction and the Climate Crisis in Africa*, Cape Town: Pambazuka Press.

Bazilian, M., Onyeji, I., Aqrawi P-K., Sovacool, B. K., Ofori E., Kammen, D. M. and de Graaf, T. V. (2013) Oil, energy poverty and resource dependence in West Africa, *Journal of Energy & Natural Resources Law*, 31(1), pp. 33–53.

Bon, R., Birgonul, T. and Ozdogan, I. (1999) An input–output analysis of the Turkish construction sector, 1973–1990: A note, *Construction Management and Economics*, 17(5), pp. 543–551.

Carmody, P. (2011) *The New Scramble for Africa*, London: Polity Press.

Chan, M. (2011) Fatigue: The most critical risk in oil and gas construction, *Construction Management and Economics*, 29(4), pp. 341–353.

Collier, P. (2008) *The Bottom Billion: Why the Poorest Countries Are Failing and What Can Be Done about It*, New York: Oxford University Press.

Corkin, L. (2011a) Uneasy allies: China's evolving relations with Angola, *Journal of Contemporary African Studies*, 29(2), pp. 169–180.

Corkin, L. (2011b) Chinese construction companies in Angola: A local linkages perspective. Discussion paper no. 2: Making the Most of Commodities Programme (MMCP), pp. 1–80 [online]. Available at: http://commodities.open.ac.uk/8025750500453f86/%28httpassets%29/9407b19720d08f308025787e0039e3fc/$file/chinese%20construction%20companies%20in%20angola.pdf [Accessed 13 December 2014].

Corkin, L. (2011c) Redefining foreign policy impulses toward Africa: The roles of the MFA, the MOFCOM and China EximBank, *Journal of Current Chinese Affairs*, 40(4), 61–90.

Corkin, L. (2012) Chinese construction companies in Angola: A local linkages perspective, *Resources Policy*, 37(4), pp. 475–483.

Damluji, M. (2013) The oil city in focus: The cinematic spaces of Abadan in the Anglo-Iranian Oil Company's Persian story, *Comparative Studies of South Asia, Africa and the Middle East*, 33(1) , pp. 75–88.

Darkwah, K. A. (2013) Keeping hope alive: An analysis of training opportunities for Ghanaian youth in the emerging oil and gas industry in Ghana, *International Development Planning Review*, 35 (2), pp. 119–134.

Davidson, D. J. and Dunlap, R. E. (2012) Introduction: Building on the legacy contributions of William R. Freudenburg in environmental studies and sociology, *Journal of Environmental Studies and Sciences*, 21, pp. 1–6.

Elbra, A. (2014) Gold mining in sub-Saharan Africa: Towards private sector governance, *The Extractive Industries and Society*, 1 (2), pp. 216–224.

El-Gamal, M. A. and Jaffe, A. M. (2010) *Oil, Dollars, Debt, and Crises: The Global Curse of Black Gold*, Cambridge: Cambridge University Press.

Fioramonti, L. (2013) *Gross Domestic Problem: The Politics behind the World's Most Powerful Number*. London: Zed Books.

Ford, N. (2011) Building Ghana one property at a time, *African Business*, November, pp. 72–76.

Frankel, E. G. (2007) *Oil and Security: A World Beyond Petroleum*, Dordrecht: Springer.

George, H. ([1879] 2006) *Progress and Poverty*, New York: Robert Schalkenbach Foundation.

Guyer, J. I. (2002) The Chad–Cameroon petroleum and pipeline development project, *African Affairs*, 101(402), pp. 109–115.

Hardin, G. (1968) The tragedy of the commons, *Science*, 162(3859), pp. 1243–1248.

Hartwick, J. M. (1977), Intergenerational equity and the investing of rents from exhaustible resources, *The American Economic Review*, 67(5), pp. 972–974.

Hays Recruiting Experts in Oil and Gas (2013) *Oil and Gas Global Salary Guide*, Manchester: Hays Plc.

Hilson A. E. (2014) Resource enclavity and corporate social responsibility in sub-Saharan Africa: The case of oil production in Ghana. PhD thesis in Management, Aston University, UK.

Hirschman, A. O. (1958) *The Strategy of Economic Development*, New Haven and London: Yale University Press.

Hirschman, A. (1984) Albert O. Hirschman: A dissenter's confession: 'The strategy of economic development' revisited, in: G. M. Meier and D. Seers (Eds), *Pioneers in Development*, Oxford: Oxford University Press, pp. 87–111.

Innis, H. (1930) *The Fur Trade in Canada: An Introduction to Canadian Economic History*, New Haven: Yale University Press.

Kofoworola, O. F. and Gheewala, S. (2008) An input–output analysis of Thailand's construction sector, *Construction Management and Economics*, 26(11), pp. 1227–1240.

Laryea, S. A. (2010) Challenges and opportunities facing contractors in Ghana, Paper presented at West Africa Built Environment Research (WABER) conference, 27–28 July, Accra, Ghana.

Lewis, T. M. (2004) The construction industry in the economy of Trinidad & Tobago. *Construction Management and Economics*, 22(5), pp. 541–549.

Mason, G. (2009) *After the Crash: Designing a Depression-Free Economy*, Chichester: John Wiley and Sons Ltd.

Mills, S. and Sweeney, B. (2013) Employment relations in the Neostaples resource economy: Impact benefit agreements and Aboriginal governance in Canada's nickel mining industry, *Studies in Political Economy*, 91, pp. 7–33.

Obeng-Odoom, F. (2013) A critique of Africa's so-called failed development trajectory. *African Review of Economics and Finance*, 4(2), pp. 151–175.

Obeng-Odoom, F. (2014a) *Oiling the Urban Economy: Land, Labour, Capital, and the State in Sekondi-Takoradi, Ghana*, London: Routledge.

Obeng-Odoom, F. (2014b) Sustainable urban development in Africa? The case of urban transport in Sekondi-Takoradi, Ghana, *America Behavioral Scientist*, DOI:10.1177/000 2764214550305.

Obeng-Odoom, F. (2014c) Africa: On the rise, but to where? *Forum for Social Economics*, DOI:10.1080/07360932.2014.955040.

Olatunji, O. A. (2010) The impact of oil price regimes on construction cost in Nigeria, *Construction Management and Economics*, 28(7), pp. 747–759.

Owusuaa, D. (2012) Gender and informality in the construction industry in Ghana's oil city Takoradi, unpublished Master's thesis, Department of Geography, University of Bergen, Bergen.

Panford, K. (2014) An exploratory survey of petroleum skills and training in Ghana, Africa Today, 60(3), pp. 56–80.

Ramsaran, R. and Hosein, R. (2006) Growth, employment and the construction industry in Trinidad and Tobago, *Construction Management and Economics*, 24(5), pp. 465–474.

Richardson, T. and Weszkalnys, G. (2014) Introduction: Resource materialities, *Anthropological Quarterly*, 87(1), pp. 5–30.

Robbins, G. (2012) Mining FDI and urban economies in sub-Saharan Africa: Exploring the possible linkages, *Local Economy*, 28(2), pp. 158–169.

Ross, M. L. (2008) Oil, Islam, and women, *American Political Science Review*, 102(1), pp. 107–123.

Saka, N. and Lowe, J. G. (2010) The impact of the petroleum sector on the output of the Nigerian construction sector, *Construction Management and Economics*, 28(12), pp. 1301–1312.

Sarbu, B. (2014) *Ownership and Control of Oil*, London and New York: Routledge.

Schlatter, R. (1951) *Private Property: The History of an Idea*, New York: Russell and Russell.

Seers, D. (1970) The meaning of development, *Revista Brasileira de Economia*, 24(3), 29–50.

Stevenson, M. (2014) Public–private partnering in natural resource extraction, *Global Environmental Politics*, 3, pp. 139–145.

Stilwell, F. (2000) Ways of seeing: Competing perspectives on urban problems and policies, in: P. Troy (ed.), *Equity, Environment, Efficiency, Ethics and Economics in Urban Australia*, Victoria: Melbourne University Press, pp. 13–37.

Sutherland, K. M. and Flin, R. H. (1989) Stress at sea: A review of working conditions in the offshore oil and fishing industries, *Work and Stress: An International Journal of Work, Health and Organisations*, 3(3), pp. 269–285.

Watkins, M. H. (1963) A Staple theory of economic growth, *The Canadian Journal of Economics and Political Science/Revue canadienne d'Economique et de Science politique*, 29(2), pp. 141–158.

3 Construction, real estate mortgage market development and economic growth in Turkey

Işıl Erol

1. Introduction

In line with the "economic sectors" definition of the Turkish Statistical Institute (2013), it is possible to define the overall real estate (RE) sector in terms of two main components, the construction industry and other RE business activities. The other RE business sector comprises establishments primarily engaged in renting or allowing the use of their own assets by others. This sector also includes establishments engaged in RE management, selling, buying, appraisal and mortgaging (Turkish Statistical Institute, 2013).

Although the other RE business activities have been growing rapidly over the past decade, especially after the 2001 financial crisis, the construction industry has been the largest component of the RE sector. The construction industry has been considered one of the engines of economic growth in Turkey from the early 1960s onwards. Through the reorientation of economic policies starting with the 1980 stabilization program, the industry has been assigned a new role as part of the export oriented growth strategy as Turkish contractors have expanded their activities abroad, especially in the Middle East and North Africa (MENA) region (Sonmez, 1982). Turkey has been one of the region's fastest developing RE markets as a result of the economic growth and favourable demographics in the 2000s (FESSUD, 2014). The recent improvements in the Turkish economy, especially the drop in the inflation rate, together with increasing domestic and foreign demand for residential units of RE have led the government to work on a raft of regulatory changes that would facilitate the legal environment for the establishment of a RE mortgage system. These efforts to develop the mortgage system resulted in an increase in the construction of new housing units, the development in mortgage products, and a significant decline in mortgage interest rates. The Turkish Parliament ratified the Housing Finance Law in March 2007 and long-term fixed rate borrowing became, for the first time ever, an available financing option for potential homeowners in Turkey (Erol and Tirtiroglu, 2011).

This chapter aims, firstly, to investigate the relationship between the construction industry and the aggregate economy in Turkey by using the Granger causality analysis. Using quarterly data for the period from 1998 Q1 to 2013 Q4, the chapter tests whether construction investments stimulate aggregate economy or aggregate economy leads the construction activities or if there exists feedback effects between construction flows and the aggregate economic activities in Turkey. Secondly, the chapter examines the lead–lag relationship between residential RE mortgage market development and economic growth in Turkey over the same period of time.

Admittedly, several studies have investigated the direction of the causal relationship between the construction industry and economic growth in developing countries such as Sri Lanka (Ramachandra, Rotimi and Rameezdeen, 2013), Korea (Kim, 2004), Hong Kong (Tse and Gnesan, 1997; Yiu et al., 2004), Ghana (Anaman and Osei-Amponsah, 2007), Singapore (Lean, 2002), China (Hongyu et al., 2002), Barbados (Jackman, 2010) and Cape Verde (Lopes, Nunes and Balsa, 2011). However, only a limited number of studies have examined the causal relationship between the growth in mortgage markets and economic growth in developing economies, and they mainly focused on the MENA countries (see, for example, Buckley, 1996; Li, 2001; Erbas and Nothaft, 2005). Thus, this study is the first attempt to investigate the lead–lag relationship between the construction industry, mortgage market development and economic growth in Turkey, which is a rapidly growing economy with significant growth potential in its construction sector. Gaining a better understanding of the general lead–lag relationship between the construction industry and mortgage market development, on one hand, and economic growth on the other requires more evidence, especially from developing economies like Turkey with considerable growth potential in the construction sector.

The chapter is organised as follows. Section 2 discusses briefly the research methodology as well as describes the data used. Section 3 provides a summary of existing research on the relationship between RE construction and economic growth, especially in developing economies. The relevant literature on the causal relationship between the growth in mortgage markets and economic growth in MENA countries is also presented in this section. Section 4 provides an overview of the size, growth and volatility of the RE sector in conjunction with macroeconomic fluctuations in Turkey between 1998 and 2013 and then discusses housing construction and residential RE mortgage market development in Turkey. Section 5 presents the Granger causality test results and discusses empirical results. Finally, the chapter ends with concluding remarks.

2. Research methodology

This study adopts the quantitative research methodology and using data collected from publicly available sources. As earlier indicated, the main goal of this research is to determine the causal relationship between the growth of the construction industry and mortgage market development, on one hand, and economic growth in Turkey on the other. More specifically, the Granger causality test, a descriptive tool for time series data, is used to investigate the causal relationship between time series of economic growth and the development of the RE mortgage market and the construction industry.

The study uses quarterly data on real and nominal production-based GDP series, construction activity and the mortgage loan outstanding balance series between 1998 Q1 and 2013 Q4. All of the data come from the Central Bank of the Republic of Turkey (CBRT) and Turkish Statistical Institute (TSI) electronic data delivery systems. Construction activity is measured using the value of construction activities in the total gross domestic production in the country, and the market value of outstanding mortgage debt is used as a proxy for the mortgage market development.

Time series data, especially economic data, is non-stationary in the level form and need to be transformed into a stationary form in order to be used for the Granger causality test (Huang, 1995; Feige and Pearce, 1979). As a preliminary step to the analysis, the order of integration of the selected variables is determined – namely, the GDP, market value of construction activity and mortgage loan balance series. Unit root tests are widely used to detect the stationary problem and transform time series into stationary forms. The present study uses the Augmented Dickey–Fuller (ADF) test that considers situations in which the white noise error terms are correlated, and thus is accepted as an improvement over the Dickey–Fuller test.

Granger (1969) proposed a time series data based approach in order to determine causality. In the Granger sense, x is a cause of y if it is useful in forecasting y. Granger causality is consistent with the notion that the cause precedes the effects but cannot be applied to the contemporaneous values of x and y (Foresti, 2006, p. 3). In other words, this test seeks to determine whether or not the inclusion of past values of a variable x do or do not help in the prediction of present values of another variable, y. If variable y is better predicted by including past values of x than by not including them, then x is said to Granger-cause y.

The present study uses a simple Granger causality test, which uses only two variables and their lags in order to test whether the construction sector growth and mortgage market development separately Granger-causes economic growth (GDP growth) and vice versa. The following vector autoregressive model of lag order m, VAR (m), is used:

$$Y_t = \alpha_1 + \sum_{i=1}^{m}\beta_{1i}X_{t-i} + \sum_{i=1}^{m}\delta_{1i}Y_{t-i} + \varepsilon_{1t} \qquad \text{Eq. (1)}$$
$$X_t = \alpha_1 + \sum_{i=1}^{m}\delta_{2i}Y_{t-i} + \sum_{i=1}^{m}\beta_{2i}X_{t-i} + \varepsilon_{2t}$$

The choice of the optimal lag length as all inferences in the VAR is naturally based on the chosen lag order, that is, the number of lags chosen in the above equations has a significant impact on the decision to reject or accept the null hypothesis. Hence, the Schwarz Information Criterion (SIC) is employed to determine the optimal lag length.

3. Causal relationship between RE construction and the macroeconomy

Given that construction output is an integral part of national output, it is expected that an expansion of a construction activity is preceded by an increase in economic output, with the initial effect felt largely within the construction sector and only subsequently on the aggregate economy (Tse and Ganesan, 1997). According to Akintoye and Skitmore (1994), construction investment is a derived demand, which is growth dependent. However, the lead–lag relationship between national output (GDP) and construction investment is not clear in the existing literature. Some studies have suggested that the construction industry influences economic growth because of its strong linkages with other sectors of the economy. That is to say, an increase in (particularly) residential RE construction is often associated with increased employment and income for workers in the housing sector and also in related sectors that provide goods and services associated with housing (Hirschman, 1958; Bon and Pietoforte, 1990; Lean, 2001; Ewing and Wang, 2005; Hosein and Lewis, 2005).

For instance, Anaman and Osei-Amponsah (2007) analysed the causality links between the growth in the construction industry and the macroeconomic growth in Ghana using data from 1968 to 2004 and showed that the growth in the construction industry Granger-caused growth in GDP, with a three-year lag. The authors concluded that although the government aims to use the agricultural sector as the major vehicle for achieving high growth rates in the aggregate economy, the construction industry needs to be considered as one of the major drivers of economic growth in Ghana. Similarly, Lean (2002) concluded that the construction sector leads other sectors' output as well as GDP in Singapore.

On the other hand, several studies have shown that the economic expansion (GDP growth) causes growth in construction output; see, for example, Tse and Ganesan (1997), Yiu et al. (2004) for Hong Kong and Lopes et al. (2011) for Cape Verde. Green (1997) performed a series of Granger causality tests to see whether residential and non-residential RE investment in the USA Granger-causes GDP and whether GDP Granger-causes each of the two types of investments. He showed that residential RE investment Granger-causes GDP, but not vice versa. These results suggest that housing markets lead the business cycle in the US economy. A similar Granger causality test was carried out by Kim (2004) using 1970–2002 quarterly Korean data and concluded that housing is not a driver of GDP but a follower of fluctuations of the wider economy. One possible explanation of the passive role of housing investment as a follower of the macroeconomy is that the government has used residential RE investment to counter business fluctuations. Similarly, a recent study by Ramachandra et al. (2013) investigated the direction of the causal relationship between construction and the economy of Sri Lanka using the Granger causality test for the period 1990 to 2009. The findings revealed that national economic activities preceded construction activities for all indicators except construction investment. The study concluded that the national economy in Sri Lanka has been inducing growth in the construction sector and not vice versa for the studied sample period. Moreover, Hongyu et al. (2002) and Jackman (2010) investigated the causal relationship between the residential RE construction and economic growth in China and Barbados, respectively and concluded that there has been bidirectional causality between housing investment and GDP for these countries. According to Wang et al. (2000), it is reasonable to believe that the cross-country differences in the market structure of construction industries affect the dynamics of construction activities.

Studies have investigated the causal relationship between economic growth and growth in RE mortgage markets, especially in developing economies. Numerous studies have addressed the question of whether affordable mortgages can stimulate additional savings and growth. Li (2001) provided a comprehensive review of the literature and empirical results for some MENA countries and argued that when borrowing constraints are very restrictive, most low and middle-income households may defer or even forgo the purchase of a home and attain a certain utility-maximizing allocation of consumption and saving where saving is lower than it would be if households had access to mortgage financing. Buckley (1996) concluded that for the selected MENA countries studied, the mortgage market development is likely to have a favourable impact on savings mobilisation as the return on housing rewards saving, ensuring positive returns and housing provides the most secure collateral against market fluctuations as well as yields a positive rate of return, at least, in the long run.

Erbas and Nothaft (2005) claim that making affordable mortgage loans available to a large cross-section of the population can result in economic growth. Widespread availability of mortgages has a beneficial impact on the quality of housing, infrastructure and urbanization, and hence improves living standards and alleviates poverty. Moreover, by increasing home ownership, mortgage availability may help improve the quality of citizenship and community life. The widespread availability of affordable mortgages may also enhance savings, promote financial market development and stimulate investment in the housing sector (Erbas and Nothaft, 2005, p. 213).

Many studies have recognized the potential for increasing employment through mortgage markets to finance housing construction in developing countries. However, on average, housing construction activities in many developing countries have tended to be capital-intensive and dominated by imported materials. Erbas and Nothaft (2005) argue that in order to develop mortgage markets and encourage construction sector growth in developing countries, it is important to use labour-intensive construction methods and increase the share of domestic materials in the construction sector. In addition, subsidization should be targeted at lower-income housing, which has a higher labour intensity and lower import component.

4. RE construction and the Turkish economy

The construction industry as a sub-sector of the RE sector has been considered an engine of economic growth in Turkey ever since the adoption of import substitution industrialization as a development strategy from the 1960s onwards. With the reorientation of economic policies that started with the 1980 stabilization program, the industry has been assigned a new role as part of an export oriented growth strategy, and Turkish contractors have expanded their activities abroad, especially in the MENA region (Sonmez, 1982). Meanwhile, the construction industry has been prominent within the domestic economy with the establishment of the Housing Development Administration (HDA) in 1984. The HDA has been instrumental in undertaking numerous projects of mass housing and landscaping from the mid-1990s onwards and has gained significant momentum since 2002. For the period from 2003 to 2012, housing starts by the HDA reached 562,000 dwelling units, accounting for nearly 11% of all national starts during the same period of time (Turel and Koc, 2014). Since 2005, Turkey has been one of the region's fastest developing RE markets as a result of the economic growth and favourable demographics in the 2000s. In May 2012, the Law on the Transformation of Areas under Natural Disaster Risk was enacted, which authorized public sector involvement in urban transformation process with an initial estimation of 6.5 million dwelling units with natural disaster risk (FESSUD, 2014).

The construction sector has had a share of 5% to 6.5% of Gross Domestic Product (GDP) over the past sixteen years from 1998 to 2013 (Figure 3.1). Figure 3.2 displays the growth rate of the construction industry in conjunction with the fluctuations in the real GDP growth rate between 1999 and 2013. The growth rate in real GDP has been significantly negative in the past three financial crises of 1999, 2001 and 2009. It is clearly seen that the 1999 and 2001 financial crises in Turkey and the global financial crisis in 2008–2009 directly resulted in negative growth rates in the sector. That is, growth of the sector has declined to –3.1% in 1999, –17.4% in 2001 and to –16.1% in 2009. Hence, the data has revealed the fragility and volatility of the construction industry during the years of economic crises. One may conclude that during periods of rapid economic expansion, construction output usually grows faster than the output of other sectors, but during periods of stagnation, the industry is the first to suffer; see, for example, Ramachandra et al. (2013) for a similar discussion.

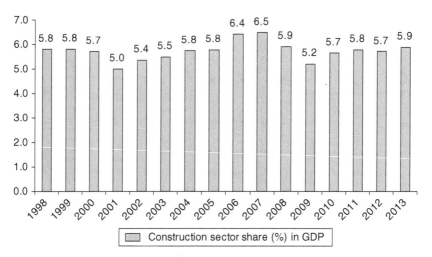

Figure 3.1 Share of construction industry in GDP between 1998 and 2013

Source: Turkish Statistical Institute (2013).

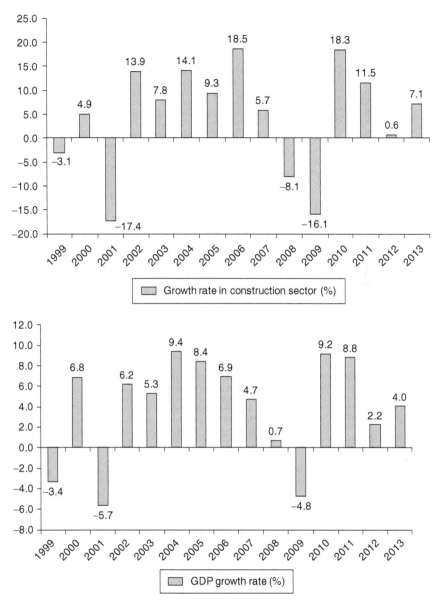

Figure 3.2 Construction industry growth and GDP growth in Turkey

Source: Turkish Statistical Institute (2013).

The share of the market value of the overall RE sector in the Turkish economy (GDP at 1998 prices) has increased from 8.3% in 1998 to 9.8% in 2013. As at the end of 2013, while the construction sector-to-GDP ratio was 5.8%, the RE renting and other business activities-to-GDP ratio was 4%.

The newly amended Reciprocity Law substantially eased foreign investment restrictions in Turkey in 2012 and accordingly, European and especially Gulf-based RE investors have turned their attention to the Turkish RE markets, and particularly to the commercial RE sector.[1] According to Jones Lang LaSalle's 2014 report, retail development in Turkey is seen as a priority market for international retailers. According to the report, as at the end of 2013, the shopping centre gross leasable area in Turkey had reached 9.5 million square metres in 352 centres, an increase of 1.15 million square metres compared with the end of 2012. Istanbul, the most developed and active office market in Turkey, operates as a regional business hub to companies that serve the MENA and Commonwealth of Independent States (CIS) regions. Lately, Turkey has been recognized as one of the developing economies in which the e-commerce sector is triggering the transformation of retail logistics. The expansion of physical retail space continues to build along with the growth in e-commerce. Furthermore, RE-based capital market instruments, including mortgages, RE Investment Trusts' shares, RE certificates and lease certificates, have been growing rapidly along with the Turkey's financial sector development.

According to a report titled "Emerging Trends in Real Estate Europe" (prepared jointly by the Urban Land Institute and PriceWaterhouseCoopers), Istanbul was ranked as the most attractive investment market in Europe. In this report, it is noted that in the "Existing Property Performance", "New Property Acquisitions", and "Development Prospects" categories, Istanbul was followed by Munich, Warsaw, Berlin and Stockholm. Moreover, according to a survey conducted by the Association of Foreign Investors in RE (AFIRE, 2012), Turkey ranks as the third most attractive RE investment destination among emerging countries in 2012 (ULI and PWC, 2012).

Figure 3.3 displays both RE sector growth and economic growth between 1999 and 2013. Noticeably, the RE sector has been significantly affected by the financial crises in the years 1999, 2001, 2008 and 2009. The decline in the RE sector growth has been much faster and bigger than that of growth in real GDP. Except for the crisis periods, the growth rate in the RE sector has been faster than economic growth. During the years 2006, 2010 and 2011, the Turkish RE sector grew significantly at 16.6%, 13.9% and 10.7%, respectively. The data set clearly shows that the economy drives the RE sector in Turkey. In an attempt to have a better understanding of this lead–lag relationship, a Granger causality test is analysed below in Section 5.

According to the Turkish Statistical Institute (2013), the distribution of new buildings and additions (construction permits) by type of usage shows that between 1985 and 2013, 85% to 92% of the construction permits belonged to residential RE usage. The following sections provide a brief discussion of the residential RE or housing market in Turkey with a focus on housing production in Section 4.1 and the residential RE mortgage market development in Section 4.2.

4.1 Housing production in Turkey between 1985 and 2013

The history of housing policies in Turkey reveals that the housing markets operate with little regulation and under highly competitive conditions. One outcome

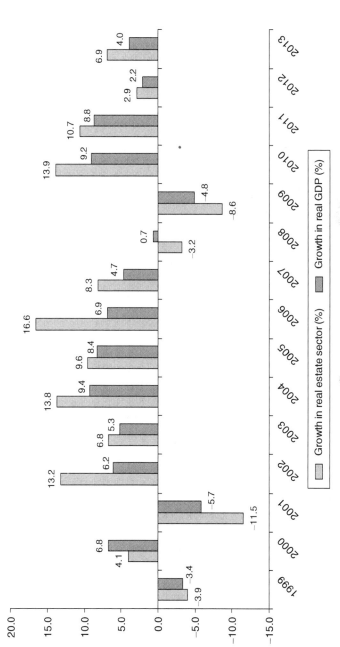

Figure 3.3 RE sector growth and GDP growth in Turkey

Source: Turkish Statistical Institute (2013).

of the less-regulated housing markets is a broad variation in the housing starts in the provinces of Turkey (Turel and Koc, 2014): housing production is increasing faster than the number of households in most provinces and cities while in some provinces it is much less (Turel and Koc, 2014).

Turkey is one of the biggest countries in Europe in terms of housing production. Over the last decade, in most years, annual housing starts have been between 500,000–600,000 dwelling units (Turel and Koc, 2014). Culturally, home ownership is the most embraced means of investment, and socially the Turkish households mostly prefer to be homeowners rather than tenants. Figure 3.4 shows the construction permits or housing starts measured by dwelling units between 1985 and 2013. Annual housing starts began to rise in 1986, exceeded 500,000 dwelling units in 1993 and remained over that level during the following two years. The fall that began in 1996 continued until 2002, when housing starts were as low as 161,920 dwelling units. This fall can be related to the effects of a very destructive earthquake that hit the north-western regions of the country in 1999, new building regulations that were introduced following the earthquake and a series of economic crisis during the 1999–2002 period. Recovery began in 2003; housing starts went over 500,000 again in 2005, remained over 500,000 during the global financial crisis of 2008–2009 and reached up to 907,451 in 2010.

The high increase in the number of housing starts in 2010 was due to an increase in the geographical coverage of building regulations from 19 to all 81 provinces. As new regulations involve additional cost in the preparation of housing projects, many house-builders aim to avoid such costs by getting construction permits for the next several years' projects before the regulations become applicable in their provinces (Turel and Koc, 2014).

Figure 3.5 shows construction permits in respect of the total number of buildings between 1985 and 2013. An examination of the housing production in terms of producer groups reveals that there are three main actors. These are the public sector, the private sector and the construction cooperatives. Total housing production increased from 71,844 buildings in 1985 to 147,033 buildings in 1993 and then declined to 36,379 buildings in 2002. As seen from the figure, housing starts began to increase substantially from 2004 onwards, mainly due to the significant increases in starts by the private sector.

Figure 3.6 shows the percentage share of producers when housing starts are measured in terms of the number of buildings. During the period from 1985 to 1998, the share of cooperative housing in the total number of buildings with construction permits was around 20% to 30%, which declined to 6% in 2005 and then to 3% in 2013. On the other hand, the share of private sector construction permits increased from 70% in the late 1980s to 90–92% after 2003. Figure 3.7 displays the percentage share of producers when housing starts are measured in terms of the market value of buildings. Once again, the private sector has the leading part in the market value of housing starts. After 2000, the share of private sector housing in the total number of buildings with construction permits was around 76% to 85%.

According to Turel (2012) private sector residential building has been dominated by small-capital builders that produce mostly apartments on single parcels.

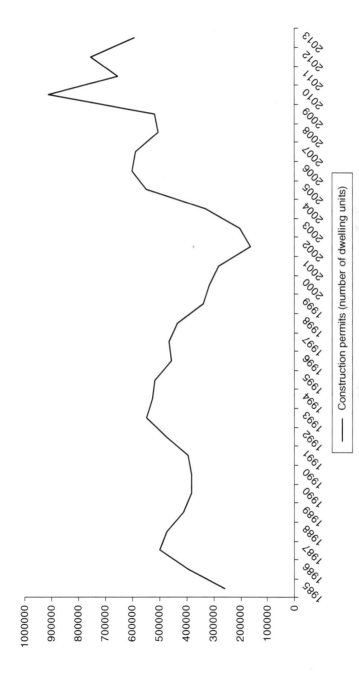

Figure 3.4 Construction permits by number of dwelling units

Source: Central Bank of the Republic of Turkey (2013).

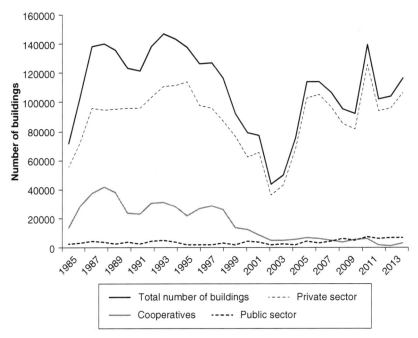

Figure 3.5 Construction permits by number of buildings and type of investor
Source: Turkish Statistical Institute (2013).

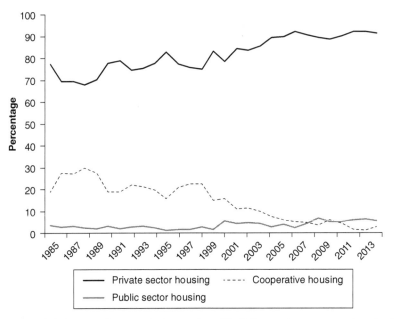

Figure 3.6 Construction permits by number of buildings and percentage share by the type
of investor

Source: Turkish Statistical Institute (2013).

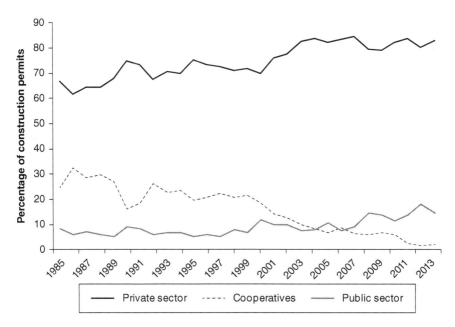

Figure 3.7 Construction permits by market value and percentage share by the type of
         investor

Source: Turkish Statistical Institute (2013).

In recent years, moderate-to-large capital domestic builders and even global construction companies have been increasing their share in housing supply as they
produce housing on large tracks of land with many on-site amenities, including
parking, sport facilities and the facilities to employ private guards. Many of these
units of RE are in the form of gated communities and supplied to the upper income
groups (Turel, 2012). Consequently, the rise in the share of the private sector has
been at the expense of cooperatives, which together with the public sector, are
regarded as non-profit producers (Turel, 2012). Construction cooperatives are also
in competition with the public sector, which is the other non-profit housing producer in Turkey. HDA had increased its involvement in housing production by the
year 2003. The public sector had a higher market share in terms of housing starts
than cooperatives between 2007 and 2013. HDA provides housing by developing
publicly owned land and for this reason, cooperatives became unable to buy land
from public institutions during this period (Turel, 2012).

4.2 Residential RE mortgage market development in Turkey

The sources of housing finance are both institutional and non-institutional in
Turkey. While cooperative housing and equity sharing agreements are the non-
institutional alternatives, project debt financing and housing loans or mortgages
are the main institutional financing instruments. Housing cooperatives, the legal

entities established to provide their members with flats or houses, were tradition-ally one of the most favoured methods of acquiring RE among middle-income households. Equity sharing agreements have also been used widely in Turkey. In this arrangement, the land owner offers his land to the contractor in return for a portion of the equity interest, that is, for half of the apartments that will be built. In project debt financing, the project developer applies to a financial institution, which agrees to provide a secure loan of appropriate maturity and terms. The most rapidly growing source of housing finance in recent years has been RE mortgages.

In spite of the historically high demand for RE, a well-organized and deep enough mortgage market did not exist in Turkey until the early 2000s. The absence of an efficient mortgage market was mainly due to a long-running process of persistently high inflation, the inability of the banks to fund mortgages from their deposit base and the lack of standardization within the title and appraisal systems (Erol and Cetinkaya, 2009). The measures, taken after the financial cri-sis of 2000–2001 have been effective in suppressing inflation, building investor confidence and attracting substantial amounts of foreign investments. The recent improvements in the Turkish economy, especially the drop in the inflation rate, have led the government to work on a draft of regulatory changes that would facilitate the legal environment for the establishment of a RE mortgage system. These efforts towards the development of a mortgage system have attracted the construction sector and related financial sectors. The result was an increase in the construction of new housing units, the development of mortgage products and a significant decline in mortgage interest rates. Eventually, in March 2007, the Turkish Parliament ratified the Housing Finance Law (No. 5582), which intro-duced a mortgage system and envisaged the possibility of the eventual securitiza-tion of RE mortgages. Hence long-term fixed rate borrowing became for the first time ever an available financing option for potential homeowners in Turkey. The main target of the mortgage system is thought to be the middle-income house-holds (Erol and Tirtiroglu, 2008).

Figure 3.8 shows that the mortgage interest rate declined sharply from 40%–50% in the years 2002 and 2003 to the 9%–15% band after 2010. According to the European Mortgage Federation's *Hypostat* report (2013), the representative annual interest rates on new residential loans in Turkey declined from 48.3% in 2002 to 12.4% in 2012.

The percentage share of outstanding mortgage loan balance in total outstand-ing consumer loans is displayed in Figure 3.9. Mortgage loans represented only 9.7% of the overall consumer loan value in 2003, whereas, in September 2013, housing loans had a share of 44.1%. Between 2005 and September 2013, out-standing mortgage loan balance was 43.7% to 49% of the overall consumer loan portfolio. In the early 2000s and especially in 2005, the RE mortgage market began to grow significantly, largely as a result of a change in the investment poli-cies of commercial banks. More specifically, as the supply of high-income gov-ernment bonds dried up, commercial banks greatly increased their involvement in mortgage loans, and by 2004, with the fall in inflation and mortgage interest rates, had become the primary source of mortgage loans. According to the Banks

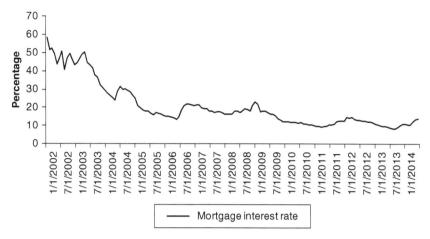

Figure 3.8 Mortgage interest rates from 1 January 2002 to 28 March 2014

Source: Central Bank of the Republic of Turkey (2014).

Association of Turkey (2013), mortgage market value jumped from 2.4 billion TL in 2004 to 12.4 billion TL in 2005 and reached up to 97.5 billion TL (USD 48 billion) in September 2013.

Unlike its European and US counterparts, mortgage lenders in Turkey are almost totally deposit banks, and investment banks do not have any role in mortgage origination. Since the mortgage market is mainly funded through a saving deposits base, the mortgage debt-to-GDP ratio is very low for Turkey: the share of mortgage loans, which had been less than 1% of GDP before 2004, had increased to 6.1% by the end of 2012. Additionally, the share of mortgage loans in the overall banking sector loans declined from 13.5% in 2007 to 10.8% as at February 2014 (Central Bank of Republic of Turkey, 2014).

Within the framework of the Housing Finance Law of March 2007, legal procedures for enforcing claims and the rules on default and foreclosure are well-defined, and principally the middle-income households are targeted for mortgage lending. The law modified the Bankruptcy and Foreclosure Law, Capital Markets Law, Consumer Protection Law, Financial Leasing Law, Mass Housing Law and some tax laws. Moreover, two bylaws (Serial III, Nos 33 and 34 by the Capital Markets Board) were put into effect for covered debt instruments (covered bonds) and securitisation instruments (asset-backed and mortgage-backed securities).

In fact, there has been limited access to mortgage financing even by middle-income households, mainly because households are constrained in their mortgage financing in two ways: an inability to afford a minimum down-payment of 25% of the house price and an inability to access capital markets for mortgage products with moderate interest rates and sufficiently long maturities. Households with some amount of accumulated wealth for down-payment can be eligible for mortgage loans. Since lenders have to bear high risks caused by maturity (or

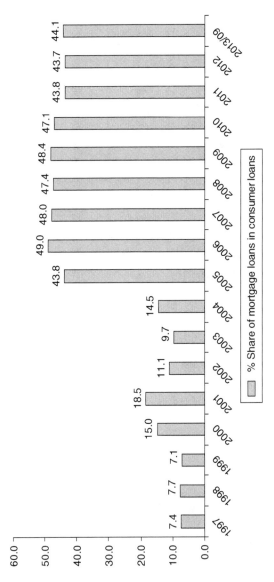

Figure 3.9 The share of housing loans in the total outstanding consumer loan balance

Source: Banks Association of Turkey (2013).

asset-liability) mismatch, banks have mainly targeted households with lower risk and high-income profiles. Thus, a wide segment of households are excluded from mortgage loans. Mortgage interest rates are still at higher levels in comparison to those in developed markets and very volatile, ranging from 21.8% in August 2006 to 8.3% in June 2013. Commonly, mortgage loans have 7 to 10 year maturities, which is remarkably short compared to the mortgages with 20 to 30 year maturities in developed markets. In addition, mortgage loans in Turkey are not tax-deductible. There are no mortgage interest rate subsidies for households who buy dwellings in which to reside themselves. However, for buy-to-let houses, the interest paid on the mortgage loan can be deducted from the part of the rent that is subject to income tax (Turel and Koc, 2014).

Along the lines of Erbas and Nothaft's (2005) study, affordable mortgages can be defined in terms of low down-payment requirements and low interest rates over a long period of mortgage maturity, to make the monthly payment a smaller portion of the household's income. Hence, it can easily be argued that mortgages are not affordable for low-income households – and even for some middle-income households in Turkey under the current circumstances.

In terms of insurance products, hazard and earthquake insurance is required by all lenders in the Turkish mortgage market. This has been a requirement since the 1999 earthquake and is provided by the Turkish Catastrophe Insurance Pool (TCIP). TCIP takes the first loss position and private insurers take the second loss position. The annual premiums due to TCIP are collected by private insurance companies from the home owners and then forwarded to TCIP. In addition, very few lenders also require a life insurance policy that would remain in effect over the term of the mortgage. Such a policy would help to cover the full repayment of the loan in the event of the borrower's death. However, mortgage default insurance products are not prevalent in Turkey. Existing sectoral studies suggest that there is no urgent need for mortgage insurance as this will increase the cost of funds for borrowers. Recently, a number of banks have started to require mortgage payment protection insurance in case of the borrowers becoming unemployed or injured. However, this product is different from the mortgage default insurance that is widely used in developed mortgage markets (Erol and Çetinkaya 2009).

In addition to the Housing Finance Law of 2007, two bylaws (Serial III, Nos 33 and 34 by the Capital Markets Board) were put into effect for the introduction of covered bonds and mortgage-backed securities. Nevertheless, there is currently no active secondary market for mortgages. In spite of these attempts to develop primary and secondary mortgage markets, especially within legal framework, some factors have restrained the growth of the Turkish mortgage market up to now. First, a number of governmental institutions, including the Capital Markets Board, the Banking Regulation and Supervision Agency, the Central Bank of the Republic of Turkey and the Under-Secretariat of Treasury, have taken different responsibilities for the development of the mortgage market. This multi-agency regulatory framework may create significant coordination problems. Second, issuances of the Turkish Treasury still have low credit ratings from external credit assessment institutions. Therefore most secondary market products would not get

high rating grades, which will possibly limit the growth potential of the secondary –
and hence the primary – mortgage market. Finally, although the banking sector
has shown strong growth in the last decade, the capital markets have been lagging behind it. This is another reason to defer the initiation of a secondary market
(trading securitised instruments) and correspondingly, to limit product variability
in the primary mortgage market (Kutlukaya and Erol, 2014).

5. Data analysis and discussion

In this section, firstly, the causal relationship between the construction industry
and economic growth is investigated, and secondly, the extent to which the RE
mortgage market growth Granger causes economic growth in Turkey and vice
versa during the last fifteen years is examined.

Table 3.1 presents the results for ADF unit root tests for the selected variables,
which shows that the null hypothesis of non-stationary series for the real and nominal GDP, construction activity (CONST) and mortgage loan balance (MORTG)
in level form with and without time trend is not rejected at all conventional levels
of significance. Hence, all time series are non-stationary in level terms. When the
series are first-differenced and the unit root tests are re-run, the ADF test statistic
rejects the null hypothesis of a unit root at 1% significance level for nominal GDP,
CONST and MORTG time series, but at different significance levels for the real
GDP and real construction series. Thus, a comparison of the critical values with
the calculated values reveals that nominal GDP, CONST and MORTG series are
stationary at first level of difference, I(1). It is important to note that nominal
series for the RE are non-stationary both at level and first difference terms. Hence,
separate Granger causality tests between the overall RE sector and GDP were not
carried out.

Table 3.1 Unit root test results for variables

Variables	At level		At first difference	
	With intercept and no trend	With trend and intercept	With intercept and no trend	With trend and intercept
Real GDP (base year 1998)	0.138 (0.966)	−0.869 (0.952)	**−3.624***** (0.008)	**−3.672**** (0.032)
Nominal GDP	2.094 (0.999)	−0.869 (0.952)	**−3.862***** (0.004)	**−4.573***** (0.003)
Real construction (base year 1998)	−0.260 (0.924)	−2.164 (0.500)	**−3.226**** (0.024)	**−3.243*** (0.087)
Nominal construction	0.633 (0.989)	−2.164 (0.500)	**−4.188***** (0.001)	**−4.305***** (0.006)
Nominal mortgage loan balance	2.307 (0.499)	0.189 (0.997)	**−4.977***** (0.001)	**−5.053***** (0.001)

Equation (2) is used to test whether construction sector growth in Turkey Granger-causes economic growth (GDP growth) and vice versa, whilst Equation (3) is used to test whether mortgage market growth Granger-causes economic growth (GDP growth) and vice versa.

$$GDP_t = \alpha_1 + \sum_{i=1}^{m} \beta_{1i} CONST_{t-i} + \sum_{i=1}^{m} \delta_{1i} GDP_{t-i} + \varepsilon_{1t}$$

$$CONST_t = \alpha_1 + \sum_{i=1}^{m} \delta_{2i} GDP_{t-i} + \sum_{i=1}^{m} \beta_{2i} CONST_{t-i} + \varepsilon_{2t} \qquad \text{Eq. (2)}$$

$$GDP_t = a_1 + \sum_{i=1}^{m} \gamma_{1i} MORTG_{t-i} + \sum_{i=1}^{m} \theta_{1i} GDP_{t-i} + \varepsilon_{1t}$$

$$MORTG_t = a_1 + \sum_{i=1}^{m} \theta_{2i} GDP_{t-i} + \sum_{i=1}^{m} \gamma_{2i} MORTG_{t-i} + \varepsilon_{2t} \qquad \text{Eq. (3)}$$

The Granger causality test results for Eq. (2) both in nominal and real time series are presented in Tables 3.2a and 3.2b respectively. In nominal terms, there has been bidirectional causality between CONST and GDP for lags one to three; that is, GDP causes construction activities and construction industry causes GDP.

Table 3.2a Granger causality test results at first difference – causality between nominal GDP and CONST

Lag length	GDP does not cause CONST		CONST does not cause GDP	
	F-statistic	Prob.	F-statistic	Prob.
1	**4.005****	0.050	**17.084*****	0.000
2	**5.584****	0.025	**7.249*****	0.002
3	**11.577*****	0.000	**4.999*****	0.006
4	**4.192*****	0.005	1.286	0.288
5	**4.040*****	0.004	0.062	0.997

Table 3.2b Granger causality test results at first difference – causality between real GDP and CONST (base year 1998)

Lag Length	GDP does not cause CONST		CONST does not cause GDP	
	F-statistic	Prob.	F-statistic	Prob.
1	**5.966****	0.018	1.630	0.207
2	**6.437*****	0.003	0.671	0.515
3	**3.971****	0.012	0.906	0.444
4	1.301	0.282	0.765	0.553
5	0.342	0.885	0.575	0.719

Note
Since the Granger causality test is very sensitive to the number of lags included in the regression, Schwarz Information Criteria (SIC) has been used in order to find an appropriate number of lags.

The F statistics (or the probabilities) for lags four and five indicate that GDP growth Granger-causes construction, but not vice versa. In order to check the robustness of results, the Granger causality tests were re-run with longer lags (six, seven and eight quarters) but the results remained the same as before. The nominal GDP growth causes growth in construction activities with a one-two year lag and the conclusion that construction is not Granger-causing GDP is robust.

Empirical results for real GDP and CONST series show that the null hypothesis of "Construction does not cause GDP growth" cannot be rejected at the conventional significance levels (1%, 5% and 10% levels) for any lags (Table 3.2b). In other words, the causal effect running from construction activities to the GDP is statistically rejected in real terms (at 1998 constant prices) for the sample.

Overall, the cause–effect analysis for construction industry and economic growth reveals that there has been a bidirectional relationship between the construction sector and the economic activities for the short term; that is, with one to three-quarter lags. Along the lines of Tse and Ganesan's (1997) discussion, expansion of construction activity is preceded by an increase in economic output with the initial effect felt largely within the construction sector and only subsequently on the aggregate economy. Granger causality test results also reveal that the economy leads the construction sector in nominal terms with a one to two year lag. In other words, during the period between 1998 and 2013, economic growth in Turkey has preceded construction activities with a one to two year lag, but not the vice versa. The result supports the findings of Tse and Ganesan (1997), Yiu et al. (2004) and Ramachandra et al. (2013) in that GDP tends to lead construction flows. Changes in GDP initially will affect demand for construction projects, then housing and credit availability, and then the level of construction output.

Turkey's economy has been moving full speed ahead, especially after the 2001 financial crisis, except for a temporary reversal in 2009 during the global financial crisis (see Figure 3.2). In the years 2010 and 2011, Turkey benefited from foreign investments and the impressive GDP growth, which reached an annual 9%, was largely driven by foreign investment, debt-fueled private consumption and RE investments by both domestic and international construction firms (TSI, 2013). The growth of the construction industry delivered the largest contribution to the GDP growth in 2010 and 2011 (see Appendices 3.A.1 and 3.A.2). The growth in the wholesale and retail trade sector, transport, storage and communication sector and manufacturing industry are the other main contributors due to their considerably higher growth rates in these two years. Immediately after this notable growth in the economy, the Central Bank introduced a tightening policy. The measures taken by the Central Bank resulted in a significant decline in debt-based domestic spending. According to Standard & Poor's Global Credit Portal (Six, 2012), the fall in domestic spending in 2012 was not the sole contributor to the low growth. Economic developments in the world also limited the flow of foreign capital into Turkey, which directly affects growth because the economy is reliant on exports.

When the contribution of sectors to the GDP growth over the sample period of 1998 to 2013 is examined, one observes that average annual growth rate of GDP was 3.9% and that it was the higher growth rates in financial intermediation sector

(7.6%), RE renting and other business activities (7.2%) and transport, storage and communication sector (6%) that affected the economic growth in Turkey (see Appendix 3.A.3). Hence, the data shows that the temporary effect of construction industry growth on the GDP growth in the years 2010 and 2011 is not sustainable for the overall sample period.

The second analysis investigates the extent to which residential RE mortgage market growth Granger-causes economic growth in Turkey and vice versa during the last 15 years. The residential RE mortgage market began to grow significantly, especially after 2005, in line with the recent improvements in Turkish economy. The share of mortgages, which had been less than 1% of GDP before 2004, increased to 6.1% as the end of 2012. However, as discussed above, there has been limited access to mortgage financing. Mortgage loans are not affordable for low-income and even for some middle-income households under the current economic environment, where main mortgage lenders are almost totally deposit banks. The banking sector offers volatile mortgage interest rates for relatively shorter maturities of 7 to 10 years.

The Granger causality test results for Eq. (3) are presented in Table 3.3. While outstanding mortgage loan balance is used as an indicator for mortgage market growth, economic growth is measured by the nominal GDP growth. The values of F statistic suggest that, historically, there has been bidirectional causality between mortgage market growth and economic growth for lags one to four. The findings reveal that over the period from 1998 to 2013, there has been feedback effect between residential RE mortgage market growth and the aggregate economy. When the mortgage market grows, more construction, a larger volume of banking sector credits, more insurance products and more appraisal works take place due to the higher demand for housing units, and the economy expands with more residential RE investment. The Granger causality test also supports the growth-driven mortgage market development hypothesis. That is to say, the housing sector is largely influenced by income and during expansionary periods, individuals appear

Table 3.3 Granger causality test results at first difference – causality between mortgage market growth (MORTG) and nominal GDP

Lag length	GDP does not cause mortgage		Mortgage does not cause GDP	
	F-statistic	Prob.	F-statistic	Prob.
1	**5.715****	0.020	**7.007*****	0.010
2	**3.626****	0.033	**10.467*****	0.000
3	5.181***	0.003	**5.952*****	0.001
4	**4.915*****	0.002	**4.538*****	0.003
5	**4.794*****	0.001	1.627	0.172

Note
Since the Granger causality test is very sensitive to the number of lags included in the regression, Schwarz Information Criteria (SIC) has been used in order to find an appropriate number of lags.

to demand more mortgage loans due to the credit availability in the economy and the income growth, while recessionary periods are generally associated with a fall in mortgage loan origination.

Granger causality tests are also run with longer lags and the finding is that the nominal GDP growth Granger-causes mortgage market development with a five to six quarters lag, but not the vice versa. However, the results did not remain the same as before for lags seven to eight quarters. There is no Granger causality between nominal GDP growth and mortgage market growth for lags seven and eight.

To sum, empirical findings indicate that the construction industry, by itself, did not lead to economic growth over the sample period between 1998 and 2013. Indeed, it was the expansion of the residential RE mortgage market that affected economic growth interactively. The mortgage market includes not only construction activities in the RE sector of the economy but also banking sector credits, insurance products, appraisal activities, etc., in the financial sector of the economy. Although the government aims to use the construction industry as the major vehicle for achieving high growth rates in the aggregate economy, the residential RE mortgage market and the related industries need to be considered as major drivers of economic growth in addition to construction activities.

6. Conclusion

This chapter provides new evidence for the lead–lag relationship between the construction industry, RE mortgage market development and economic growth in Turkey, a rapidly growing economy with a significant growth potential in its construction sector. Using quarterly data between 1998 and 2013, the present study employs the Granger causality analysis in order to test whether construction sector growth and mortgage market development separately Granger-causes economic growth (GDP growth) and vice versa.

After the 2001 financial crisis, Turkey's economy has been moving full speed ahead, except for a temporary reversal in 2009 during the global financial crisis. As noted earlier, during periods of rapid economic expansion, construction output usually grows faster than the output of other sectors, but during periods of stagnation, the construction industry is the first to suffer (see Figure 3.2). The main hypothesis that "the economy drives the construction industry in Turkey" is supported by the Granger causality analysis, which revealed that the Turkish economy leads the construction industry in nominal terms, with a one to two year lag. Over the period of 1998 through 2013, economic growth in Turkey Granger-caused the construction activities with up to a two-year lag, but not the vice versa. Hence, unlike the widespread belief that the construction plays a crucial role in Turkey's economic development, this study concludes that the construction industry is not a driver of GDP but a follower of fluctuations in the macroeconomic environment in Turkey. Indeed, similar results were obtained for Sri Lanka (Ramachandra et al., 2013), Korea (Kim, 2004), Hong Kong (Tse and Gnesan, 1997; Yiu et al., 2004) and Cape Verde (Lopes et al., 2011).

The empirical findings also reveal that, over the period from 1998 to 2013, there has been bidirectional causality between mortgage market development and economic growth with one to four quarters lags. The causality running from residential RE mortgage market growth to the aggregate economy can be explained as follows. When the mortgage market grows, more construction, a larger volume of banking sector credits, more insurance products and appraisal works take place due to higher demand for housing units and the economy expands with more residential RE investment. The Granger causality test also supports the growth-driven mortgage market development proposition. That is to say, the housing sector is largely influenced by income; during expansionary periods, individuals appear to demand more mortgage loans due to the credit availability in the economy and the income growth while recessionary periods are generally associated with a decline in mortgage loans. Moreover, nominal GDP growth Granger-caused mortgage market development with five to six quarters lag, but not the vice versa.

It is a fact that in the years 2010 and 2011, short-term remarkable growth in the construction industry, supported by foreign investment and debt-driven private consumption, significantly affected economic growth. However, the results suggest that this temporary effect could not be sustainable in a longer period of time; that is, between 1998 and 2013. This study concludes that the construction industry, by itself, does not lead to economic growth in the long run. Indeed, it is the expansion of the residential RE mortgage market that affects economic growth interactively. The mortgage market includes not only construction activities in the RE sector of the economy but also banking sector credits, insurance products, appraisal activities, etc., in the financial sector of the economy. Although the government aims to use the construction industry as the major vehicle for achieving high growth rates in the aggregate economy, residential RE mortgage market and the related industries need to be considered as major drivers of economic growth in addition to construction activities. Finally, making affordable RE mortgage loans available to a large cross-section of the Turkish population and also developing the secondary market for RE mortgage loans will deepen the financial sector and enhance economic growth in Turkey.

Note

1 Articles 35 and 36 of the Land Registry Law Numbered 2644 are the two main articles which regulate the foreigners' right to acquisition of RE in Turkey. Article 35 stipulates that: "With reservation of reciprocity and compliance with legal restrictions, foreign natural persons can acquire real property in Turkey for the purpose of using it as residence or as business place, provided that such real properties are allocated and registered in the implemented development plans or localised development plans for these purposes."

References

Akintoye, A. and Skitmore, M. (1994) Models of UK Private Sector Quarterly Construction Demand, *Construction Management and Economics*, 12(1), pp. 3–13.

Anaman, K. A. and Osei-Amponsah, C. (2007) Analysis of the Causality Links between the Growth of the Construction Industry and the Growth of the Macro-Economy in Ghana, *Construction Management and Economics*, 25(9), pp. 951–961.

Banks Association of Turkey (2013) *Statistical Reports: Deposits and Loans by Provinces under the Scope of Official Statistical Programme – 2013*, Istanbul: Banks Association of Turkey.

Bon, R. and Pietoforte, R. (1990) Historical Comparison of Construction Sectors in the United States, Japan, Italy and Finland Using Input–Output Tables, *Construction Management and Economics*, 8, pp. 233–247.

Buckley, R. M. (1996) *Housing Finance in Developing Countries*, London: MacMillan Press.

Central Bank of Republic of Turkey (2014) *Annual Report 2014: Central Bank of Republic of Turkey*, Ankara: Central Bank of Republic of Turkey.

European Mortgage Federation (2013) *Hypostat: A Review of Europe's Mortgage and Housing Markets*, Brussels: European Mortgage Federation.

Erbas, S. N. and Nothaft F. E. (2005) Mortgage Markets in the Middle East and North African Countries: Market Development, Poverty Reduction, and Growth, *Journal of Housing Economics*, 14(3), pp. 212–241.

Erol, I. and Cetinkaya, O. (2009) Originating Long-Term Fixed-Rate Mortgages in Developing Economies: New Evidence from Turkey, *METU Studies in Development*, 36(2), pp. 325–362.

Erol, I. and Tirtiroglu, D. (2008) The Inflation-Hedging Properties of Turkish REITs, *Applied Economics*, 40(20), pp. 2671–2696.

Erol, I. and Tirtiroglu, D. (2011) Concentrated Ownership, No Dividend Pay-Out Requirement and Capital Structure of REITs: Evidence from Turkey, *Journal of Real Estate Finance and Economics,* 43, pp.174–204.

Ewing, B.T. and Wang, Y. (2005) Single Housing Starts and Macroeconomic Activity: An Application of Generalised Impulse Response Analysis, *Applied Economics Letters*, 12(3), pp. 187–190.

Feige, E. L. and Pearce D. K. (1979) The Causal Relationship between Money and Income: Some Caveats for Time Series Analysis, *The Review of Economics and Statistics*, 61(4), pp. 521–533.

Financialisation, Economy, Society & Sustainable Development) (FESSUD) (2014) Comparative Perspective on Financial System in the EU: Country Report on Turkey, The EU 7th Framework Program Research Project, FESSUD Studies in Financial Systems, No. 11.

Foresti, P. (2006) Testing for Granger Causality between Stock Prices and Economic Growth, Paper No 2962, Munich: Munich Personal RePEc Archive (MPRA).

Granger, C. W. J. (1969) Investigating Causal Relations by Econometric Models and Cross-spectral Methods, *Econometrica*, 37(3), pp. 424–438.

Green, R. K. (1997) Follow the Leader: How Changes in Residential and Non-Residential Investment Predict Changes in GDP, *Real Estate Economics*, 25(2), pp. 253–270.

Hirschman, A. O. (1958) *The Strategy of Economic Development*, New Haven CT: Yale University Press.

Hongyu L., Park, Y. W. and Siqi, Z. (2002) The Interaction between Housing Investment and Economic Growth in China, *International Real Estate Review*, 5(1), pp. 40–60.

Hosein, R. and Lewis, T. M. (2005) Quantifying the Relationship between Aggregate GDP and Construction Value Added in a Small Petroleum Rich Country: A Case Study of Trinidad and Tobago, *Construction Management and Economics*, 23(2), pp. 185–197.

Huang, B. N. (1995) Do Asian Stock Market Prices Follow Random Walks? Evidence from the Variance Ratio Test, *Applied Financial Economics*, 5(4), pp. 251–256.

Jackman, M. (2010) Investigating the Relationship between Residential Construction and Economic Growth in a Small Developing Country: The Case of Barbados, *International Real Estate Review*, 13(1), pp. 109–116.

Jones Lang LaSalle (2014) *On Point: Turkey Real Estate Overview*, February.

Kim, K. H. (2004) Housing and the Korean Economy, *Journal of Housing Economics*, 13, pp. 321–341.

Kutlukaya, M. and I. Erol (2014) Housing Finance in Europe: New Empirical Evidence, Unpublished working paper.

Lean, C. S. (2001) Empirical Tests to Discern Linkages between Construction and Other Economic Sectors in Singapore, *Construction Management and Economics*, 10(4), pp. 355–363.

Lean, C. S. (2002) Responses of Selected Economic Indicators to Construction Output Shocks: The Case of Singapore, *Construction Management and Economics*, 20(6), pp. 523–533.

Li, X. (2001) Mortgage Market Development, Savings and Growth, Unpublished Working Paper, WP/01/36, Washington DC: International Monetary Fund.

Lopes, J., Nunes, A. and Balsa, C. (2011) The Long-Run Relationship between the Construction Sector and the National Economy in Cape Verde, *International Journal of Strategic Property Management*, 15(1), pp. 48–59.

Ramachandra T., Rotimi J. O. B. and Rameezdeen, R. (2013) Direction of Causal Relationship between Construction and the National Economy of Sri Lanka, *Journal of Construction in Developing Countries*, 18(2), pp. 49–63.

Six, J.-M. (2012) Economic Research: Turkey's Dynamic Growth Sets It Apart, but Reliance on Foreign Investment Is a Vulnerability, *Standard & Poor Global Credit Portal*, 1 May.

Sonmez, M. (1982) Türkiye Ekonomisinde Bunalım: 1980 Sonbaharından 1982'ye, İkinci Kitap, İstanbul Belge Yayınları.

Tse, R. T. C. and Ganesan, S. (1997) Causal Relationship between Construction Flows and GDP: Evidence from Hong Kong, *Construction Management and Economics*, 15(4), pp. 371–376.

Turel, A. (2012) High Housing Production under Less-Regulated Market Conditions in Turkey, Paper presented at the 19th Annual European Real Estate Society (ERES) Conference, 13–16 June, Edinburgh, Scotland.

Turel, A. and Koc, H. (2014) Housing Production under Less-Regulated Market Conditions in Turkey, *Journal of Housing and the Built Environment*, DOI: 10.1007/s10901-014–9393-6.

Turkish Statistical Institute (2013) *Turkey's Statistical Yearbook 2013*, Ankara: Turkish Statistical Institute, Printing Division.

Urban Land Institute (ULI) and PriceWaterhouseCoopers (PWC) (2012) *Emerging Trends in Real Estate Europe*, Washington DC: Urban Land Institute and PriceWaterhouseCoopers.

Wang, K., Zhou, Y., Chan, S. H. and Chau, K. Q. (2000) Over-Confidence and Cycles in Real Estate Markets: Cases in Hong Kong and Asia, *International Real Estate Review*, 3(1), pp. 93–108.

Yiu, C.Y., Lu, X. H., Leung, M.Y. and Jin, W. X. (2004) A Longitudinal Analysis on the Relationship between Construction Output and GDP in Hong Kong, *Construction Management and Economics*, 22(4), pp. 339–345.

APPENDICES TO CHAPTER 3

GDP growth in Turkey and growth rates by sector

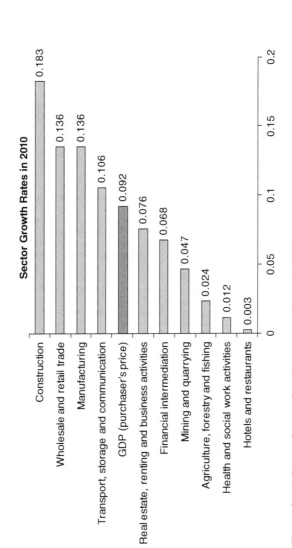

Figure 3.A.1 Economic growth and sector growth rates in 2010

Source: Turkish Statistical Institute (2013): main statistics, national accounts, gross domestic product by production approach, gross domestic product in constant prices by kind of economic activity, 1998–2013.

Figure 3.A.2 Economic growth and sector growth rates in 2011

Source: Turkish Statistical Institute (2013): main statistics, national accounts, gross domestic product by production approach, gross domestic product in constant prices by kind of economic activity, 1998–2013.

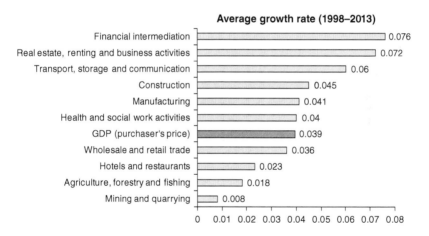

Figure 3.A.3 Average growth rates for the period 1998 to 2013

Source: Turkish Statistical Institute (2013): main statistics, national accounts, gross domestic product by production approach, gross domestic product in constant prices by kind of economic activity, 1998–2013.

4 A causal analysis between construction, real estate, and economic growth

A case study of Slovenia

Maruška Šubic Kovač

1. Introduction

Many authors point out that the construction industry plays a vital role in economic development. It promotes growth, accumulates capital formation, constitutes a source of employment, and provides critical backward and forward linkages to the rest of economy (Wells, 1984; Kirmani, 1988; Chia et al., 2014). The performance of the construction sector both affects and is influenced by general economic conditions. The global economic and financial crisis has affected the relations between construction and economic development. Analysis in many European countries (Sun et al., 2013) shows that construction shares (as a percentage of GDP) varied greatly before and during the recent economic crisis. However, the construction shares increased, some to very high levels, during the boom period. During 2000–2008, the real estate (RE) boom in Ireland, Spain, and Cyprus was synonymous with the construction boom, which boosted growth. In emerging Europe, similar overheating also took place in the Baltic countries, and to a lesser extent in Croatia. Higher construction activity prior to the crisis was associated with a lower unemployment rate. The same study (Sun et al., 2013) found that in recent years, Europe has had one of the largest variations in construction shares in the world. Changes in construction shares revolve around a norm that is determined by country-specific characteristics – the country's geography, demographics, and economic conditions.

Construction is only a portion of economic activity associated with RE. RE activities constitute an important part of the RE sector. In the extremely dynamic circumstances of the RE market brought about by economic and financial crisis, greater attention should be paid to the relationship not only between construction and economic development, but also between construction, economic activity, and economic development.

The purpose of this chapter is to establish, based on statistical analysis, the strength of correlation and the causative–resultant interconnection between particular factors (indicators of the state) in the economy, in construction, and in RE activities in the period 2000–2013 in the Republic of Slovenia. The factors that are taken into consideration include:

- Gross Domestic Product (GDP);
- Unemployment (UNEMP);

- Value of construction put in place (VALCONS);
- Construction cost (or producer prices) of newly built dwelling (COST);
- Value added by RE activities (ADDRE); and
- House prices of existing dwellings (PRICE).

In order to evaluate the role and function of construction and RE activities in the process of economic development, it is important to identify the nature and directions of that causal relationship. What is of most interest is how well the theoretical macroeconomic model of Granger causality illustrates the developments in the socio-economic environment in Slovenia, which, on account of its specificity, has not been included in certain studies (Sau et al., 2013) of the factors behind pro-cyclical but widely varying construction shares (as percentage of GDP) across countries, with a strong focus on European countries.

The structure of this chapter is as follows. First, the research methodology adopted for the study is described, after which the relationship between economic growth, construction, and RE activities is presented. Following on from this, the basis for analysing the relationship between construction, RE, and economic growth in Slovenia is considered, along with demand and supply issues in the RE market in Slovenia. The data is then analysed and discussed before the discussion concludes.

2. Research methodology

A quantitative research methodology is used for this study, in which data for the factors under discussion was collected from the publicly accessible web portals of the Statistical Office of the Republic of Slovenia (SORS), the Bank of Slovenia, and the Employment Service of the Republic of Slovenia (ESS). The time series comprises the period 2000–2013, and data was collected by quarters of each year. Considering local conditions in Slovenia, this is a relatively long time series that may bring about a certain variation in data collection procedures applied, which may lead to incomparability between the particular intervals of time. Thus, in cases where analysis confirmed that the relative relationships presented by the chain index were relatively constant irrespective of the procedure applied, such data was incorporated into the time series in spite of the different data collection procedure.

Seasonal influences are relatively high in the activity of construction as well as in RE activity. For this reason, the chain indices (I) are computed as value ratio in a particular year and quarter of the year (Qi, where I means 1st, 2nd, 3rd, 4th quarters) with regard to the preceding year and identical quarter (Qi), as for instance:

$$I(2001Q_1) = \frac{V(2001Q_1)}{V(2000Q_1)} \, 100,$$

Where:

- I . . . index
- V . . . value

The above indices were computed from:

- Absolute data for the gross domestic product (GDP);
- Rate of registered unemployment (UNEMP);
- Real value index of construction put in place and contracts made in construction, deseasonalized, mean 2010 = 100 (VALCONS);
- Absolute data for the value added by RE activities (ADDRE); and
- Price index of existing dwellings, mean 2010 = 100 (PRICE).

The indices were denominated by relevant factors, that is, the entire factors above, plus COST, which are briefly defined as follows.

- GDP is value added at basic prices by activities, increased by taxes on products and reduced by subsidies on products. Thus, it is the sum of value added at basic prices of all domestic (resident) production units and net taxes on products (taxes less subsidies on products).
- UNEMP is the percentage share of registered unemployed persons among the active population.
- VALCONS, which is value of construction put in place, covers the value of work done in the reporting period, irrespective of whether or not it has been paid. It is given at current prices and without value added tax. It includes the value added and the construction costs. It is a type of manufacturing value that is monitored in Slovenia in the construction sector, based on its own methodology. For this reason, the value added was not included in the model.
- COST is construction cost and refers to the construction cost of new dwelling.
- PRICE is the price of existing dwellings. Fixed assets (new and existing) in gross investments are valued at purchase value, which is composed of purchase price plus eventual taxes (for example, VAT), costs of delivery and other direct costs (for example, transport, assembly, etc.). As the intention was to analyse only the RE prices, without any supplements, the RE prices (PRICE) as gross investments were included in the analysis.
- ADDRE is the value added by RE activities. Among the twenty activities, the Statistical Office of the Republic of Slovenia also keeps data on the value added for RE activities.

With the exception of the data on UNEMP, which was sourced from ESS, the data on the other factors/variables was compiled from the records of SORS in 2014.

First, the correlation for the time series of chain indices (hereinafter referred to as index/indices) was computed. The correlation between the individual factors was computed using Spearman's correlation coefficient. In case of weak correlation of a certain factor with the other factors, or where the Spearman's coefficient in all the cases was less than 0.3, such a factor was excluded from further analysis.

Where the Spearman's coefficient was more than 0.3, the existence of correlation between the two factors under discussion was confirmed but no causal relationship was found. In other words, the existence of a relationship between variables does not prove the causality or direction of influence. For the time series data, the succession of onset of individual events is specific. In such a case, the causal relationship between factors may be determined based on the Granger causality test. The Granger causality test assumes that the information relevant to prediction of the respective variables is contained solely in the time series data on these variables (Gujarati, 2004).

The test involves estimating the following pair of regressions (Granger, 1969; Gujarati, 2004):

$$A = \sum_{i=1}^{n} \alpha_i B_{t-i} + \sum_{i=1}^{n} \beta_j A_{t-j} + u_{1t}$$
$$B = \sum_{i=1}^{n} \lambda_i B_{t-i} + \sum_{i=1}^{n} \delta_j A_{t-j} + u_{2t},$$

. . . where it is assumed that the disturbances u_{1t} and u_{2t} are uncorrelated. Since there are two variables, A and B, bilateral causality is what one is dealing with.

There are three situations in which the Granger causality test can be applied:

- Simple Granger causality test: there are two variables and their lags;
- Multivariate Granger causality test: there are more than two variables included because it is assumed that more than one variable can influence the results; and
- Granger causality test in a VAR (vector auto regression) framework: the multivariate model is extended in order to test for the simultaneity of all included variables.

The empirical results presented in this study are calculated pairwise within the Granger causality test in the VAR framework, so as to test whether the variable A Granger-causes the variable B, and vice versa. The Granger causality test is conducted using the following steps (Gujarati, 2004):

- Checking whether the variables and their time series (for example, A and B) included in the model are stationary. Time series, especially economic data in level form, is non-stationary and most statistical methods, including the Granger causality test, require that time series be transformed into stationary form. Therefore, the theory behind the Granger causality test is based on a stationary time series. In this study, stationarity is detected using the Augmented Dickey–Fuller (ADF) unit root test.
- Determining the number of lagged terms to be introduced in the causality test using the "VAR lag order selection criteria" (sequential modified LR test statistic, final prediction error, Akaike information criterion, Schwartz information criterion, Hannan-Quin information criterion). We checked that the error terms entering the causality test were uncorrelated, using the Lagrange Multiplier (LM) statistic.

In cases where autocorrelation was not established, the Granger causality was tested, where:

Ho (null hypothesis): A does not Granger-cause B; and B does not Granger-cause A.

The Granger approach (Granger, 1969) to the question "of whether causes" is to see how much of the "current" can be explained by "past values of" and then to see whether adding "lagged values of" can improve the explanation. A is said to be Granger-caused by B if B helps in the prediction of A or equivalently if the coefficients on the lagged are statistically significant. The two-way causation is frequently the case: A Granger-causes B, and B Granger-causes A.

It is important to note that the statement "A Granger-causes B" does not imply that it is the effect or the result of Granger causality, which measures precedence and information content, but does not by itself indicate causality in the more common use of the term.

3. Causal relationships between construction, RE, and the national economy

The construction sector contributes directly to the GDP of any country by entering the national accounts as a component of investment (Aligadide, 2012). It is part of aggregate demand and so it enters the circular flow of income and determines output movements in the short run. By augmenting the nation's stock of productive assets, investment is central to the determination of long run economic growth. The construction sector is critical as any government's policy of stimulating the economy works through spending on physical infrastructure. The sector can thus be used for governments' counter-cyclical macroeconomic policy. Increasing the level of capital stock through improving physical assets in a recession would counter the effects of the fall in output, smooth economic cycles and put the economy in steady state growth.

In dealing with the above stated, the country or regional level of development expressed by per capita income level is significant as well. Alagidade (2012) focused on the economy and the construction sector in West Africa which, according to the World Bank's income classification, belongs to the low income economies. He analysing stylized features of the construction sector; the drivers of demand in the sector and suppliers of construction services; the financial system and their role in providing funds and allocating investments; and construction risks. This is an extensive analysis in which Aligadide endeavoured to highlight the different impacts on the construction sector in the specific economic conditions of West Africa. In conclusion, Alagidade (2012) notes that GDP growth does not solve the problems of the area. He argues that although West Africa has achieved significant growth rates over the past few years, many socio-economic challenges remain and these would require bold public sector intervention.

In mathematical models, authors normally restrict themselves only to analysing the impact of a limited number of factors and therefore a certain risk attaches to use of the results. Chia et al. (2014) analysed economic development and construction productivity in Malaysia, which is classified as an upper middle income country according to the World Bank's classification of countries. They took as their basis a hypothesis that the productivity of the construction industry had a significant effect on national economic growth, and examined the relationship between construction productivity and economic fluctuations using the partial correlation method, so as to establish the underlying factors driving the change in construction productivity. The results of analysis conducted for three significant construction cycles showed that fluctuations in construction cycles were more pronounced than in the general business cycle.

Construction work involves long-term investment and long-term risks. Though governments frequently intervene in construction investment through fiscal and monetary policies in order to regulate the economy, it is unlikely that this will lead to perfectly synchronized economic and construction fluctuations. The empirical analysis of statistical data between 1970 and 2011 found a significant positive correlation between construction productivity and economic fluctuations in the 1985–1998 and 1998–2009 construction cycles. The underlying factors driving the change in construction productivity in the former cycle did amplify construction activity and increased construction employment associated with the supply side of the economy. The change in construction productivity in the 1998–2009 cycle was mainly energised by the demand generated.

In order to define the interconnection between national economy, construction, and RE activity, it is important to define RE activities. According to EUROSTAT (2013), RE activities are divided into three separate groups and include: (1) buying and selling of RE; (2) renting (to third parties) and operating owner-occupied or leased residential and non-residential RE, including both furnished and unfurnished RE – the development of building projects for one's own operation is also included; (3) appraising RE, providing RE agency services as an intermediary, and managing RE as an agent. This definition of activity shows that the field of RE activities is directly connected to the field of construction and also to the field of economic activity.

The output in RE activities is significantly impacted by the circumstances in the RE market. A decrease in RE sale more or less brings about a decrease in RE prices and thereby a devaluation of all RE, whether on the market or not. This may lead to a reduced purchasing power of the inhabitants, which in turn may have severe consequences for the economy, triggering higher unemployment, thereby impacting on the income of the country and, as a result, yielding an even lower purchasing power of the inhabitants – thus concluding the vicious circle.

Many studies so far have dealt with the interconnection between national economy and construction or with the interconnection between national economy and RE. Research on interconnections between all these sectors have been rather infrequent; there are certain empirical studies on the links between construction output and property prices. Researchers such as Di Pasquale and Wheaton (1992) found that construction output depended on price level, which in turn depended on interest rates and fundamental economic factors. Zheng (2012), who confirmed a distinct long-run causal flow from RE price to construction output in Hong Kong, arrived at similar conclusions. There are certain assertions that the two sectors are inseparably interlinked (Ball, 1981). Other researchers, such as Tse and Ganesan (1997), argue that the effects of price level and the financial cost of construction output may be reflected in the change of GDP. Additionally, the ratio between RE price and construction cost (Hofmann, 2004) should play an important role, meaning that a rise in RE price would increase construction activity.

4. Basis for analysing the relationship between construction, RE, and economic growth in Slovenia

On attaining independence in 1991, a period of market economy began in the Republic of Slovenia. In 2004 the country attained full EU Membership and in 2007 it entered the Eurozone. As assessed by the number of inhabitants (approximately two million) and its surface area (approximately 21,000 km²), Slovenia is one of the smaller EU member states. In 2013, among the 28 EU member states, Slovenia was:

- Seventeenth by the annual GDP growth level; the GDP growth level (–1.1%) was below the EU28 mean (0.1%);
- Fourteenth by the current deficit of the state sector; the deficit amounted to –3.8% of the GDP in 2012, which is close to the EU28 mean (–3.9%);
- Fifteenth (in the third quarter of 2013) by the level of unemployment of persons aged between 15 and 64; the unemployment level (9.5%) was below the EU28 mean (10.5%) (SORS, 2014c).

Whilst the RE business activity is a relatively novel activity in Slovenia that began developing more only after 1991, construction has had a longstanding tradition. Construction in general constitutes one of the more significant activities impacting on economic growth. The multiplicative effects of construction play an important role in the development of other activities. Špacapan (2008) finds that the multiplier in Slovenia ranges between 2 and 2.5. Construction is important for Slovenia also from the point of view of employment as, according to data from ESS (2014), construction is the fourth greatest activity by number of employees (following wholesale manufacturing, repair of motor vehicles, and education).

In Figure 4.1, the vertical axis (left and right) represent the number of persons employed in the construction sector (in 0'000) and RE sector (in percentages) respectively whilst the horizontal axis represents years. Construction reached its pinnacle in 2008, employing approximately 98,000 persons. During the ensuing economic crisis, this number rapidly decreased and dropped in 2011 to a level similar to that of 2006. The value of construction activities accomplished was decreasing and employers began adapting the number of employees accordingly. At the outset of the crisis, definite term employment contracts were mainly not prolonged and redundant workers were discharged permanently. By the end of 2010 and in 2011, several notorious declarations of bankruptcy by major Slovenian construction companies took place.

The share of persons employed in RE business activities is relatively low, but it does grow from year to year, and in the period under discussion it ranged between 0.32% (June 2000) and 0.54% (June 2009 and 2010). An increase in the number of employees in this activity is a result, in particular, of the fact that this is a

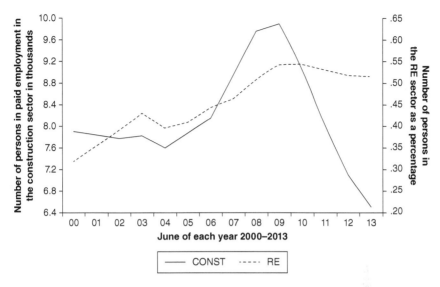

Figure 4.1 Share of persons in paid employment in construction activities and RE
         activities in June of each year in Slovenia (2000–2013)

Source: SORSb (2014).

relatively novel activity in Slovenia that was legally regulated for the first time
only in 2003 with the passing of the RE Agencies Act 2003.

Despite the relatively great differences in employment, the construction and
RE sectors do contribute a relatively great percentage to GDP.

The vertical (left and right) and horizontal axes in Figure 4.2 represent added
value shares (in percentages) and years respectively. In the period 2000–2013, the
added value share in construction moved in the interval between 4.6% (in the second quarter of 2013) and 7.9% (in the third quarter of 2008) and in RE activities
in the interval between 5.9% (in the second quarter of 2007) and 7.7% (in the first
quarter of 2010). With the latter, the variability of the share in the period under
discussion was lower in construction.

The economic crisis in Slovenia also affected the number of building permits
and number of transactions completed. Based on date sequence of the number of
building permits and of transactions completed, the situation prior to and following the beginning of the economic crisis in Slovenia may, in part, be illustrated.
Figure 4.3 represent the number of building permits (PERMIT) and number of
dwellings transactions (TRANS) (vertical axis) as against years (horizontal axis).
Data on PERMIT and TRANS is recorded by the Statistical Office of the Republic
of Slovenia. The number of building permits comprises all the building permits
(excluding the engineering works) for new construction, extension, conversion-
improvement, reconstruction, investment maintenance operations, or regular
maintenance. By 2011, the facilities were classified according to the old CC SI

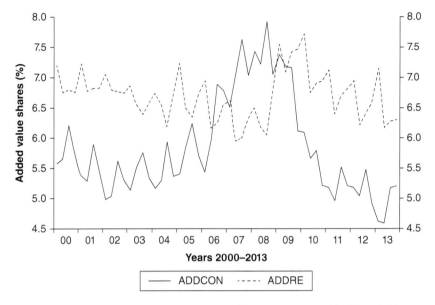

Figure 4.2 Added value shares (%) in construction (ADDCON) and in RE activities
(ADDRE) in Slovenia (2000–2013)

Source: SORSb (2014).

classification (*Official Gazette of the Republic of Slovenia*, No. 33/2003). Data
was limited to dwellings as data on all the construction facilities was not recorded
during the entire period under discussion. The number of apartment transactions
(TRANS) comprises only the data on existing dwellings (buildings). Data on other
RE types was not recorded during the entire period under discussion (2000–2013).

The lowest number of building permits (493) was issued in the first quarter of
2012 and the highest number in the second quarter of 2006 (1,276). Somewhat different was the situation with the number of transactions. The lowest number of
transactions was concluded in the first and second quarter of 2009. The highest
number of transactions coincides with the last year prior to introduction of the Euro
and the onset of the economic crisis; that is, the fourth quarter of 2006 (2,185).

Though the relative shares of construction, RE, and economic activities have
declined over recent years, they are still of high importance for the Slovenian
economy. The indices for the development of construction output constitute an
important tool for the European Central Bank and the National Central Bank in
monitoring and analysing economic development. Production in construction is
one of the so-called "Principal European Economic Indicators" (PEEIs), which
are used for monitoring and steering the economic policy in the EU and in the
Eurozone (EUROSTAT, 2014).

Slovenia is at present in a financial and economic crisis, which is reflected
in investments in RE: in a smaller scope of investment in RE, in a greater scope

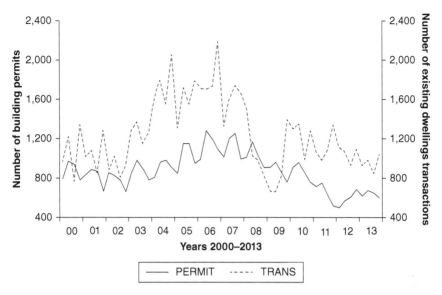

Figure 4.3 Number of building permits (PERMIT) and number of existing dwellings transactions (TRANS) in Slovenia (2000–2013)

Source: SORSb (2014).

of uninhabited and non-finalised RE, and in a weak RE market. Slovenian economists such as Fabjančič et al. (2013) are sure that Slovenian economy had – irrespective of the global financial and economic crisis – already been exhausted and broached at the time of the crisis' emergence: the global crisis only accelerated and exposed all the weaknesses brought upon the country during the transition period that officially ended with Slovenia's accession to the European Union. Similar things apply to the RE construction sector. The many bankruptcy declarations by national companies and scandals in the construction sector have contributed to a relatively low level of overall citizen confidence in the construction sector.

SORS computes the confidence indicators per particular activities, more precisely, Economic Climate (ECLIM) and level of confidence in Industry (IND), Wholesale (WSALE), Consumers (CONS), Services (SERV) and Construction (CONST). The economic climate indicator is based on surveys of business tendencies and on consumer opinions. The data on this is illustrated in Figure 4.4 where the vertical axis represents the confidence level whilst the horizontal axis represents years.

From Figure 4.4, we see that the level of confidence in construction in the period 2007–2010 constantly decreased and in 2010 it reached the bottom line. Although in 2011 and in 2012 the level of confidence in construction among the citizens increased somehow, it remained the lowest of the activities. In order to improve the situation in construction, these circumstances first need to be analysed and, in light of that analysis, relevant measures taken.

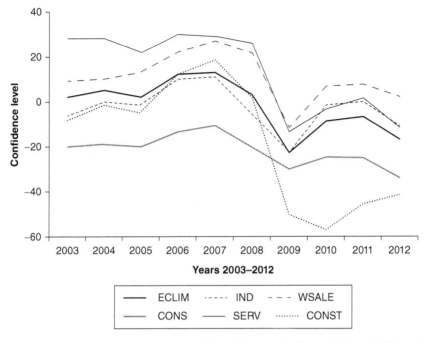

Figure 4.4 Confidence indicators per activity and economic climate indicator in Slovenia (2003–2012)

Source: SORSa (2014).

5. Demand and supply in the RE market in Slovenia

This study comprises the different indicators that represent the national economy and the construction and RE sectors. Based on the factors/variables defined in the research methodology section, what is of particular interest is whether the research results regarding the impact and impact direction of such factors/variables can be integrated into a real economic, civil construction, and RE environment in Slovenia. Therefore, we need to consider the more significant factors of demand and supply in the Slovenian RE market. This will provide the appropriate context for the discussion of results below in Section 6.

Demand and supply factors have been considered by Pavlin (2011) and reviewed as follows.

The increasing numbers of households consisting of a single person or a single-parent family have led to a poorly developed market of rental apartments and created part of the demand for residential RE. Following privatisation, a great share of residential RE passed into private RE (approximately 84% of residential RE is privately owned), and some households in privately owned apartments have, in furtherance of their desire to improve standards in residential RE, sold old and purchased new apartments.

The demand for residential RE has increased by the concentration of inhabitants in the central region of the country. Based on the new role of the city of Ljubljana as the capital of a newly emerged country, the demand for residential RE in Ljubljana and its surroundings increased and exceed the national average. The Housing Fund of the Republic of Slovenia prepared the National Housing Savings Scheme, which constitutes a savings-credit system in cooperation with the business banks and, as from 2004, moderately supported demand for residential RE, which, at fixed supply, impacted the increase in residential RE prices.

In addition, the influence of psychological factors in the Slovenian residential RE market is not negligible either. On account of news of a possible increase in VAT on the sale of new residential RE in 2007, the demand for new apartments increased in 2007; in 2008, following confirmation of existing VAT levels being retained for new residential RE, the prices of new residential RE remained at the same level or began falling. Following the accession of Slovenia to the EU the demand for residential RE by foreigners increased. In 2007, they purchased 4.2% of residential RE whilst later this share decreased and now does not have a significant impact on demand for residential RE in Slovenia.

Following the change of socio-economic system in Slovenia and Slovenia's independence as a state, Slovenian construction companies at least temporarily lost the former Yugoslav market and were faced with a drastic reduction in commissions; this was followed by numerous bankruptcies and an increase in redundancies in the construction sector. The formerly publicly owned construction companies were privatised and their proprietary and business consolidation took more than a decade. Simultaneously, many smaller construction companies were established, which on the residential RE market mainly functioned as investors and focused in particular on smaller projects. After 1991 (following Slovenia's new independence as a state), a residential RE funding system was not yet in place. At first, the state granted loans only to households, but not to companies – investors. National banks were undercapitalised and the emerging foreign banks advanced more and more credits to investors into the residential RE sector, thereby increasing the supply in the residential RE market.

Credit conditions for households and investors were relatively favourable on account of the planned inflation minimisation during Slovenia's process of accession into the EU in 2004 and into the Eurozone by 2007. Thereafter inflation increased and, concurrently, the interest rates on loans. Completion of the privatisation process (restitution of RE that had been nationalised after the Second World War back to its rightful proprietors) contributed to an increase in supply in the period 2003–2013 and therefore supply on the residential RE market rather increased. The number of transactions made in 2008 was substantially lower than in 2007, and towards the middle of 2008 the scope of turnover decreased by approximately 40% as compared to the preceding quarters of the year. A decrease in turnover and the falling prices of used apartments continued through 2009, reaching the lowest annual price fall to date (-8.3%). By the end of 2009 and in the beginning of 2010, the turnover in used residential RE reanimated and prices increased to the level of the last quarter of 2008 and were maintained at that approximate level through 2010.

In 2010, as a result of the relatively low level of encumbrance of Slovenian households and low interest rates for housing loans, Slovenian households, by purchasing residential RE, contributed to a mitigation of the contraction of the residential RE market and of the decline in prices. Thus, the decline of the residential RE market at the end of the decade was less dramatic in Slovenia than in several other European countries.

6. Data analysis and discussion

6.1 Analysis

Based on the factors (variables) defined in the research methodology section, the time series of such factors (input data) are presented in Figure 4.5, where the horizontal axis represents years whilst the vertical axis represents the chain indices of the factors considered.

By computing the Spearman's correlation coefficient, the linear relationship between two continuous variables is described. Table 4.1 displays the correlation coefficients for the factors/variables.

In Table 4.1 the maximum Spearman's correlation coefficient exists between PRICE and GDP (0.667**); it is somewhat smaller in the relationship between both COST and GDP (0.658**) and COST and PRICE (0.620**); and minimum between ADDRE and PRICE (0.261). There is a significant positive correlation between these variables. A relative high but negative is the Spearman's correlation coefficient between the UNEMP index and the PRICE (−0.713**), and between the UNEMP index and the GDP (−0.679**). There is a significant negative correlation between the two variables.

The circumstances of Slovenia in the period 2000–2013 confirm the obtained result. In the period 2000–2007, relatively favourable socio-economic conditions prevailed, with the following characteristics (see Figure 4.5):

Table 4.1 Spearman's correlation coefficient for factors/variables in Slovenia (2000–2013)

	GDP	UNEMP	VALCONS	COST	ADDRE	PRICE
GDP	1.000	−0.679**	0.421**	0.658**	0.571**	0.667**
UNEMP	−0.679**	1.000	−0.445**	−0.580**	−0.469**	−0.713**
VALCONS	0.421**	−0.445**	1.000	0.342*	0.524**	0.374**
COST	0.658**	−0.580**	0.342*	1.000	0.321*	0.620**
ADDRE	0.571**	−0.469**	0.524**	0.321*	1.000	0.261
PRICE	0.667**	−0.713**	0.374**	0.620**	0.261	1.000

Notes
* Correlation is significant at the 0.05 level (2-tailed).
** Correlation is significant at the 0.01 level (2-tailed).

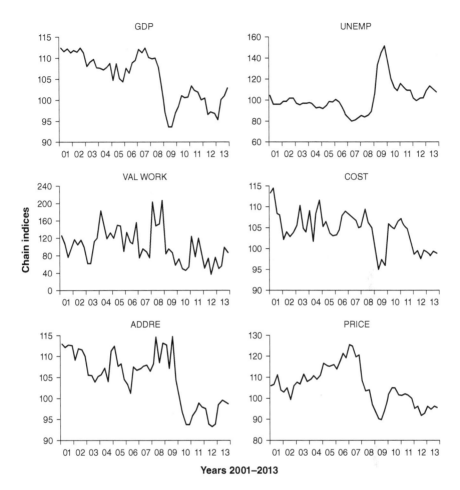

Figure 4.5 Time series of factors (variables) (processed input data) in Slovenia (2000–2013)

Source: SORSb (2014) and ESS (2014).

- relatively constant GDP growth and increasing activities in civil construction, thereby lessening unemployment;
- increasing purchasing power of the inhabitants and increasing demand for residential RE exceeded the supply, and for this reason residential RE prices were constantly increasing;
- in the long term, the accession of Slovenia to the EU did not essentially impact conditions in civil construction or on the residential RE market; and
- RE purchasing may be financed via a commercial bank loan, a loan on savings under a national savings scheme, which is a project of the Slovenian government, or under a leasing contract; in this respect, the income capacity of the borrower – RE buyer – constituted no real problem, given the relatively low unemployment rate.

The civil construction and RE sector in Slovenia is relatively small and rather inaccessible: most work is done by domestic civil construction companies and RE agencies and most buyers and sellers are Slovenian nationals. For this reason, they rather quickly adapted to the changed circumstances.

In the period 2008–2013, the socio-economic circumstances deteriorated. This means that (Figure 4.5):

- GDP began falling, construction activity diminished, and unemployment increased;
- the purchasing power of inhabitants fell and supply exceeded the demand for RE and thus residential RE prices began falling; and
- the purchasing of residential RE under the circumstances of growing unemployment grew more and more limited; loans were granted by commercial banks only, or under leasing contracts, and loan approval conditions were stricter so that only few buyers were in a position to obtain loans.

Results obtained on the basis of the linear relationship between two continuous variables do not indicate any causal relationship between factors. Thus, the response to the question, "Which is the driver and which is the driven?" was subsequently answered using the Granger causality test. Stationarity was analysed on the basis of the Augmented Dickey–Fuller (ADF) test statistics (Table 4.2).

Taking into account the results in Table 4.2, it can be concluded that all-time series of variables are non-stationary series in the level form but taking the 1st difference (GDP, VALCONS, COST, ADDRE, PRICE) or 2nd difference (UNEMP) of the variables makes them stationary.

A Johansen cointegration test was conducted for the time series and for each pair of time series. It was established that the "unrestricted cointegration test" indicated no cointegration at the 0.05 level. Then, for every pair of time series the optimum number of lags was defined. To test causality, the results were validated using the Lagrange multiplier (LM) test. The LM statistic is useful in identifying serial correlation not only of the first order but of higher orders, as well. Where in certain cases the serial autocorrelation was identified, the Granger causality was not determined. Results of the pairwise Granger causality test in the VAR framework are shown in Table 4.3.

In Table 4.3, only those time series are defined for which the Granger causality and their lags lengths have been established. The probability value for the null hypothesis that Factor A does not cause Factor B for all the lags indicates that it cannot be rejected at the 5% significance level if probability is more than 0.05.

In the interpretation of these results, one should bear in mind that the method used may have certain inadequacies. For this reason, one needs to exercise great caution in using the Granger methodology as it is most sensitive to the lag length applied in the model. Even though a variable A Granger may cause a variable B, that does not mean that A is exogenous (Gujarati, 2004).

6.2 Discussion

Section 5 above provides the appropriate context for the discussion of results as follows.

Table 4.2 The Augmented Dickey–Fuller unit root test results for variables

Time series	Specification		ADF t-statistic	Test critical values (t-statistic):		
				1% level	5% level	10% level
GDP	Level	None	−0.686	−2.611	−1.947	−1.613
		Constant	−1.676	−3.565	−2.920	−2.598
		Constant, Linear trend	−3.144	−4.157	−3.504	−3.182
	1st difference	Constant	−6.526	−3.568	−2.921	−2.598
		Constant, Linear trend	−6.486	−4.152	−3.502	−3.181
UNEMP	Level	None	−0.116	−2.616	−1.948	−1.612
		Constant	−2.626	−3.581	−2.926	−2.601
		Constant, Linear trend	−2.964	−4.171	−3.511	−3.185
	1st difference	Constant	−3.706	−3.568	−2.921	−2.598
		Constant, Linear trend	−3.647	−4.152	−3.502	−3.181
	2nd difference	Constant	−6.879	−3.571	−2.922	−2.599
		Constant, Linear trend	−6.801	−4.157	−3.504	−3.182
VALCONS	Level	None	−0.711	−2.615	−1.948	−1.612
		Constant	−1.956	−3.578	−2.925	−2.601
		Constant, Linear trend	−4.317	−4.161	−3.506	−3.183
		Constant, Linear trend (lag=1)	−3.224	−4.152	−3.502	−3.181
	1st difference	Constant	−10.947	−3.568	−2.921	−2.598
		Constant, Linear trend	−10.831	−4.152	−3.502	−3.181

(continued)

Table 4.2 (continued)

Time series	Specification	ADF t-statistic	Test critical values (t-statistic):		
			1% level	5% level	10% level
COST	Level				
	None	−0.730	−2.611	−1.947	−1.613
	Constant	−3.272	−3.565	−2.920	−2.598
	Constant, Linear trend	−3.903	−4.148	−3.500	−3.180
	1st difference				
	Constant	−8.874	−3.568	−2.921	−2.598
	Constant, Linear trend	−8.809	−4.152	−3.502	−3.181
ADDRE	Level				
	None	−0.705	−2.611	−1.947	−1.613
	Constant	−2.066	−3.565	−2.920	−2.598
	Constant, Linear trend	−2.818635	−4.148	−3.500	−3.180
	1st difference				
	Constant	−8.979	−3.568	−2.921	−2.598
	Constant, Linear trend	−8.888	−4.152	−3.502	−3.181
PRICE	Level				
	None	−0.440	−2.615	−1.948	−1.612
	Constant	−1.549	−3.581	−2.927	−2.601
	Constant, Linear trend	−2.480	−4.171	−3.511	−3.185
	1st difference				
	Constant	−6.822	−3.568	−2.921	−2.598
	Constant, Linear trend	−6.796	−4.152	−3.502	−3.181

Table 4.3 Results of Granger causality test

Lag length	Chi-sq	Probability	Chi-sq	Probability
	GDP does not cause VALCONS		VALCONS does not cause GDP	
1	6.401	0.011	4.091	0.043
	GDP does not cause COST		COST does not cause GDP	
1	5.814	0.016	0.100	0.752
	GDP does not cause ADDRE		ADDRE does not cause GDP	
1	6.254	0.012	0.471	0.492
	COST does not cause UNEMP		UNEMP does not cause COST	
2	0.040	0.980	3.829	0.147
	ADDRE does not cause UNEMP		UNEMP does not cause ADDRE	
2	1.331	0.514	19.197	0.000
	COST does not cause ADDRE		ADDRE does not cause COST	
2	3.760	0.153	6.269	0.043
	ADDRE does not cause PRICE		PRICE does not cause ADDRE	
4	4.564	0.335	13.005	0.011
	COST does not cause VALCONS		VALCONS does not cause COST	
1	2.190	0.139	0.083	0.773
	PRICE does not cause VALCONS		VALCONS does not cause PRICE	
3	7.176	0.066	0.710	0.871
	UNEMP does not cause VALCONS		VALCONS does not cause UNEMP	
3	7.154	0.067	1.729	0.630
	PRICE does not cause UNEMP		UNEMP does not cause PRICE	
3	10.887	0.012	8.063	0.045

6.2.1 Causality between GDP and VALCONS

The results show that taking into account the given input data and the methodology described, there is a bi-directional relationship in Slovenia between the GDP in VALCONS (lag = 1). This means that changes in GDP impact the changes in VALCONS and vice versa. Similar research results were obtained by Bon et al. (1999), Hongyu et al. (2002), Khan (2008), and Jackman (2010).

Table 4.4 Causality between national economy (GDP, UNEMP), construction
(VALCONS, COST) and RE (ADDRE, PRICE)

Factor A		Factor B	Lag length
GDP	does Granger-cause	VALCONS	1
VALCONS	does Granger-cause	GDP	1
GDP	does Granger-cause	COST	1
GDP	does Granger-cause	ADDRE	1
UNEMP	does Granger-cause	ADDRE	2
ADDRE	does Granger-cause	COST	2
PRICE	does Granger-cause	ADDRE	4
PRICE	does Granger-cause	UNEMP	3
UNEMP	does Granger-cause	PRICE	3

Slovenian economists such as Tajnikar (2010) are convinced that changes in civil construction in the past period significantly impacted the decrease in the GDP: "The current crisis in the construction sector is pulling the entire carriage backwards and for this reason the economic growth is not promising." He believes that changes in the construction sector could at present most advantageously be stimulated by the state, which should accelerate residential RE construction, commence investment into certain infrastructure facilities, and assist in obtaining business abroad.

Nevertheless, the current state of the construction sector is more or less on account of an inappropriate economic policy and of inadequate decisions by the construction companies. In the years prior to the financial crisis, Slovenia belonged to those countries in which the construction sector had an above average share of value added in GDP within the Eurozone. The value-added share culminated in the third quarter of 2008 when it reached 8.5% of GDP, compared to 6.4% on average in Eurozone countries. This situation was brought about by economic policy that, by decreasing the interest rates of loans, stimulated the increased financing of loans to companies. The offer of easily obtainable money on the market was enormous and unrestricted spending ensued. In such a situation, the state should have decreased its investment expenditure, which did not happen. As from 2006, the extensive private accumulation of debt and investments was joined by that of the state. With the onset of financial crisis, the financing of loans stopped and contractors and buyers were left stranded (without liquid assets). In the period prior to financial crisis, the construction companies built houses on low price loans and without stable funds for the operation of working capital, thus becoming over-indebted, and managers additionally encumbered the companies with the tycoon court proceedings.

Jackman (2010) points out that results obtained on the basis of the Granger causality test should be interpreted with a certain amount of reserve, in particular in the case of a small developing country where the consequences of improper decisions may be particularly devastating. This applies also to Slovenia where the

the variability of VALCONS is particularly high. Taking into account the chain indices by quarter, GDP growth in Slovenia in the period 2001–2008 was positive, the GDP increased from quarter to quarter (in the interval between 12.4% in the first quarter of 2001 and 3.4% in the fourth quarter of 2008). After 2008, the GDP growth was smaller and also negative. In conformity with these changes in the GDP there are the relatively extensive deviations in VALCONS, which is an extremely important piece of information relevant to the creation of economic policy and/or creation of the policy in construction. Any decision-making that will impact changes in the GDP or changes in VACONS should be based on relevant analysis and simulations of decisions in the real environment.

6.2.2 Causality between GDP and ADDRE and causality between GDP and COST

Changes in the GDP have an impact on the changes in ADDRE (lag = 1) and COST (lag = 1). This is a unilateral interconnection. In the period prior to the year of onset of the financial crisis, GDP growth favourably impacted the RE market: more sales and purchases of RE at higher prices, more leases, more valuations and, as a consequence, a higher ADDRE. On account of a relatively low supply of residential RE in the beginning of the period under discussion, the supply of new dwellings was increasing which, given a high demand for residential RE, facilitated the realisation of the relatively high COST. After 2008, the situation changed completely. The fall of the GDP brought about the fall of ADDRE and of COST.

6.2.3 Causality between UNEMP and PRICE and causality between UNEMP and ADDRE

The analysis has established Granger causality in both directions between UNEMP and PRICE (lag = 3) and unilateral Granger causality exists also between UNEMP and ADDRE (lag = 2), which may be explained as follows.

Residential RE prices in Slovenia were increasing in the period from the privatisation of social apartments in the 1990s up to 2008. Residential RE purchasing was also seen as capital investment, an alternative to purchases on the stock market or bank deposits. Circumstances for such purchases matured in 2003–2004 where the gains on global stock markets decreased due to falling stock value in the information technology sector. On account of decreasing inflation, bank loan interest rates were rapidly decreasing, and bank deposit interest rates decreased considerably, as well. For this reason, RE became an interesting area for the realisation of higher capital gains: thus demand increased.

In addition, the traditional conviction of Slovenians that investment in RE was a safe investment contributed to maintaining the spiral of constant and increasing demand and of increasing the prices of residential RE in Slovenia. In this particular period, the level of unemployment was decreasing, which together with favourable credit conditions led to an increase in demand which then, because of its relatively fixed supply, led to a short term increase in prices. The expectation that prices would continue to rise led buyers to pay far more for RE than they would have otherwise paid.

Residential RE supply in the relevant period was constantly increasing in Slovenia and after 2007 already greatly exceeded demand, leaving more and more unfinished and/or unoccupied residential buildings. The decline in home prices caused the 2008 and 2009 financial crisis. However, some economic studies, for example that of Amadeo (2007), show that residential RE prices declines of 10–15% are enough to eliminate equity and create a snowball effect. And what was the residential RE price decline in Slovenia? Taking into account the chain indices by quarter in the period 2001–2003, prices of used apartments were moderately increasing where the annual growth rates did not surpass 7.7%. The period 2004–2007 was a period of exceptional price increase of used apartments: in the first quarter of 2004, the prices increased by 9.4% and in the first quarter of 2007 by 25.4%. There was only an 8.0% increase in the first quarter of 2008, a decrease in the fourth quarter of 2009 up to 2.9%, a temporary increase in 2010 and in 2011 (in an interval from 1.0% to 2.0%) and a decrease in 2012 and in 2013.

A decline in the prices and value of residential RE reduces the amount of home equity loans the homeowners can get, which reduces consumer spending. Reduction in consumer spending contributes to a downward spiral in the economy. This results in further unemployment and reduction in income and consumer spending.

6.2.4 Causality between PRICE and ADDRE and Causality between ADDRE in COST

UNEMP Granger-causes ADDRE (lag = 2), PRICE Granger-causes ADDRE (lag = 4) and ADDRE Granger-causes COST (lag = 2). RE activities (ADDRE) are directly connected to conditions in the residential RE market, as the payment for a service is conducted according to the Real Estate Agencies Act (*Official Gazette of the Republic of Slovenia*, No 42/2003; 72/2006; 49/2011), at least regarding the buying and selling of RE, renting and operating of owner-occupied and leased residential and non-residential RE, which is directly defined by a percentage of the price of RE sold, and/or by the RE rental. Most of the RE activities are realised on the secondary market concerned with existing residential RE, though some, such as RE developers for owner-occupiers, are active in the primary market and are, therefore, closely connected to the construction sector.

Thus, one may conclude that changes in RE prices do have an impact on the changes of ADDRE. It is evident from Figure 4.5 that these changes are not as explicit as changes in RE prices. This may be explained, in particular, by the reorientation of the activities of RE agencies in the period 2007–2013 from the field of mediating purchases and/or sales of residential RE to the field of RE valuation for the purposes of foreclosure by the courts and of value adjustments of credits at commercial banks.

Changes in ADDRE have an impact on the COST of new construction. The solution to the so-called "land issue" of how to acquire in due time a piece of land that is appropriate for construction by its location, price, and size, constitutes a major problem in Slovenia. Despite the fact that spatial plans envisage relatively extensive surfaces for construction, such land is in fact not disposable for construction. The fragmented ownership structure and undeveloped land intended for building construction only increase the relative limitation of land for construction and decrease the actual supply of land, causing land development costs to increase and thereby increasing the cost of new residential RE construction.

7. Conclusion

In this chapter, the strength of correlation and the causative–resultant interconnection between particular factors in the economy, construction, and RE activities in the period 2000–2013 in the Republic of Slovenia, a country belonging to the emerging economies, has been examined. These factors include: GDP, unemployment, the value of construction put in place, construction cost, the prices of existing dwellings, and the value added by RE activities. The results of the Granger causality test in the VAR framework show the causality between the time series of the above factors and thus clarify the situation in these sectors in Slovenia.

However, it is important to note that the results and their usability are based on several assumptions.

Among such assumptions is that a mathematical model as such this illustrates well enough neither the developments in the economic, civil construction, and RE areas in Slovenia nor the interrelations between them. Issues that remain open are: Does the linear VAR model used in the study suffice? Are the impacts perceived direct impacts on particular factors? How do substantial changes within a particular factor favourably have an impact on a change in another factor? And, in particular: Is the GDP the appropriate criterion by which to judge the progress and wellbeing of humans? Or does economic growth lead to a significant improvement in the standard of living in the long run?

Weeks (2014), in *Economics of the 1%*, "analyses how the mainstream economics serves the rich, obscures the reality and distorts the policy". In his opinion, there is a widespread public perception that the economics discipline is bankrupt – based on faulty assumptions, using ineffective methods, unable to explain what is happening, and incapable of either accurate prediction or sensible policy prescription. He argues that the current construction of economic analysis is not scientific at all, it is ideological. Thus, the existing ways of thinking and models should be changed in order for them to be used in economics for the 99% of inhabitants of this planet who are living in completely different circumstances.

Most recent literature critically discusses the expected constant GDP growth. Piketty (2014) outlines the consequences of various possible scenarios for the dynamics of wealth distribution. He finds that growth can create new forms of inequality. Even today, many people believe that the period of crisis is going to end soon, like a bad dream, and that things will once again be as they were before. However, this is not the case. The current high GDP growth should be accepted as temporary only and much lower GDP growth should be expected as a target or desirable target in the future.

In particular, the GDP as the criterion by which to judge the progress and wellbeing of humans contains many deficiencies. For this reason, alternative indicators (for example, the Human Development Index, Genuine Progress Indicator, Index of Economic Wellbeing, Better Life Index and Happiness Index) are applied at present. Today, one does not need growth without improved wellbeing. One needs green, intelligent, and all-inclusive growth. However, without changes to the criteria for progress and wellbeing, no changes to the way one thinks are to be expected, let alone changes to one's actions.

Acknowledgement

The author would like to thank Franklin Obeng-Odoom for his helpful comments on an earlier draft of the chapter.

References

Alagidede, P. (2012) The economy and the construction sector in West Africa, in: *Construction in West Africa*, Accra: EPP Book Services Ltd, pp. 1–32.

Amadeo, K. (2007) How does real estate affect the U.S. economy? [online]. Available at: http://useconomy.about.com/od/grossdomesticproduct/f/Real_estate_faq.htm [Accessed 20 November 2014].

Ball, R. (1981) Employment created by construction expenditures, *Monthly Labour Review*, 104, pp. 38–44.

Bon, R., Birgonul, T. and Ozdogan, I. (1999) An input–output analysis of the Turkish construction sector, 1973–1990: A note, *Construction Management and Economics*, 17, pp. 543–451.

CC SI classification (Classification of Types of Construction), *Official Gazette of the Republic of Slovenia*, No. 33/2003.

Chia, F. C., Skitmore, M., Runeson, G. and Bridge, A. (2014) Economic development and construction productivity, *Construction Management and Economics*, 32(9), pp. 874–887, DOI: 10.1080/01446193.2014.938086.

Dickey, D. A. and W.A. Fuller (1979) Distribution of the estimators for autoregressive time series with a unit root, *Journal of the American Statistical Association*, 74, pp. 427–431.

DiPasquale, D. and Wheaton, W. C. (1994) Housing market dynamics and the future of housing price, *Journal of Urban Economics*, 35(1), pp. 1–27.

ESS (2014a) *Trg dela/Labour market*, Zavod Republike Slovenije za zaposlovanje/ Employment Service of Slovenia (ESS) [online]. Available at: http://www.ess.gov.si/ [Accessed 20 April 2014].

EUROSTAT (2013) *Real estate activity statistics: NACE Rev. 2* [online]. Available at: http://www.revijakapital.com/kapital/nepremicnine.php?idclanka=5702 and http://epp. eurostat.ec.europa.eu/statistics_explained/index.php/Real_estate_activity_statistics_-_ NACE_Rev._2 [Accessed 21 November 2014].

EUROSTAT (2014) *Construction production* [online]. Available at: http://epp.eurostat. ec.europa.eu/statistics_explained/index.php/Construction_production_(volume)_ index_overview) [Accessed 20 April 2014].

Fabjančič, Z., Grahek (2013) *Bi bila Slovenija v krizi tudi brez globalne krize?* [online]. Available at: http://www.siol.net/novice/gospodarstvo/2013/10/bi_bila_slovenija_v_krizi_tudi_brez_globalne_krize.aspx [Accessed 20 April 2014].

Granger, C. W. J. (1969) Investigating causal relations by econometric models and cross-spectral methods, *Econometrica*, 37, pp. 424–438.

Gujarati, D. (2004) *Basic econometrics*, fourth edition, New York, NY: McGraw-Hill.

Hoffmann, B. (2004) The determinants of bank credit in industrialized countries: Do property prices matter? *International Finance*, 7(2), pp. 203–234.

Hongyu L., Park, Y.W. and Siqi, Z. (2002) The interaction between housing investment and economic growth in China, *International Real Estate Review*, 5(1), pp. 40–60.

Jackman, M. (2010) Investigating the relationship between residential construction and economic growth in a small developing country: The case of Barbados, *International Real Estate Review*, 13(1), pp. 100–116.

Khan, R. A. (2008) Role of construction sector in economic growth: Empirical evidence from Pakistan economy, First International Conference on Construction in Developing Countries (ICCIDC–I), 4–5 August, Karachi, Pakistan.

Kirmani, S. S. (1988) *The construction industry in development: Issues and options*, Infrastructure and Urban Development Department, Washington, DC: World Bank.

Pavlin, B. (2011) *Značilnosti trga in gibanje cen stanovanjskih nepremičnin v Sloveniji v obdobju 2003–2011, Statistični dnevi* [online]. Available at: http://www.stat.si/StatisticniDnevi/Docs/Radenci2011/Pavlin-Trg_stanovanjskih%20_nepremicnin-prispevek.pdf [Accessed 20 April 2014].

Piketty, T. (2014) *Capital in the twenty-first century*, Cambridge, MA: Harvard University Press.

SORS (2014a) *Poslovne tendence in mnenja potrošnikov/Business tendencies and consumer opinions*, Statistical Office of the Republic of Slovenia [online]. Available at: http://www.stat.si [Accessed 20 April 2014].

SORS (2014b) Statistical Office of the Republic of Slovenia [online]. Available at: http://www.stat.si [Accessed 20 April 2014].

SORS (2014c) *To je Slovenija, naše prvo desetletje v EU, 2014/This is Slovenia, our first decade in the EU*, Statistical Office of the Republic of Slovenia [online]. Available at: www.stat.si/pub.asp [Accessed 20 April 2014].

Špacapan, B. (2008) Analiza panoge – Gradbeništvo: Panoga, ki cveti/Branch Analysis – Construction: A branch that flourishes [online]. Available at: http://www.revijakapital.com/kapital/nepremicnine.php?idclanka=5702 [Accessed 20 April 2014].

Sun, Y., Mitra, P. and Simone, A. (2013) The driving force behind the boom and bust in construction in Europe, International Monetary Fund [online]. Available at: https://www.imf.org/external/pubs/ft/wp/2013/wp13181.pdf [Accessed 20 November 2014].

Tajnikar, M. (2010) Za upad BDP je krivo gradbeništvo, Finance [online]. Available at: http://www.finance.si/281057/Tajnikar-Za-upad-BDP-je-krivo-gradbeni%C5%A1tvo [Accessed 20 November 2014].

Tse, C.Y.R. and Ganesan, S. (1997) Causal relationship between construction flows and GDP: evidence from Hong Kong, *Construction Management and Economics*, 15, pp. 371–376.

Weeks, J. (2014) *Economics of the 1%*, London and New York: Anthem Press.

Wells, J. (1984) The construction industry in the context of development: A new perspective, *Habitat International*, 8(3/4), pp. 9–12.

Zheng, X., Chau, K. W., Hui, E.C.M. (2012) The impact of property price on construction output, *Construction Management and Economics*, 30(12), pp. 1025–1037.

5 The ups and downs of the real estate market and its relations with the rest of the economy in China

Meine Pieter van Dijk

1. Introduction

During the last decade, the real estate (RE) sector in China expanded more than tenfold. However, the slump in fixed asset investments in China in 2014 was due to a slowdown in RE investments, which led to less work for the construction sector and reduced demand for building materials and the commodities of which they are made, thus emphasizing the linkages between the RE and construction sectors. Residential RE construction in China has always been a source of demand for other commodities in China and worldwide, both in the construction industry and outside the industry, and consequently is an engine of growth for the economy generally.

According to the *Financial Times* (2014c, p. 4), the current picture for the RE sector is not rosy as there are too many towers where no lights burn. In the summer of 2014, the RE market showed all the signs of a collapse. Headlines in the financial press ranged from "Property slowdown fuels China fears" to "China property correction would be painful, but salutary" (*Financial Times*, 2014e, p. 3). Housing demand has been increasing due to higher incomes, rapid urbanization and China's rural urban migration strategy. However, most of these migrants cannot afford the type of houses being built at the moment.

Experts predicting a collapse of the Chinese economy have always pointed to bubbles in the construction sector or industry as a whole. The construction sector has played an important role in China's economic development process and it is true that RE prices were booming almost continuously until 2007. A bubble has been identified very often and experts are watching what kind of landing the Chinese economy in general, and the RE market in particular, will make.

Pelosky (2014) argues that a RE crash in China would be socially explosive. He points to the fact that in the Chinese system of housing loans it often takes three to four individuals' savings to meet the 30% deposit requirements for buying an apartment with a loan. If the value of the bought apartment strongly decreases, all these people would see their savings evaporate. One could argue that they have chosen to invest in housing since the official government banks only pay 2% interest in a country where inflation is often twice as high. It is not certain, however, that people realized the risks they were taking, and in the summer of 2014 Chinese citizens went on strike against the decreasing value of their RE due to declining house prices.

Using China as a case study, the questions addressed in this chapter are: (a) what are the factors behind the development of the housing market and how is this major emerging economy managing its RE market, and (b) what will be the effect of a downturn in the housing market and in turn its impact on the construction industry and the real economy as a whole? Regarding how the rest of the chapter is structured, after a section on the theoretical and methodological framework, rapid economic growth and urbanization is examined, followed by the treatment of a booming RE market under which the mechanisms that fuel RE boom and bust, the consequences of rapidly expanding cities for the demand for rural land, the consequences for the Chinese economy of a slowdown of the RE market, the role of government in the RE sector, and the issue of promoting low cost housing are considered, in that order.

2. Theoretical and methodological framework

Housing investment has increasingly been considered a contributor to economic growth. Harris and Arku (2006) find that investment in housing affects economic development through its impact on employment, savings, total investment and labour productivity. Chen and Zhu (2008) study the direction of the relationship in China. Is increased investment in housing (leading to increased construction and supply of houses) affecting economic development, or is it a consequence of variations in economic growth (driven by fluctuating demand)? They use quarterly province-level panel data and show a clear support of a stable long-run relationship between housing investment, non-housing investment and GDP in China. However, there is bidirectional Granger causality between housing investment and GDP in both the short run and long run for the whole country, while the impacts of housing investment on GDP are strikingly different in the sub-regions of China studied by these authors.

Table 5.1 shows the major factors influencing the demand and supply of housing in China. The development of household income is an important factor, just

Table 5.1 Major factors influencing demand and supply of housing in China

Demand for housing	Supply of housing
The rapid development of per capita income and the availability of housing loans.	National policies, for example the low rate of interest on loans for developers and mortgages; and regulation such as an obligatory 30% cash down-payment.
Increased urbanisation reinforced by current rural urban migration policies.	Local level policies, for example making available land to lease; and regulation, such as local taxes and rules concerning house ownership and the ability to incorporate rural land.
Quality and price: the right size and price for very rich or very poor people in cities.	The role of RE developers and their ability to mobilize funds.

Source: Richardson (1976) and Chen and Zhu (2008).

like the rate of urbanization and the quality and price of what is offered. On the supply side, national and local policies are important, just like the role of the RE developers who have been able to make good returns on housing investments in the past. In the past, supply side factors have been very important and this has led to an oversupply of houses. Whether the current crisis means a change to demand driven development of the housing market and what the consequences are for the growth of the Chinese economy will be analysed. The second step is to link the increased housing (whether supply or demand driven) to the development of the economy. Do they contribute to the same extent to rapid economic growth?

The research is based on secondary data published particularly by the National Bureau of Statistics (Ministry of Commerce, 2014). Other secondary data is from relevant journals on the latest developments concerning the RE market and government policies to mitigate the negative effects of RE crisis; for example, the work of van Dijk (2011a) on "Three Ecological Cities, Examples of Different Approaches in Asia and Europe", in which he interviewed various project developers and government officials on the role of project developers and local government. Also, Chen and Zhu (2008) have studied the relation between housing investment and economic growth before the crisis and help to understand the implications of the recent developments in the market. However, given the recent changes in the market, some primary data was also collected through interviews with key stakeholders who were selected based on their expertise. Various researchers and government officials were interviewed on the functioning of land markets and their role in RE crises at a conference on land held at Beijing University on 10th June 2013. Two specialists (an urban environmental specialist and a RE market analyst) were also interviewed on 8th September 2014 at a seminar on urban issues in Beijing for HOVO, Beijing University of Civil Engineering and Architecture. The issues covered in these interviews related mainly to the possibility of another RE crises.

3. Rapid economic growth and urbanization in China

Economic reforms in China started after 1978 and can be described in four stages, the waves of reforms caused by Deng (*The Economist*'s *China Survey*, 1992) to create a market economy (for an overview of the whole period, see Yusuf et al., 2006). The first stage ran from the 3rd Plenum of the 11th Congress in 1978 to the 3rd Plenum of the 12th Congress in 1984. In this phase, there was rapid development of the township and village enterprises (TVEs) (Chen, 2000) and the beginning of the state-owned enterprises (SOEs) reform. In the second phase, there was an emergence of a banking sector. A third phase started in 1992 with Deng Xiaoping's famous southern tour and the emerging stock market. Systematic market reforms were undertaken and Liang (2004) noted that there was a rapid development of joint-stock companies, foreign invested companies (FIEs) and private-owned enterprises (POEs). The private sector has become more important in China and is now responsible for three-quarters of economic output and employment. The fourth phase of the reform process started after joining the World Trade Organization in 2001 (Brahm, 2002).

The Chinese economy has grown very fast since 1978, particularly in the big cities in the eastern part of the country, fueled by the Special Economic Zones located closely to these cities. During the last decade, the RE sector expanded more than tenfold. Western provinces have grown more slowly in the past and many workers from these provinces migrated to the eastern provinces, sending back a substantial part of their revenues, money partially invested in building houses.

Drivers such as rapid economic growth, migration and low productivity in the agricultural sector have contributed to rapid urbanization in China. Already 43% of China's population (559 million) lived in cities in 2005 (CSP, 2006). China was more than 50% urbanized in 2014 and 64% of the population is supposed to live in cities by 2025. In some Chinese provinces, this percentage is much higher. City states like Shanghai (89%), Beijing (84%) and Tianjin (75%) are leading. Table 5.2 gives an impression of the increased demand for housing between 1978 and 2010 due to urbanization. Taking the average value of column one, this means almost half a billion people in just over 30 years came to live in cities in China, or about 15 million per year. Assuming three people per housing unit, that would require, every year, five million new apartments.

The central government has launched urbanization plans aiming to allow another 100 million migrant workers to achieve urban residence status (Hukou) by 2020. This means that they will be eligible for subsidized housing, medical care and education, but it also means that the demand for housing will continue to increase, although it may not be equally spread over cities of different size. In fact, government policy is to promote the "smaller" cities and it can be seen in Table 5.2 that the biggest increase has effectively taken place in the cities with fewer than five million inhabitants.

China has become the workplace of the world (although it still produces only about 7% of all industrial products worldwide) and it is the result of far-sighted planning. In 1820, China's GDP was ten times that of Japan (Maddison, 2003, p. 170). A century later it was still 3.5 times bigger. Only in 1961 did Japan have a bigger GDP than China. In 2013, China became the second economy in the world in terms of GDP (World Bank, 2004) and in 2014 the biggest in purchasing power terms (IMF World Economic Outlook, 2014).

Table 5.2 Increased demand for housing between 1978 and 2010 due to urbanization

Million inhabitants	2010	1978	Increase in no. of cities
10+	6	0	6
5–10	10	2	8
3–5	21	2	19
1–3	103	25	78
0.5 to 1	138	35	103
Less than 0.5	380	129	251

Source: Ministry of Commerce (2014).

As van Dijk (2006b) notes, the reforms in the urban areas, just like in the rural areas, started with the introduction of the so-called responsibility system which gave managers of enterprises more autonomy. Managers would sign a contract with local authorities or the ministries concerning the profits to be made or the taxes to be paid. They would also sign a contract specifying how profit would be shared. Production above the agreed quotas could be marketed at floating rates between the minimum and maximum fixed by the state. The township and village enterprises would act as independent legal entities with specified rights and obligations. Because production above the quotas could be marketed at floating prices, the earnings depended on fulfilling the quotas. Mandatory planning changed into "guidance planning" and the introduction of market forces.

Economic growth in China is driven by these reforms and the rapid growth of credit and of labour productivity. There has been an emphasis on exports, although there is also a strong domestic demand fed by low prices and a demand for low cost housing. China's lower prices are not just due to lower wages but also to a good infrastructure, lower taxes, lower cost of capital and the fact that the productivity of workers is higher than, for example, in India, depending of course on the sector (van Dijk, 2006a). The most important success factors according to van Dijk (2006b) are a clear vision and strategy where the government wants to go: foresighted planning with a role for private sector and systematic reforms focusing on making the economic system function well. Part of the strategy has been stimulating investments in infrastructure and housing and using the huge savings of the population for that purpose. This high savings rate and experimentation with reforms on a small scale (for example, concerning land rights) were important success factors.

The rapid expansion of manufacturing activities in China, which began in the 1990s, is based on increased exports and had very positive effects on the Chinese economy. In fact, during the last decade, Chinese industry has become one of the most competitive in the global market. China's per capita GDP as at 2014 was around US$3,583 (*Financial Times*, 2014b, p. 1). The Chinese economy has usually grown two to three times faster than that of India since 1990, particularly if measured in per capita terms (van Dijk, 2010). The growth of manufacturing implied a gradual shift away from agriculture towards industry. This is suggested by the theory of structural transformation (Todaro, 1989), which points to a gradual shift away from agricultural to non-agricultural activities and at a later stage a move from industry to services, the latter development starting in China only in the nineties. The process is also reflected in the different rates of growth of the manufacturing, services and agricultural sectors.

Foreign capital played an important role in China's rapid economic development. In 20 years China attracted US$336 billion in foreign investment compared with India's mere US$18 billion; foreign direct investment in India increased more than 100% in 2006 to reach US$5.3 billion but in China FDI increased to US$63 billion in 2006 (van Dijk, 2006b). China's export increased fourfold in a decade; while India's export tripled, but from a much lower base: China is now the largest exporter to the USA and Japan (van Dijk, 2006b). Foreign funds are also channeled into the housing sector.

Economic growth is the most important determinant of poverty reduction. Poverty is much lower in China than in India, if only because the average income is twice as high. In China's case, poverty reduction is very clearly linked to economic liberalization and the country's export-oriented policies. The effect of high economic growth on the income distribution is less clear (van Dijk, 2011b). It depends on the social and investment policies of the government and its ability to stimulate the development of different parts of the country through income transfers or additional incentives for economic activities. However, even the European Union with high amounts spent on backward regions has not been very successful in spreading development equally. Like in Europe, it is seen that in India and China the poor tend to migrate to the more developed regions and cities. Jacobs (1970) notes that the relation between the city and its hinterland can be very fruitful if properly managed, with the urban manager as the key actor. These cities provide ideas, technology, products and markets for the rural areas and in this way can contribute to their development. It would not be wise to isolate the two systems too much. More can be gained from urban development than from rural development, as has been shown for India and China.

The strategy in which infrastructure investments and the development of the RE sector was a major engine of macroeconomic growth is replaced by a strategy where ongoing and increasing urbanization will be a major driver for economic growth and transformation. It will allow increasing productivity in agriculture and keep wages low in the urban area. There is great potential in what is called the "in-between" cities. According to van Dijk (2010), in-between cities are good for about 70% of GDP in China and 54% in India with the potential to increase their contribution to GDP from the present 2–2.5 times the average to 3 times the average, just like the mega-cities in China and India. China is the only country in the world which has changed from a policy restricting urbanization to a policy of encouraging urbanization. This type of urbanization is new, according to Lan Xinzhen (2014, cited in *Beijing Review*, 2014, pp. 34–37), because it focuses on people's rights: "the plan emphasizes the integrated development of new-type urbanization, new-type industrialization, IT application and agricultural modernization", and thinks in terms of urban clusters while taking cultural continuity and ecological factors into account. As such, the integrated development of new-type urbanization contrasts with the city level approaches, which often focus on one objective to be achieved. China expects that particularly the smaller and medium size cities still have huge economic potential and could absorb most of the migrants from the planned urbanization process. It fits in the government policy that many of these cities will be developed in the western part of the country, and the government has also shifted investments to western China (van Dijk, 2011b).

The southern provinces took up the suggestions made by Deng quickly. The Jiangsu Provincial Committee of the Chinese Communist Party and Provincial institutions, for example, became responsible for urban development in Nanjing, the capital of the Jiangsu province. Each city has a municipal government that has a number of districts. These districts have their own layer of government, the district authorities, and there are about ten in Nanjing with each about half a million inhabitants.

The importance for this study is the emphasis on decentralization of control, the introduction of market forces and an open door policy which meant opening up to foreign capital and technology – also seen in the housing sector where, for example, Hong Kong based project developers played an important role. In collaborating with local government they were actively building housing complexes one after another and contributed in this way to higher economic growth.

4. A booming RE market

The construction and supply of housing in China is also driven by the abundant availability of finance, which is mainly from Hong Kong but also involves western investors. According to the *Financial Times* (2014b, p. 5) foreigners own US$48 billion bonds and another 6 billion renminbi in yuan-denominated bonds in the RE sector. RE developers account for 20% of all non-financial bonds. Thus, many of these bonds have been bought by foreigners because they were floated in Honk Kong, New York or London. Better-off Chinese citizen prefer to invest their savings directly in housing by buying two or more apartments, which means that they may suffer heavy losses from a housing price decline. The potential crisis can be seen in the stock market, where shares of developers are down one third (*Financial Times*, 2014f, p. 22) and bonds are at 88 or 89 cents to the dollar. Hence some developers have delayed bond sale, but a complete cut-off from global debt markets would put financial pressure on the sector.

The picture in 2013 was still one of a booming RE market: "China's property boom shows no sign of cooling", according to the *Financial Times* (2014h, p. 5): the Chinese government intends "to rein in rising property prices and resurgent investment binge in the real estate market". It was expected that this trend would continue for some time. Hence the government also took policy measures in 2013 to rein in further price rises: in February 2013 a RE transaction sales tax of 20% was suggested:

- the government has raised mortgage requirements during the last three years;
- the central bank was required to raise down-payment requirements and increased interest rates for second mortgages;
- local governments have been supported to tighten home purchases limits; and
- a property tax has been introduced in cities like Shanghai and Chongqing.

These measures were supposed to prevent bubbles, but interventions also hamper market developments. The fear was that the building up of a bubble in the housing market would continue and would spell disaster. Three indicators were generally provided for the crisis in the RE market in the summer of 2014:

1   the volume of sales of apartments, which was -7.8% in renminbi terms in the first four month of 2014 compared to 2013 or -5.7% in terms of floor space;
2   the number of transactions declined: fewer people were willing to buy at current prices, which made the RE market look shaky. Prices were going down,

which was unsatisfactory according to the UNDP (2014) because it could trigger a downward spiral; and

3 the number of newly started construction projects was down by 22.1% or even one third (*Financial Times*, 2014f, p. 22).

China is a big country and the situation differs from region to region and from city to city. According to the *China Daily* (2014, p. 8), recent reports on price cuts for RE in Beijing, Hangzhou and Shenzhen, where RE sales used to be very strong, have sent a chill down the spine of some RE developers. The question is whether a collapse of RE prices will pose a systematic threat to the overall Chinese economy and even have consequences for the slow recovery of the world economy. Many assume the Chinese government will intervene but it will be argued that the margins for government interventions have become smaller, given the increased debt of central and local governments and of households who have taken mortgages in the past that now need to be paid back. Total debt would currently be 250% of GDP and it is estimated that the debt servicing ratio is 17% (the percentage of income which is necessary for paying back debts); this is at the expense of people's income or investments (*Financial Times* Special Report, 2014k, p. 1).

The evidence provided by the Ministry of Commerce (2014) is that the supply, in terms of square metres of floor space under construction, is a multiple of the amount sold in the past 12 months. What is currently being built or constructed would take over five years on average to clear, and in some provinces (Inner Mongolia) developers are building 14 years' worth of housing! These are all indicators of supply driven construction projects contributing to economic growth. Also, floor space sold showed a 16.3% decline between June and July (Kynge and Wildau, 2014, cited in *Financial Times*, 2014j, p. 1). Another way to indicate the seriousness of the current crisis is to look at the number of years of annual income necessary to buy a house. In Shenzhen, for a 70 m² house, 38 times the disposable income is required, but in Xiamen it is only 32 times. This illustrates the big regional differences but is also reflected in the decline of the prices. Xiamen was, in September 2014, the only city out of 70 being tracked by the government that had a month-on-month increase in housing prices (+0.2%), showing that demand is still bigger than supply. The average decline for the other cities was –0.9%, while big cities like Beijing and Shenzhen faced –1.2% decline and Shanghai and Guangzhou even –1.3%. The International New York Times (2014, p. 16) concludes that there may not be one RE crisis in China, but that "a conflagration among regional problems could be just as bad as one big national one". One has to be prepared for many RE bail outs. The *International New York Times* adds that many RE markets are local, but since "funding, confidence and guidance come mostly from Beijing, there is lots of room to get it wrong" (p. 16).

4.1 Mechanisms fueling RE boom and bust

Understanding the mechanism behind the RE boom is important. The availability of finance has contributed to it. Gorton and Ordonez (2014) argue that the 2008 worldwide crisis should be considered a collateral crisis. There was no major external shock explaining it, but it was preceded by a decline in the value of mortgage backed assets. The value of mortgage backed assets is difficult to assess and largely depends on trust in the issuing organization. Providing detailed information about these assets is costly and hence all parties benefit from the mechanism of supplying new capital for investment (and hence economic growth) based on selling the mortgage backed assets. Dissonant information may lead to a shock because the trust on which the transaction is based diminishes. The dilemma Gorton and Ordonez (2014) point to is that there is too little information before the crisis but the regulators ask for too much information once the system has failed, increasing the cost and slowing down the process of recovery.

The *Financial Times* (2014e, p. 4) notes that the Chinese banking sector, which provided most of the loans for the RE boom, is at risk although it has a strong ability to withstand economic shocks if only given the huge reserves of foreign exchange owned by the Chinese government. However, shareholders of Chinese banks are not convinced. Since April 2014, Chinese bank stocks have continued to lose value. Direct lending by the banks listed in Hong Kong to RE related companies is only 17% of the total loans (in the category of mortgages); however, there is also 6% of the loan portfolio going to RE developers and another 3% to construction companies (Credit Suisse figures). This adds up to 26%, but the percentage increases if the non-standard credit products are added (Noble, 2014, cited by *Financial Times*, 2014f, p. 22). This refers to the activities of the non-bank sector. The increased indebtedness is part of the general rise in debt in China. Furthermore, mortgages are collateral for 40% of the bank loans and another 10% are backed by land – this could trigger a negative spiral if these underlying values (the collateral) lose value and force the owners to reduce their activities or repay their loans (ibid., p. 22).

Government officials used to be judged by their contribution to the economic growth of their city or province and so any investment is welcome. An important aspect of local governments is that they are not allowed to raise directly money from capital markets. However, they are able to make a lot of money by converting rural land into urban land. Once the local government has obtained rural land, it tries to find project developers to build on it.

4.2 Consequences of rapidly expanding cities for the demand for rural land

Currently 50% of the population are living in cities in China, but according to estimates this will be 64% of the population by 2025. Land is not only very important for agricultural development but it also plays a major role in urban development. Li (2012) points to the innovative land policies that aim to reduce rural poverty in China.

The rapidly growing urban population and economic growth result in high demand for land. It is often noted that the speed at which cities expand, although important for the economy, contributes to inflation and the over-supply of land and houses, particularly because there are so many sources of credit available in China (Kamsma and van Dijk, 2015).

According to Kamsma and van Dijk (2015), lack of transferable land rights make it difficult for rural residents to move to China's cities. Land is essential for the development of the rural areas; however, it is also a primary asset for urban development. Urban land is in the hands of the state in China while rural land is owned by the village collectives. Land users (for example, farmers, house owners, developers or industrialists) are granted land use rights for a fixed period of time. In the early 1980s, the allocation of these land use rights was one of the first steps away from the Maoist "planned economy model" towards a more "socialist market economy model". In present-day China, the fast pace of urbanization has raised a lot of problems. Collectively, owned rural land has to be converted to state-owned land to fulfill the demand for urban land. However, rural and urban land markets are strictly separated and rural land conversion is only possible through official expropriation. This makes the land acquisition process inefficient and urban development unsustainable.

Since rural land is owned collectively, farmers do not have the right to actually buy or sell their land (whether it is farmland or construction land) or houses. Local governments have the power to appropriate and sell land and make money from these land sales. The compensation for the farmers is low and illegal land grabs and forced evictions cause social unrest in the rural areas, since peasants are aware that the land they till today could be taken away tomorrow. The absence of a clear and equitable land use rights transfer system for rural land has contributed to a widening of the urban-rural income gap, since farmers are not allowed to transfer the land use rights of their land to non-members of the collectives. Rural residents do not have a fair share of the benefits coming from rapid economic growth, urbanization and industrialization (Kamsma and van Dijk, 2015).

Most registered conflicts in China take place in the urban periphery and concern acquisition of land for the expansion of cities. The functioning of the land market in the urban areas is a real issue in China's development process. China's expanding cities together need about 200,000 hectares of rural land for new industrial zones, infrastructure and housing projects every year. In China the government owns all land but the government provides user rights to farmers, usually for a period of 30 years (Kamsma and van Dijk, 2015). There is a lot of negative publicity about land issues in China, as illustrated by a recent article in *The Wall Street Journal* (2013, pp. 14–15): "Tensions mount as China snatches farms for homes." The fact that number of registered revolts increased to 120,000, of which 65% were related to land issues, is an indication of the seriousness of the issue. These "mass actions" are what are called organized conflicts. Of the 70,000 registered conflicts in 2005, two-thirds were related to land and probably half of that number concerned conflicts in the periphery of the rapidly growing cities. The mechanism is that the compensation for farmers for giving up their user rights is usually paid through local government structures, which tend to be political and do not always transfer the full compensation to the farmers, who then start to revolt.

According to Amnesty International (2012), an important reason for too much land acquisition is that cash-strapped local governments are not allowed to raise money directly from capital markets. The mechanisms are that the municipalities generate high revenue by acquiring the land at low cost and leasing it out at a higher price. Mayors are judged on their ambitious plans and saving their "face" (status) is very important. At the same time, one sees insufficient implementation of national level policies and control at the local level, which fuels illegal practices. Finally, there is a lack of transparency, which leads to many cases of not compensating the farmers properly for the transfer of their user rights.

It is concluded that land acquisition for urban expansion is problematic because the absence of a clear system of transfer rights for rural land has led to a widening urban–rural gap and inefficient land use. The acquisition of too much land for urban purposes concerns two million farmers each year. Many of these farmers do not receive (equitable) compensation for their land that is expropriated. The expropriation is sometimes done in very unreasonable, non-transparent or even illegal ways, which shows poor governance. However, China's agriculture is inefficient and small-scale due to the small scattered land plots, and consolidation may help increase yields.

A further distinction is made between agricultural and rural construction land. Jin and Deininger (2009) rightly remark that the strict separation between rural and urban land markets with expropriation as the only way of trading between them makes land acquisition inefficient and urbanization unsustainable. CBF (2008, p. 23) notes that the rural construction land can bypass the land acquisition by the government and directly enter the market of industrial and commercial land: "The farmers who own those lands can enjoy the legal income from the land appreciation." Hulshof and Roggeveen (2011) analyze the issue in more journalistic terms. In particular, the case of Shijiazhuang (the capital of Hebei province) is interesting because it deals with farmers who build for city dwellers and with "informal architecture", showing that this is one of the ways in which rural inhabitants deal with the threat of the expanding city. The results are called "urban villages" in the literature and pose a problem for the authorities and RE developers who like to obtain the land and get rid of such "urban villages" in their modern city centres; other documented examples concern Shenzhen and Shanghai.

China has taken a few steps in the direction of a real functioning land market. The approach to a better functioning land market decided at the national level has to be elaborated by the provinces, which have opted for different solutions. To what extent are these land policies effective? Land sales are responsible for 38% of the local government revenues and too much land has been converted into urban land (Anderlini, 2014, cited by the *Financial Times*, 2014c, p. 14). Too much credit fueled the RE crisis. In global value chain terms, the Chinese construction sector was responsible for buying a lot of commodities abroad and contributed to the price increase of these raw materials.

4.3 Consequences of a slowdown in the RE market for the Chinese economy

RE alone in China is 13% of GDP, but if including the construction materials themselves it would be 16%. In investment terms, RE is one-third of all fixed asset investments and one-fifth of all commercial bank loans. Since the financial crisis, big Chinese RE developers have started to issue bonds in renminbi and dollars to finance their construction projects. Financial returns in the past have assured great interests in their bonds.

Experts do not agree on the effects of the RE crisis on China's economic development. The pessimists take the view that increasing debts (not just in the RE sector) will eventually cause a systemic crisis that would really slow down the rate of China's economic growth substantially – sometimes the cyclical nature of the RE market is emphasized by reference to an earlier decline in the prices of apartments after the crisis of 2007 followed by prices booming again in 2013 (*Financial Times*, 2013, p. 2). More optimistic analysts, such a Louis Kuijs, chief economist of RBS in Hong Kong, point to the big buffers of Chinese state banks and the fact that buyers have to pay 30% cash for their first house and 50% for their second. He does not expect a system crisis in particular, since China's new president and first minister have started a series of reforms and are less focused on economic growth and more on solving China's other problems such as corruption, pollution and a weak pension and health system (NRC, 2014). The optimists will say that the current bubble is already deflating through price reductions and declining transaction volumes. Clearly, the construction sector will be less booming in the near future and international demand for construction related commodities will also fall, while demand for low cost housing will remain huge.

The fate of the RE market is critical to the health of the overall economy. There seems to be severe over-capacity, not just in apartments built but also in terms of steel, cement and glass produced, and other construction related activities. However, the margins for government intervention have become smaller. As the *China Daily* (2014, p. 8) notes, Chinese policy makers are unlikely to sacrifice macroeconomic control and economic restructuring to save the decade-old surge in house prices (in 2013) or decline (in 2014). For the moment, the result is that developers have scaled down their investments, which leads economists to predict further financial defaults and a slowing down of economic growth in the second half of 2014. Given that all construction related activities make up 16% of the Chinese GDP, it is estimated that the decline of the RE sector may cost the country up to 1.5% of its annual growth, estimated to be 7.5% for 2014.

4.4 Role of government in the RE sector

The Chinese government has many different objectives and creates the institutional arrangements facilitating the functioning of land and housing markets (Jin and Deininger, 2009). China privatized home ownership in 1998 and prices for land and housing have been increasing ever since. During the boom the rules concerning housing were, for example, a down-payment requirement of 30% before getting a mortgage and a tax that had to be paid if a second or third home was sold. The functioning of RE markets in China is highly regulated, but influenced by a number of factors, which are not all fully under the control of the national government:

1   economic growth and its consequences for household income;
2   increasing urbanisation;
3   the large number of rural migrants coming to the cities in the next 10 years and the resulting demand for low cost housing;
4   an increased demand for better housing for successful urban dwellers; and
5   the active role of local government trying to convert rural land into urban land as an important source of revenue.

The direct role of the national government in the Chinese economy has helped to restructure state-owned enterprises (SOEs), which has allowed the private sector to take more initiative in the RE and construction sectors. Through its policies and through the control of the financial sector and the SOEs, the government still plays an important role in the economy. It has to decide whether to support the economy and especially the fledging RE sector.

In particular, in smaller Chinese cities, RE developers are struggling to survive and local officials must decide whether to bail them out or let them collapse. The *Financial Times* (2014b, p. 5) gives the example of huge building complexes in Qinhuangdao, which were supposed to house about 10,000 people: construction work stopped when the developer ran out of money, although some of the units were sold already. These developers have benefited from government support in the past and also receive loans at very low rates of interest – an average of 2.7% according to the *Financial Times* (2014b). Due to saving surpluses in China, many people invest in apartments. The annual price increase (the rate of interest on these investments) used to be much higher than the 2% interest they receive in the official banks, which is eaten up by inflation anyway. Now legislation has limited the number of apartments somebody can own, although it is possible to put a second or third home in the name of a partner or family member.

Another reaction to the RE crisis is the introduction or the changing of existing institutional arrangements. For example, the People's Bank of China and the China Banking Regulatory Commission have announced on their website (2014) a relaxation of the mortgage rules. Now banks are accepting that those who have fully repaid mortgages on previously bought homes will be treated as first time home buyers when it comes to financing new houses. This means they pay less interest and is meant to increase demand.

4.5 Promoting low cost housing

Finally, there is a mismatch in the Chinese housing market. There are too many units constructed and too few are affordable for low income people, for example, the more than 200 million migrants arriving in the cities during the next ten years who are looking for work and housing. Some authors have suggested that the new administration should build more social housing to beat supply (*Business Times* of Vietnam, 2013, p. 26). The government indeed wants to embark on special programs for low cost housing. In the meantime, the RE sector is in a decline and may face a crisis with consequences for economic growth – although it may be a regional crisis in which the smaller and less important cities are more vulnerable than the big mega cities.

China's State Council published, in March 2014, a draft regulation on subsidized urban housing. The *China Daily* (2014, p. 8) hopes subsidized urban housing will give China's cooling RE market a shot in the arm. The hope is that by explicitly including 100 million migrant workers in government-subsidized urban housing schemes, the new regulation will assure home builders of sufficient demand and give migrants an opportunity to live in the cities where their cheap labour is necessary. Other measures announced concern stepping up infrastructural investments and loosening purchase restrictions as well as making it easier for home buyers to access earmarked housing funds (*Financial Times*, 2014i, p. 6). In the meantime, developers try to sell their RE with heavy discounts and to increase their share capital by bringing out claims' issues.

5. Conclusion

The housing market in China is changing from a supply driven to a demand driven market. However, given incomes are not growing as fast anymore, and people have to spend more money on repaying their debts, demand for housing is currently lower and more directed to low cost housing, which is often not available. In future, slower growth of the national economy is expected and this will even have international repercussions.

The theory of the declining value of collateral seems relevant in the Chinese situation, because that is what is currently happening. It will have huge consequences for the construction sector and the economy. However, most of the financial consequences will be borne by the house owners, the Chinese banking sector (which is largely government owned) and the foreign and the private local and international share and bond holders of these institutions and of project developers.

In terms of institutional economics, the current policies concerning land and housing are not so much focusing on regulating the market (although a policy discouraging the buying of a second or third apartment would be an example of such an institutional arrangement) but rather making a choice to leave more to the market. Local government officials sometimes intervene to solve the problems of developers (as in Qinhuangdao). However, the national government seems to have chosen to let the market do its work, as can be seen by some recent bankruptcies in other sectors of the economy.

Chen and Zhu (2008) showed there is bidirectional Granger causality between housing investment and GDP for the whole country, while the impacts of housing investment on GDP are strikingly different in the sub-regions of China. This chapter also indicates big differences, but there are currently few cities where the prices of RE are not declining. For that reason it is feared that given the oversupply in a number of cities and developers scaling down their investments, the result will be a negative effect on the real economy of the construction firms and their suppliers.

This chapter has argued that the current crisis means a change from supply driven to demand driven development of the housing market in 2014. The consequences are a slowing down of the growth of the Chinese economy because demand is lower at the moment, and effective demand only in certain cities and for certain types of (mainly low income) houses.

References

Amnesty International (2012) Standing their ground, thousand face violent evictions in China, London: Amnesty International, International Secretariat [online]. Available at: http://www.amnesty.org/en/library/asset/ASA17/001/2012/en/976759ee-09f6-4d00-b4d8-4fa1b47231e2/asa170012012en.pdf [Accessed 12 July 2013].

Brahm, L. J. (2002) *China after the WTO*. Beijing: Intercontinental Press.

CBF (2008) *China Business Focus*, No. 118, November.

Chen, H. (2000) *The institutional transition of China's township and village enterprises*. Aldershot: Ashgate.

Chen, J. and A. Zhu (2008) The relationship between housing investment and economic growth in China, A panel analysis using quarterly provincial data. Uppsala: Uppsala University, WP 17.

CSP (2006) *China statistical yearbook 2006*, Beijing: China Statistical Press.

Dijk, M. P. van (2006a) Can China remain competitive? The role of innovation systems for an emerging IT cluster in the Jiangsu province capital Nanjing, HOVO of the Erasmus University in Rotterdam.

Dijk, M. P. van (2006b) Different effects of globalization for workers and poor in China and India, Comparing countries, clusters and ICT clusters? *Journal of Economic and Social Geography, Dossier Globalization and Workers*, 97(5), pp. 463–470.

Dijk, M. P. van (2010) *The contribution of cities to economic development, An explanation based on Chinese and Indian cities*. Saarbrucken: Lap, 61 pp.

Dijk, M. P. van (2011a) Three ecological cities, examples of different approaches in Asia and Europe, in: Wong, T.-C. and Yuen, B. (Eds), *Eco city planning*, Berlin: Springer, pp. 31–51.

Dijk, M. P. van (2011b) A different development model in China's western and eastern provinces? *Modern Economy*, 2(5), pp. 757–768.

Gorton, G. and Ordonez, G. (2014) Collateral crisis, *The American Economic Review*, 104(2), pp. 343–378.

Harris, R. and Arku, G. (2006) Housing and economic development: The evolution of an idea since 1945. *Habitat International*, 30(4), pp. 1007–1017.

Hulshof, M. and Roggeveen, D. (2011) *De stad die naar meneer Sun verhuisde*, Nijmegen: Sun.

International Monetary Fund (IMF) (2014) *World Economic Outlook* (October), Washington DC: IMF.

Jacobs, J. (1970) *The economy of cities*. New York: Vintage.

Jin, S. and Deininger, K. (2009) Land rental markets in the process of rural structural transformation: Productivity and equity impacts from China, *Journal of Comparative Economics*, 37, pp. 629–646.

Kamsma, L. and Dijk, M. P. van (2015) Experimenting with changing land use rights in China, the importance of the Chengdu case? A study of the land reform process experiment in Chengdu, University of Leiden and Erasmus University Rotterdam. Paper presented at the World Bank conference on Land and Poverty 2015, Washington, DC, 23–27 March.

Li, L. (2012) Land titling in China: Chengdu experiment and its consequences, *China Economic Journal*, 5(1), pp. 47–64.

Liang, G. (2004) *New Competition, FDI and industrial development in China.* Rotterdam: Erasmus University.

Maddison, A. (2003) *The world economy, historical statistics*, Paris: OECD.

Ministry of Commerce (2014) *Chinese statistical yearbooks* (see also their website).

Pelosky, J. (2014) China's boom is over, but Beijing will avoid bust. FT.com, May 12 [online]. Available at: http://www.ft.com/cms/s/0/2061006c-d5cf-11e3-83b2-00144feabdc0.html#axzz3jS1Ab0a7 [Accessed 21 August 2015].

Richardson, H.W. (1976) *Urban economics*, Middlesex: Penguin.

Todaro, M. P. (1989) *Economic development in the third world*, New York: Longman.

UNDP (2014) *China National Human Development Report 2013: Sustainable and liveable cities: Towards ecological civilization.* Beijing: UNDP.

World Bank (2004) *Making services work for poor people*: *World development report 2004*. New York: Oxford University Press.

Yusuf, S., K. Nabeshima, and D. H. Perkins (2006) *Under new ownership, Privatizing China's state-owned enterprises.* Washington: World Bank.

Newspapers

Beijing Review (2014) 3 April, pp. 34–37.

Business Times (Vietnam) (2013) 8 March, p. 26.

China Daily (2014) 1 April, p. 8.

CBF (a Chinese weekly newspaper).

Financial Times (2013) 19 September, p. 2.

Financial Times (2014a) 17 March, p. 4. *Financial Times* (2014b) 3 April, p. 5.

Financial Times (2014c) 17 April, p. 4.

Financial Times (2014d) 30 April, p. 1.

Financial Times (2014e) 14 May, p. 3.

Financial Times (2014f) 15 May, p. 22.

Financial Times (2014g) 14 May, p. 4.

Financial Times (2014h) 22 May, p. 5.

Financial Times (2014i) 24 June, p. 6.

Financial Times (2014j) 19 August, p. 1.

Financial Times (2014k) Special Report, 4 November, p. 1.

The International New York Times (2014) 19 September.

Nieuwe Rotterdamse Courant (NRC) (2014) 27 March (a Dutch leading newspaper).

The Economist, China Survey (1992).

The Wall Street Journal (2013) 15 February, pp. 14–15.

Part II

Construction and economic development

Part II disaggregates the relationship between real estate and construction issues and drills into the field of construction and its impact on economic development to unearth new insights.

6 Managing the cultural challenges in construction projects in emerging economies

Edward Ochieng

1. Introduction

The global construction industry has been under pressure to evolve into a sector that is constantly changing to fit the needs of the broader context in which the operations are executed. Attitudes towards working have changed dramatically in recent years and there is currently much more emphasis on multicultural team working. As multinational construction organisations define more of their activities as projects, the demand for multicultural team working grows, and there is increasing interest in reforming the project delivery process. For construction organisations in emerging economies, there is an increasing need to get groups of project managers from different nationalities to work together effectively, either as enduring management teams or to resource specific projects addressing key business issues. Many construction organisations in emerging economies have found that bringing such groups of project managers together can be problematic and performance is not always at the level required or expected. In addressing the issues relating to developing effective multicultural construction project teams, it appears that the following areas should be well thought-out: communication techniques; smoothness of handover; teamwork; issue resolution; joint decision making; and people selection and prioritisation.

As discussed below, the cultural weight that each contractor brings to a project is more often than not unconscious. Part of people's culture may be conscious and explainable to others. However, few people are completely aware of how other people's actions and ways of thinking are dictated by more hidden or (in fact) unconscious values relating for example to attitudes towards authority, approaches to carrying out tasks, concern for efficiency, communication patterns, and learning styles. It is significant that cultural norms and values are passed on from generation to generation. No one culture is right and another wrong but within each cultural grouping, whether organisational or ethnic, there is a shared view of what is considered right or wrong, logical and illogical, fair and unfair. These norms do affect the ways project teams communicate and behave within project environments. Based on the studies of Hofstede (1980), human interaction does not occur in a vacuum or isolation but instead takes place in a social environment governed by a complex set of formal and informal values, norms, rules, codes of conduct, laws, regulations and policies, and as well as in a variety of organisations. Shaping and being shaped by these governing mechanisms is something that people refer to as culture. Cultures materialise and evolve in response to social cravings for answers to a set of problems common to all groups (Hofstede, 1991). In order to survive and to exist as a social identity, every project group, regardless of its size, has to come with solutions to these problems.

The growing trend in engineering design and construction is giving rise to a need to develop effective multicultural teams. Construction companies are able to move resources to almost any location worldwide and have the capacity to work on a global scale; for many organisations future opportunities to work will entail thinking more clearly, more overtly and more systematically about cross-cultural issues and developing an understanding of multicultural team working. Although much can be achieved by working with multicultural teams, the truly success-ful construction firms are likely to be those which embed integrated changes to cross-cultural team selection, joint decision making, communication, teamwork, effective people selection and project selection. What does this mean for project leaders and international construction organisations operating in emerging econo-mies? They must actively promote multicultural team working as the means of addressing poor performance in managing people and cultural issues on construc-tion projects. In particular, if organisational change is to be effectively introduced in emerging economies, multinational construction organisations will have to ensure that their key decisions are being informed by the knowledge and experi-ence of local or indigenous managers.

Many construction organisations in emerging economies have found that mul-ticultural team integration can be problematic and at times performance is not always at the level required or expected. With an ongoing increase of multicul-tural construction project teams in emerging economies, project leaders in mul-tinational construction organisations must be aware of cultural diversity issues in order to function effectively and achieve high levels of team performance. It is crucial for the construction research community to strengthen the debatable assumption that culture is an organisational variable which is subject to conscious manipulation. A more nuanced understanding of culture and team integration are required if cultural complexity is to be accurately understood and responded to.

The rest of the chapter is structured as follows. The next section treats culture and economic development followed by a section that looks at cultural challenges in construction projects. Subsequent sections examine multiculturalism; amal-gamating complexity, uncertainty and project change in teams; project complex-ity, uncertainty and change; multicultural project teams; and the way forward – in that order. The last section contains some concluding remarks.

2. Culture and economic development

According to Maxwell (2014), economic progress in emerging markets is happen-ing at an accelerated pace due partly to advances in technology, sound economic policy making, and reduction in poverty as a result of health, education and other social reforms. As noted by Maxwell (2014), from 1996 to 2010 emerging mar-kets countries grew at more than twice the rate of developed countries – about

5% versus 2% annual GDP growth respectively. What surprises governments and international policy makers, though, is the precise nature of this growth. Much of it will result from what is called South–South commerce as opposed to the more familiar North–South commerce, which is when advanced countries invest in developing countries to make products cheaply and then export predominantly to developed countries (Maxwell, 2014). Those construction organisations that thrive in emerging markets often tend to be the pioneers and their long-term success will depend on how they integrate social, economic and cultural factors in their projects. For instance, multinational construction organisations entering Africa will need to perform a detailed analysis of political risk, operational risk, cultural risk, social risk and economic risk.

The way knowledge, technology and best practice is transferred from one country to another may be problematic without considering the cross-cultural implications. This entails understanding the different levels of cross-cultural interaction, cultural convergence and organisational hybridisation. In this chapter, the rationale of culture and economic development in emerging economies is used to understand the variation of cultural factors which influence the success of construction projects and organisational hybridisation.

3. Cultural challenges in construction projects in emerging economies

The continual need for improved speed, cost, quality and safety, together with technological advances, environmental issues and fragmentation in emerging economies have contributed to the increased complexity of construction projects (Ochieng et al., 2013). How to deal with complex projects is of increasing concern. The few complexity-related studies that exist have tended to focus on either uncertainty and complexity in non-specific terms, or the experiences of individual organisations within the context of developed countries (Thompson, 1967; Meredith and Mantel, 1995; Baccarini, 1996; Cleden, 2009). According to Cleden (2009), uncertainty can be grouped into two main classes: variability uncertainty, which is behind a number of common problems in construction projects; and indeterminate uncertainty, which always leads to indistinctness in construction projects. For example, uncertainty stemming from environmental issues and the design and financial aspects of construction projects should at least be alleviated or ideally eradicated. Construction clients are intensely interested in realising some form of breakthrough that will lower team uncertainty on projects (Ochieng, 2008).

The cultural weight that each contractor brings to a project is more often than not an intended action, but some aspects of culture may be conscious and can be explained to others. However, few individuals are fully aware that their actions and ways of thinking are dictated by more hidden or indeed unconscious values, for example, attitudes towards authority, approaches to carrying out a task, concern for efficiency, communication patterns and learning styles (Tirmizi, 2008; Johnson et al., 2009). In most cases, it is not a situation of one culture being right and another wrong within a group, but rather there is a shared view of what is considered right or wrong, logical and illogical, fair and unfair.

It is significant that cultural norms and values are passed on from generation to generation (Dekker et al., 2008; Mattarelli and Tagliaventi, 2010). Cultural norms can affect the way construction project teams in emerging economies communicate and behave within project environments. According to the studies by Hall (1966), Hofstede (1972 and 1980), Trompenaars (1993) and Brett et al. (2006), human interaction does not occur in a vacuum or isolation; instead it takes place in a social environment governed by a complex set of formal and informal values. This view is supported by Brenton and Driskill (2011), who suggested that organisations with strong cultures contain surface cultural elements that are tied into employee beliefs and assumptions. Shaping and being shaped by these governing mechanisms is something that people often consider as one of the attributes of culture. Cultures materialise and evolve in response to social cravings for answers to a set of problems common to all groups (Hofstede, 1991; Brett et al., 2006; Connaughton and Shuffler, 2007). In order to survive and exist as a social identity, every construction project group, regardless of its size, has to develop solutions to these problems.

According to Ochieng et al. (2013), the construction industry throughout the world faces similar problems and challenges. However, in emerging economies these are exacerbated by a general situation of socio-economic stress and a general inability to address the key issues within the industry. The problems have become more severe in recent years. The nature of complexity within construction engineering projects has been of increasing interest, especially since the Engineering and Physical Sciences Research Council (ESPRC) research group was set up in 2003 (Winter et al., 2006). However, it could be argued that the concept of cross-cultural complexity has been neglected: there have been no empirical studies on it. Given the supposed severity of the impact of cross-cultural complexity and the obvious failings of the industry's approach towards the management of people, it is reasonable to assume that such an issue would provide a focus for research to improve practice in emerging economies. Based on the previous studies on organisational culture, Gajendran et al. (2012) created an integrated model to evaluate organisational culture. This model aligned differing perspectives on organisational culture research with some existing paradigms to understand organisational culture. The last paradigm, which is fragmentation, confirmed that some cultural values, beliefs and scripts are not shared by all stakeholders.

4. Multiculturalism in emerging economies

Multiculturalism in emerging economies has become an important focus in debates in construction management research (Ochieng and Price, 2010; Ochieng et al., 2013). An extensive literature review demonstrated that multiculturalism is often an indistinctly used term which has a diverse range of meanings, and very few empirical studies have been undertaken on its role in emerging economies. It could, therefore, be suggested that any construction project in which contractors have different assumptions about working norms (either in design engineering or team behaviour) is a multicultural project. Even when all contractors are from one country, the construction project manager may still have to deal with cultural diversity. Within the context of this chapter, some team differences are strictly cultural, while others stem from varied management styles and strategies.

Managing a multicultural construction project team presents new challenges and opportunities to harness new skills, in particular language and cultural knowledge. For instance, effective communication is an essential skill in emerging economies.

It requires clients and construction project managers to acquire and promote knowledge and understanding of cultures and attitudes. Cultural differences and a lack of management talent can make it difficult for multinational construction organisations to attain their global business objectives. According to Connaughton and Shuffler (2007), organisations working with multicultural teams face a three-fold multicultural challenge: to enable a mixed group to work towards a common goal; to maximise the contribution of each project team member; and to ensure fair treatment for all, irrespective of background.

Whether the multicultural character of an organisation arises from its operation in various countries or from the mixed backgrounds of a workforce in a single location, the client must address this cross-cultural diversity if it is to achieve its goals. Every multinational construction organisation has a strategic choice in how it will face this challenge: whether to adopt a fundamentally defensive approach, or one that develops the individual and the group.

5. Amalgamating complexity, uncertainty and project change in teams

Rethinking Construction (Egan, 1998) and more recent reports have reflected on the causes of negative teamwork. Their general conclusions highlight the individualistic and competitive focus of the industry. Some of the proposals for change by the construction reports from the UK (see for example, Egan [1998]), such as integrating teams, collaboration in the supply chain and improving the industry's commitment to people, clearly indicate the need to address cultural complexity and uncertainty in construction project teams. The inference is that project teams in the current climate accept that their existence is temporary and focus solely on the outcomes of a particular project. The single-project outlook pervades every aspect of culture and is arguably the primary cause of the industry's problem in emerging economies (Ochieng et al., 2013).

Blismas et al. (2004) identified ten main factors influencing project delivery on construction projects, which were grouped under four headings: environmental influences; client influences; third-party influences; and planning influences. Environmental influence factors exacted the greatest overall influence on project delivery, constituting five out of the ten factors. Two of the characteristics that aggravated the effects of these factors were uncertainty and immitigability. These two factors are by nature unpredictable and, therefore, uncertain in action and effect. In addition, they are generally beyond the control or influence of the organisations. When considering client influences, Blismas et al. (2004) found that a number of decisions taken by clients have ongoing effects on construction project teams, but there were two main influencing factors originating from the client body: indecisiveness and non-uniformity. They found that non-uniformity of the client resulted in inter-departmental differences and constant changes to the projects. Changes were made as a result of misunderstandings between departments, indecision, insufficient information or altered circumstances. Lack of client leadership and internal communication emerged as causes of uncertainty and change in projects.

When investigating studies under the third heading, Blismas et al. (2004) established that third parties had no vision of project planning and therefore usually applied enormous pressure on individual project phases. While a number of authors have argued the case for improved management practices that could lead to better-unified team integration across the different tiers of the construction supply chain (Bresnen and Marshall, 1999; Mitullah and Wachira, 2003; Briscoe et al., 2004; Ochieng and Price 2009; Ochieng and Price, 2010; Baiden and Price, 2011), in practice it is difficult to attain this. Commonly, clients appear to distrust their main contractors, who in turn maintain an arm's length relationship with their subcontractors and suppliers. Construction projects in emerging economies are treated as a series of chronological and mainly separate operations in which the individual players have very little stake in the long-term success of the resulting building or structure, and no commitment to it. It has been widely suggested that supply chains can exist in a number of different forms and can vary significantly in their complexity and diversity (Cox, 1999).

Construction supply chains on larger construction projects in emerging economies involve hundreds of different organisations supplying materials, components and a wide range of construction services (Ochieng et al., 2013). A continued dependence on a disjointed and largely subcontracted workforce has arguably amplified the complexity of this supply network. In reality, since the 1970s, the industry's reformation has given rise to the creation of what now appears to be an institutionally entrenched low-skill, poorly equipped and labour intensive sector (International Labour Organisation, 1972; Bosch and Philips, 2003). A number of contractors operate as flexible firms, exemplifying the hollowed out structure characterised by extensive outsourcing (Atkinson, 1984; ILO, 2001). This demonstrated a problematic context for attaining the integrated delivery of the industry's projects and processes. Despite the normal complexity that the industry faces, it is vital that it expands its supply-chain practices to deliver value to the client, rather than simply seeking to generate short-term cost savings.

If countries in emerging economies want to accomplish better practice, then thought must be given to what constitutes best practice. Fox (1999) and Cleland and Bidanda (2009) suggested that best practice included:

- standardisation of procedures and methodologies;
- efficient interface management;
- ethical actions;

- communication between government and contractors;
- attention to organisation culture;
- government understanding of the construction industry; and
- recognition of location differences in language, terminology and culture.

6. Project complexity, uncertainty and change in emerging economies

The escalating complexity of multi-purpose construction project management emerged from the growing demands of clients and the increase of multi-disciplinary teams that deliver multi-million projects. According to Cleland and Bidanda (2009), project complexity has amplified exponentially since the late 1980s. Evidence shows that the increase in project complexity has many facets: designs that approach the physical limits of construction materials and equipment; construction in remote sites; partnerships; data integration requirements; various project delivery systems; and contracting strategies (Cleland and Bidanda, 2009). Interestingly, globalisation and outsourcing also contribute to project and cross-cultural complexity. As multinational construction organisations invest in emerging economies, geographical, social and political factors add to the complexity of their projects. Managing construction projects in emerging economies requires construction organisations to ascertain how to work within an intercontinental environment with multicultural project teams.

For instance, the East–West motorway project in Algeria was one of the significant national projects undertaken by the Algerian Government. In the entire Eastern region of Algeria, there were 660 Algeria staff from Algeria and 90 Kajima Construction Corporation employees from Japan. The manual construction workers consisted of 10,500 Algerians and 6,000 from other nations including South Asia. According to the Project Director (Minoru Ishida), the people of Algeria had confidence in Japanese technology and quality. The scale of the project was big and the duties of employees were also greater. Though it was challenging to overcome cultural and national differences while executing project tasks, Minoru noted that he had to apply "5As": awatezu (calmness), aserazu (patience), atenisezu (self-reliance), anadorazu (respect) and akiramezu (persistence) (Kajima Construction Corporation, 2009).

Multicultural construction projects are invariably complex and have become more so. Baccarini (1996) argues that the construction process could be considered to be the most complex undertaking in any industry. However, Morris et al. (2000) emphasised that the construction industry has experienced great difficulty in handling the ever-increasing complexity of major construction projects. It is essential to assert that the concept of project complexity and cultural complexity has received little in-depth attention in project management literature. A review of the literature shows that certain project characteristics provide a basis for shaping the appropriate managerial actions needed to complete a project successfully (Turner, 1998; Winter et al., 2006; Ochieng et al., 2013).

Complexity is one such significant project dimension. As Bennett (1991) claimed, practitioners habitually portray their projects as simple or complex when they are discussing management issues. This suggests a practical acceptance that complexity makes a difference to the management of construction projects. As confirmed in the reviewed literature, the magnitude of complexity to the project management process is widely accredited, for example:

- Complexity is a key decisive factor in the selection of an appropriate project organisational form (Vidal and Marle 2008; Cicmil et al., 2009).
- Complexity is often used as a criterion in determining a suitable project procurement arrangement (Wozniak, 1993; Maylor et al., 2008).
- As Rowlinson (1988) and the Chartered Institute of Building (CIOB, 1991) claimed, complexity affects project, cost, time, quality and objectives. Generally, the higher the project complexity the greater cost and the time.
- Project complexity hampers the clear identification of objectives and goals of construction projects (Morris et al., 2000; Austin et al., 2002).

A common way of defining complexity is to quantify it in various dimensions, such as the number of stakeholders, units and resources. It is essential to underline that when project complexity is considered, multicultural project managers have to spell out to which of the project's dimensions (organisational or technological) they refer. Gidado (1996) showed how differentiation and interdependency transpires through technological and organisational complexity. Baccarini (1996) countered this argument by suggesting that organisational complexity based on differentiation can be either vertical or horizontal. Horizontal differentiation is determined by the number of organisational units and task structure – project job and specialisation – while vertical differentiation is the depth of the organisation's hierarchical structure. The other feature of organisational complexity in construction projects is the degree of interaction between organisational elements and operational independencies.

Team integration complexity by differentiation is determined by the array of outputs, inputs and tasks and the number of specialities involved in a project. As noted in a report by the *Strategic Forum for Construction* (2002), large construction projects are typically characterised by the engagement of diverse contractors and project teams. This leads to the formation of a temporary multicultural project structure to manage the construction. A project structure can be presented in two dimensions. The first feature is based on a relationship between complexity and uncertainty. As noted above, the second feature involves the work of Baccarini (1996). Williams (1999) claimed that there is uncertainty in the instability of circumstances and assumptions on which a project is based. Evidence has shown that as a project matures in real time, the uncertainty, and hence project complexity, is minimised (Meredith and Mantel, 1995). Uncertainty can make project situations appear weighed down with danger. External factors can be looked at as the driving force for uncertainty in projects.

The issue of complexity is the prime focus in today's construction project management literature. Project complexity can be found in three environments:

outside the project; inside the project; and in the environment outside the project (Ochieng, 2008). One of the key reasons why complexity varies is that clients have different goals, interests and expectations. A second reason is that in different project phases there are different driving factors. As a result, one could suggest that traditional forms of hierarchical team formation and leadership are being replaced by self-organising agents, self-directed and self-managed team concepts (Thamhain and Wilemon, 1996). In today's fast-paced, highly competitive world, change is inexorable. Construction organisations operating in emerging economies must respond to change to remain competitive and client focused.

The problem is that communications for implementing change often come from various sources and in many different formats. Projects are created to facilitate change, but by their nature are themselves incessantly changing. It is now thought that when considered in the context of construction projects, change occurs in two places, internally or externally. Organisational change is being considered here and in this context a construction project organisation is the same as any other organisation. The next section details issues surrounding multicultural project teams in emerging economies.

7. Multicultural project teams in emerging economies

Getting multicultural project teams to work effectively across international boundaries has become a major issue (Earley and Mosakowski, 2000; Weatherley, 2006; Brett et al., 2006). The trend is likely to continue and the future of business will increasingly depend on doing projects effectively in different cultural environments (Earley, 1993 and 1994; Peterson et al., 1995; Weatherley, 2006; Ochieng, 2008). It has been widely recognised that multicultural project teams have been common in recent years. Contemporary literature in international management has identified the management of multicultural teams as an important subject in human resource management. Most of the studies have focused on the positive effects of using multicultural teams. Earley and Mosakowski (2000) stated that multicultural teams are used because of a belief that they out-perform monocultural teams, especially when performance requires multiple skills and judgement. Multicultural team integration is a particular problem for clients and project managers in emerging economies. Once they are established, multicultural teams are perceived to outperform monocultural teams, in areas such as problem identification and resolution, by the sheer strength of its diversity (Marquardt and Hovarth, 2001).

The basic values, concepts and assumptions differ with each culture; understanding these and enabling a "settling-in" by recognising the cultural complexity is a required skill of a manager (Vonsild, 1996; Kang et al., 2006). Choosing not to recognise cultural complexity limits the ability to manage it. The fragmentation of project delivery has been blamed on the cultural complexities that exist. Project managers of multinational organisations in emerging economies often make the common assumption that cultural differences are unimportant when individual members belonging to different divisions of the same organisation are brought together as a team. The original research (Hofstede, 1980) that suggested that 80% of the differences in employee attitudes and behaviours are influenced by national culture still has resonance today.

Cultural differences reflect different expectations about the purpose of the team and its method of operation, which can be categorised into task and processes. The task area relates to the structure of the task, role responsibilities and decision making. The processes relate to team building, language, participation, conflict management and team evaluation. Culture is an issue with many different dimensions. Both Hofstede (1991) and Trompenaars and Hampden-Turner (1997) discussed different levels of culture. The former mentions gender, generation, social class, and regional, national and organisational levels. The latter presents national, corporate and professional levels of culture.

It has been widely recognised that organisational culture is important in construction project management (Kandola and Fullerton, 1998; Meek, 1998; Barthorpe et al., 1999 and 2000). For example, contractors working in emerging economies are usually drawn from a number of organisations, each with its own organisational culture. To work as a team efficiently, it is essential to have some degree of cohesion of organisational culture. Most multinational construction organisations in emerging economies have a cultural history and set ways of getting things done that can help or, in some cases, hinder a project. It is essential, therefore, to institute organisational background and culture of all the contractors involved in the project from the outset.

Typically, leadership in construction projects being delivered in emerging markets is complex and critical to success in multicultural team environments. For example, Weatherley (2006) affirmed that if management is getting the team to do what is required, project leadership involves motivating the project team in such a way that they want to do what is required. In this context, project leadership is not so much about telling the project team what to do as leading by example and developing trust and confidence in the team to take the project forward. Indeed, this kind of team leadership is not just accidental but can be developed, and is a required skill for successful multicultural project teams (Earley and Mosakowski, 2000).

In a number of emerging countries, different ethical standards apply. This affects attitudes towards the law and, indeed, national laws can be very different in different territories. Weatherley (2006) claimed that in some nations bribery and corruption have become institutionalised and are the only means by which some local officials can earn a living. There may be an unwarranted predilection given to local suppliers or contractors and planning laws and approvals can be very officious. In some emerging economies, the government approval procedure can become the critical path on the project programme. The project financing of multicultural teams can be difficult. Differences in currency rates as well as variations in different nations can play havoc with the cost management of a construction project. Owing to the above, a number of construction organisations in emerging economies have difficulty in repatriating revenue from project work done outside their home country. Ochieng (2008) established that there is also more hidden cost linked with multicultural projects such as customs, import duties, shipping, logistics and agent fees.

A number of construction organisations in emerging economies now use low-cost design centres for the completion of detailed or standardised design and this can lead to a considerable reduction in costs due the low currency rates in these countries. Emmitt and Gorse (2003 and 2007) noted that communication between the main project office and the low-cost design centre needs to be of high quality if complexity is to be reduced. It is essential that these are used with care since, although the rates may be low, productivity is often also low and so the cost savings are much less attractive than might be thought. This undoubtedly impedes integration since error rates tend to be higher for work farmed out in this way (Weatherley, 2006). This presents a problematic context for achieving the integrated delivery of the project since there can be a hostile response from project workers if there are different rates of pay for the same work.

Smith (1999) asserted that risk is present in all projects but becomes more definite in construction projects where there are often new risks, predominantly if the project is being constructed in a country where security is an issue. In some emerging economies, contract law is not well instituted by other nationalities. Emmitt and Gorse (2007) found that risks in communication and risks emerging from misunderstandings and misinterpretation are much greater. There is also a danger of the expatriate project team "going native" and becoming isolated from the project and pursuing their own project goals rather than focusing on the overall project aim and objectives (Langford and Rowland, 1995).

Another key issue that managers of multicultural project teams face is assessing the skills and competencies of the project team (Ochieng et al., 2013).

For example, in a number of emerging countries, the training and education standards and the relative value of qualifications can be very different. Weatherley (2006) also highlighted that job methods can be different because of specific local conditions such as working in heat, earthquake risk or local trade practices. Ochieng et al. (2013) established that cultural differences are usually significant when managing project teams in emerging economies. Hofstede (1991) and Trompenaars and Hampden-Turner (1997) classed the differences as national characteristics, ethnic differences, organisational culture and professional differences.

Indeed, if not addressed these differences can lead to major divergence of working practice and can severely affect the project conclusion. For example, in the Middle East status and hierarchy are very important. This can lead to a lack of empowerment of more junior project staff. Religious observance can also be very significant in getting the project completed successfully. Setting aside time for prayers during the working day might be required. In a number of countries there are religious festivals, fasts and feast days that are classed as non-working days (Reva and Ataalla, 2002). Language is another factor that affects multicultural project teams operating in emerging economies. Brett et al. (2006) noted that trouble with accents and fluency; direct versus indirect communication; differing attitudes towards hierarchy; and conflicting decision making norms can cause destructive conflicts in a team. The trouble with accents and fluency can occur when individuals who are not fluent in the team's dominant language may have difficulty sharing their knowledge. Direct versus indirect communication can transpire when some project workers use direct, explicit communication while others are indirect, for example asking questions instead of highlighting problems with a project leader.

Brett et al. (2006) further argued that team members from hierarchical cultures expect to be treated differently according to their status in the organisation. With conflicting decision making norms, project team members vary in how quickly they make decisions and in how much analysis they may require beforehand. Brett et al. (2006) asserted that an individual who prefers to make decisions quickly may grow frustrated with those who need more time. It is essential for the project manager to set out a common language so as to ensure a common understanding (Emmitt and Gorse, 2007). This arguably delimits non-native speakers who are working in their second or third language with a substantial loss of efficiency, as well as increased risks or misunderstanding.

The realisation of integrated multicultural teamwork as a single unit is a challenge in emerging economies. The various parties within the delivery team continue to face cultural issues. Egan (2002) stated that integrated teamwork is the key to construction projects that personify good whole-life value and performance. Fully integrated teams operating in emerging economies deliver greater process efficiency and by working together over time can help drive out the old style adversarial culture and provide safer projects using a qualified, trained workforce. Teams that only construct one project team at the client's expense would never be as efficient, safe, productive or profitable as those that work repeatedly on similar projects (Egan, 2002). This, particularly in emerging economies, has proven to be a long and complex process (Ochieng et al., 2013). However, partnerships and co-operatives are being formed, and integration and collaboration are becoming generally accepted needs for individuals and companies to survive.

Despite the above difficulties, it is vital for multinational construction organisations operating in emerging economies to improve multicultural team integration. It is possible to get project teams from different countries and organisations to work together effectively. The task for senior project leaders is to understand cultural issues and the secret of success so that more multicultural construction project teams can be managed effectively to the benefit of the clients in a way that properly rewards organisations involved in that delivery. In this way, they enable the key requirements for efficient multicultural team integration to be better identified.

8. Culture in construction projects in emerging economies: the way forward

This chapter has revealed that both internal and external cross-cultural communication provides the invisible glue which can hold a dislocated multicultural project team together. Effective communication is the key to managing expectations, misconceptions, and misgivings on multicultural project teams operating in emerging economies. As confirmed, good communication strategies are primary in establishing, cultivating and maintaining strong working relationships on construction projects in emerging economies. Trust is fragile, intangible and generally difficult to quantify but it is essential to the success of multicultural teamwork. Trust can be cultivated where there are good interpersonal skills and mutual respect between project leaders and team members.

In order for a multicultural project team to be fully integrated, all team members need to trust and understand each other. Communication in multicultural teams is a significant factor in the successful completion of construction engineering projects. It is essential for senior construction project leaders in emerging economies to ensure that the nature of the interactions do not affect the strength of the relationships between project teams and their ability to transfer knowledge and information required to complete project tasks successfully. Project leaders need to implement a clear and robust procedure of resolving conflicts that might arise. What needs to be well understood is that the effective structure of multicultural teamwork depends on a well interconnected communication system, between the client, the project manager and the project team.

Addressing cross-cultural communication in construction engineering projects can be viewed as a principal enabler for improving the sector in the future. What this chapter does highlight is the need for considerably more research into multicultural project teams operating in emerging economies. What it did uncover suggests that one needs a better understanding of multicultural project teams in such economies. With the growth in globalisation, construction project managers in emerging economies will need to work on culturally diverse project teams. The good news is that multicultural project teams in emerging economies will bring fresh ideas and new approaches to problem solving. The challenge is that they will also bring understanding and expectations regarding team dynamics.

As established in this chapter, the growing trend in engineering design and construction is giving rise to a need to develop effective multicultural teams. Now that construction companies are able to move resources to almost any location worldwide and have the capacity to work on a global scale, for many organisations in emerging economies future opportunities to work entail thinking more clearly, overtly and systematically about cross-cultural issues and to develop an understanding of multicultural teamwork.

As suggested in this chapter, this requires the integration of thinking and practice related to cross-cultural management. Although much can be achieved by working with multicultural teams, the truly successful construction firms in emerging economies are likely to be those that embed integrated changes to leadership style; to the team selection and composition process; and to the cross-cultural management of the team development process; and to cross-cultural communication, cross-cultural collectivism, cross-cultural trust, cross-cultural management, and cross-cultural uncertainty. In applying the above, the value of multicultural teamwork can be captured at many levels in the organisation, be they project based or permanent, and furthermore will allow project teams to reach high performance levels consistently.

It is worth noting that a number of issues have been identified that have not been discussed in the construction management literature. The issues that have been identified relate to cross-cultural teams in emerging economies. This chapter shows that the contemporary construction project manager in emerging economies is hardly ever afforded the luxury of retrospection. The coercions posed by cross-cultural complexities are real and immediate, and the stakes on a project are often high. As established in this chapter, the real challenges facing many of today's construction contractors who want to expand and become true players in emerging economies are to transform their business models and discover effective ways to integrate diverse multicultural construction teams.

The nature of delivering construction projects has changed. In today's global economy people have shifted away from the monochromic make-up of construction teams to those that are coloured by contractors from different locations. With this new multicultural make-up come differences in cross-cultural complexity, which in turn bring differences in leadership styles and many other cultural issues. Cultural awareness is now important if multicultural construction teams in emerging economies are to capitalise on their potential. The literature reviewed reaffirmed the importance of cultural awareness in organisations (Hofstede, 1980; Trompenaars, 1993; Hall, 1966; Brett et al., 2006). Although cross-cultural issues do not always result in problems, their more delicate manifestations can lead to poor cultural integration.

It has been shown that nothing ever remains static in project work, and a multicultural project team works in environments filled with uncertainty and complexity where nothing exists in a permanent state. In order to manage cross-cultural issues, project leaders need to be flexible in everything that they undertake. It is worth highlighting that cross-cultural uncertainty is one of the principal casual factors underlying project team intercultural integration. An uncertain situation could be dealt with by collecting more data about the project to minimise the cultural and information gap that might emerge. Uncertainty could have various impacts on a project. This accords with an earlier observation made in the chapter that uncertainty can make project situations appear weighed down with danger. Although uncertainty arises in many levels of the project, the art of managing uncertainty depends hugely on how much global project leaders are able to understand the realities of project situations.

The literature reviewed reaffirmed that uncertainty can be grouped into two main classes: variability uncertainty, which is behind a number of common problems in construction projects; and indeterminate uncertainty, which always leads to indistinctness in construction projects (Cleden, 2009). While evidence from this chapter suggests that uncertainty results in challenges in projects, among other internal and external factors, some participants argued that some projects could benefit from uncertainty. It is worth noting that many levels of uncertainty will never undergo conversion to create an unexpected outcome. The aim is to focus on adequate effort in the areas of uncertainty that signify most risk and have the highest chance of developing severe problems. When one comes across low uncertainty project teams, they demand less project structure, but with high uncertainty project teams they expect consistency and clear articulation to be in place. There is, therefore, a need for further work in this area to establish the influence of project leaders in managing multicultural project teams that have low and high levels of uncertainty.

In managing cross-cultural complexity, effective project leaders in emerging economies should be able to understand the type of leadership style preferred by the multicultural project team so the project leader's authority is respected. It is the responsibility of the project manager to set up a supportive and positive project culture. Effective leaders should be fair and consistent when dealing with project team members and this can be achieved by not showing any favour or partnership in the way they behave.

Cultural empathy can be used to give project team members confidence in carrying their daily project tasks. If project leaders are to manage global multicultural construction project teams smoothly, they must be aware of their own personality and characteristics. It is vital for project leaders to learn how to control their characteristics and, most importantly, to use them selectively. Two basic methods could be applied: push and pull. The push method entails forming an opinion and then arguing about its merits, whilst the pull technique depends on seeking the opinion of others. A project leader applying the pull technique needs to form an opinion first, and then use skilful questioning to encourage the team to form the same view. The pull technique can take longer to lead to a decision, but a skilful project leader needs to use both methods, depending on the team or project situation.

Project leaders of construction project teams in emerging economies must understand the culture and environment they are working in. This finding is consistent with those of other studies, such as those mentioned in the literature reviewed by Dekker et al. (2008), Mattarelli and Tagliaventi (2010) and Brenton and Driskill (2011), which suggested that knowing the relationship between values and cultures can assist organisations to better understand intercultural values in teams. A clear message is that multinational construction organisations must cultivate work locations that support their project managers and teams. This chapter provides a good foundation for understanding the influential cultural factors that affect international multicultural heavy construction engineering projects in emerging economies. Multinational construction organisations operating in emerging economies cannot afford to ignore or overlook cultural issues in construction projects. A better understanding of managing cultural issues will help minimise cross-cultural complexity in construction project teams. For a multicultural team to be effective, senior managers must acknowledge cultural issues openly and work around them.

9. Conclusion

As confirmed in this chapter, managing cultural diversity in global construction projects will be a challenge. It is worth noting that the nature of project delivery is changing with accelerating economic activity from Western economies to emerging markets in Africa, Asia and Latin America. Dewhurst et al. (2012) suggested that the ten fastest growing economies will all be in emerging markets. No single project delivery model will be best for construction organisations handling the realities of rapid growth and cultural diversity in emerging economies. That is partly because the opportunities and cultural challenges facing construction organisations will vary. Although individual multinational construction organisations are responding differently to the new opportunities in emerging economies, the findings in this chapter suggest that most will face different level of cultural challenges. For instance, a number of companies will find it difficult to be locally flexible and adaptable as they broaden their global footprint (Ochieng et al., 2013). At the same time, many global construction organisations will find deploying and developing talent in emerging economies to be a major challenge.

Efforts to standardise the common elements of essential functions such as procurement, legal work and taxation can clash with local needs. The key for international construction organisations will be to find the right mix of global and local in their operations. Cultural integration will depend on how international construction organisations find the best local and global combination that works. To understand cultural diversity and what it means for senior construction managers working in emerging economies, there must be a broader, more conceptualisation of cultural integration and culture identity developed at strategic, operational and project levels. To fully comprehend the implications of cultural integration and culture identity within the three levels, it is important to understand the basic concept of cultural diversity in organisations. Rather than considering cultural diversity as a problem with which one must cope, senior construction managers working in emerging economies can take this new understanding of reality to develop special multicultural skills that will help them deal with cultural issues and handle differences in sensitive and synergistic ways.

References

Atkinson, J. (1984) *Emerging UK work patterns in flexible manning: The way ahead*, IMS Report No. 88, Brighton: Institute of Manpower Studies.

Austin, S., Newton, A., Steele, J. and Waskett, P. (2002) Modelling and managing project complexity, *International Journal of Project Management*, 20(3), pp. 191–198.

Baccarini, D. (1996) The concept of project complexity: A review, *International Journal of Project Management*, 14(4), pp. 201–204.

Baiden, B. K. and Price, A. D. F. (2011). The effect of integration on project delivery team effectiveness, *International Journal of Project Management*, 29(2), pp. 129–136.

Barthorpe, S., Duncan, R. and Miller, C. (1999) A literature review on studies in culture: A pluralistic concept, in: Ogunlana, S. O. (Ed.), *Profitable Partnering in Construction Procurement*, London: Spon, pp. 533–542.

Barthorpe, S., Duncan, R. and Miller, C. (2000) The pluralistic facets of culture and its impact on construction, *Property Management Journal*, 18(5), pp. 335–351.

Bennett, J. (1991) *International Construction Project Management: General Theory and Practice*, Oxford: Butterworth-Heinemann.

Blismas, N. G., Sher, W. D., Thorpe, A. and Baldwin, N. A. (2004) Factors influencing project delivery within construction clients' multi-project environments, *Engineering Construction and Architectural Management*, 11(2), pp.113–125.

Bosch, G. and Philips, P. (2003) Introduction, in: Bosch, G. and Philips, P. (Eds), *Building Chaos: An International Comparison of Deregulation in the Construction Industry*, London: Routledge, pp. 1–23.

Brenton, A. L. and Driskill, G. W. (2011) *Organisational Culture in Action: A Cultural Analysis Workbook*, Thousand Oaks, CA: SAGE.

Bresnen, M. and Marshall, N. (1999) Achieving customer satisfaction? Client–contractor collaboration in the UK construction industry, in: Bowen, P. and Hindle, R. (Eds), *Proceedings of CIB W55/65 Joint Triennial Symposium on Customer Satisfaction: A Focus for Research and Practice in Construction, Cape Town, 5–10 September.*

Brett, J., Behfar, K. and Kern, M. C. (2006) Managing multicultural teams, *Harvard Business Review*, 84(11), pp. 85–91.

Briscoe, G., Dainty, A. R. J., Millett, S. and Neale, R. (2004) Client led strategies for construction supply chain improvement, *Construction Management and Economics*, 22(2) pp. 193–201.

Cicmil, S. J. K., Cooke-Davies, T. J., Crawford, L. H. and Richardson, K. A. (2009) Exploring the Complexity of Projects: Implications of Complexity Theory for Project Management Practice, Newtown Square, Pennsylvania: Project Management Institute.

Chartered Institute of Building (CIOB) (1991) *Procurement and Project Performance*, Occasional Paper No. 45, Ascot: Chartered Institute of Building.

Cleden, D. (2009) *Managing Project Uncertainty*, Surrey: Gower Publishing Company.

Cleland, D. I. and Bidanda, B. (2009) *Project Management Circa 2025*, Newtown Square, Pennsylvania: Project Management Institute.

Connaughton, S. L. and Shuffler, M. (2007) Multinational and multicultural distributed teams: A review and future agenda. *Small Group Research*, 38(387), pp. 387–412.

Cox, A. (1999) A research agenda for supply chain and business managerial thinking, *Supply Chain Management*, 4(4), pp. 209–211.

Dekker, D. M., Rutte, G. C. and Berg, V. D. (2008) Cultural differences in the perception of critical interaction behaviors in global virtual teams, *International Journal of Intercultural Relations*, 32(5), pp. 441–452.

Earley, P. C. (1993) East meets West meets Mideast: Further explorations of collectivistic and individualistic work groups, *Academy of Management Journal*, 36(2), pp. 319–348.

Earley, P. C. (1994) Self or group? Cultural effects of training on self-efficacy and performance, *Administrative Science Quarterly*, 39(1), pp. 89–117.

Earley, P. C. and Mosakowski, E. (2000) Creating hybrid team cultures: an empirical test of trans-national team functioning, *Academy of Management Journal*, 43(1), pp. 26–49.

Egan, J. (1998) *Rethinking Construction*, London: Department of the Environment, Transport and the Regions.

Egan, J. (2002) *Accelerating Change*, London: Strategic Forum for Construction.

Emmitt, S. and Gorse, C. A. (2003) *Construction Communication*, Oxford: Blackwell Publishing Limited.

Emmitt, S. and Gorse, C. A. (2007) *Communication Construction Teams*, Oxford: Taylor and Francis.

Fox, P. W. (1999) *Construction Industry Development: Exploring Values and Other Factors from a Grounded Theory Approach: Proceedings of CIB W55 and W65 Joint Triennial Symposium, Cape Town, September*, CIB Publication 234 (ISBN 0-620-239441-1), pp.121–129.

Gidado, K. I. (1996) Project complexity: the focal point of construction production planning, *Construction Management Economics*, 14(3), pp. 213–225.

Hall, E. T. (1960) The silent language in overseas business, *Harvard Business Review*, 3, pp. 87–96.

Hall, E.T. (1966) *The Hidden Dimension*, New York: Doubleday.

Hofstede, G. (1972) *The Game of Budget Control*, London: Tavistock Publications.

Hofstede, G. (1980) *Culture's Consequences: International Differences in Work-Related Values*. London: Sage Publications.

Hofstede, G. (1991) *Cultures and Organisations: Software of the Mind, Intercultural Co-Operation and Its Importance for Survival*, New York, NY: McGraw-Hill.

International Labour Organisation (ILO) (2001) *The Construction Industry in the Twenty-First Century: Its Image, Employment Prospects and Skill Requirements*, ILO, Geneva.

Johnson, S. K., Bettenhausen, K. and Ellie Gibbons, E. (2009) Realities of working in virtual teams: Affective and attitudinal outcomes of using computer-mediated communication, *Small Group Research*, 40(6), pp. 623–649.

Kajima Construction Corporation (2009) Japanese technology at work in Africa: A visit to Algeria's east-way motorway project [online]. Available at: www.kajima.co.jp/news_events/special_features/vol2/vol2-1.html [Accessed 2 February 2009].

Kandola, R. and Fullerton, J. (1998) *Diversity in Action: Managing the Mosaic*, London: Institute of Personnel and Development.

Kang, B. G., Price, A. D. F., Thorpe, A. and Edum-Fotwe, F. T. (2006) Ethics training on multi-cultural construction projects, *CIOB*, 8(2), pp. 85–91.

Langford, D. A. and Rowland, V. R. (1995) *Managing Overseas Construction Contracting*, London: Thomas Telford.

Marquardt, M. J. and Hovarth, L. (2001) *Global Teams: How Top Multinationals Span Boundaries and Cultures with High-Speed Teamwork*, Palo Alto, CA: Davies-Black.

Mattarelli, E. and Tagliaventi, M. R. (2010) Work-related identities, virtual work acceptance and the development of glocalised work practices in globally distributed teams, *Journal of Industry and Innovation*, 17(4), pp. 415–443.

Maylor, H., Vidgen, R. and Carver, S. (2008) Managerial complexity in project-based operations: A grounded model and its implications for practice, *Project Management Journal*, 39(1), pp. 15–26.

Maxwell, J. (2014) Beyond the BRICS: How to succeed in emerging markets (by really trying) [online]. Available at: http://www.pwc.com/us/en/view/issue-15/succeed-emerging-markets.jhtml [Accessed 16th January 2015].

Meek, V.L. (1998) Organisational culture: Origins and weaknesses, *Organisation Studies*, 9(4), pp. 453–473.

Meredith, J. R. and Mantel, S. J. (1995) *Project Management: A Managerial Approach*, New Jersey, NJ: John Wiley and Sons Inc.

Mitullah, W. V. and Wachira, N. I. (2003) Informal labour in the construction industry in Kenya: A case study of Nairobi, Working Paper No. 204, International Labour Office, Geneva.

Morris, P. W. G., Patel, M. B. and Wearne, S. H. (2000) Researching into revising the APM project body of knowledge, *International Journal of Project Management*, 18(3), pp.155–164.

Ochieng, E. G. (2008) Framework for managing multicultural project teams, unpublished PhD thesis, Loughborough University, Loughborough.

Ochieng, E. G. and Price, A. D. F. (2009) Framework for managing multicultural project teams, *Engineering Construction and Architectural Management*, 16(6), pp. 527–543.

Ochieng, E. G. and Price, A. D. F. (2010) Managing cross-cultural communication in multicultural construction project teams: The case of Kenya and UK, *International Journal of Project Management*, 28(5), pp. 449–460.

Ochieng E. G., Price, A. D. F and Moore, D. (2013) *Management of Global Construction Projects*. Hampshire, UK: Palgrave Macmillan's Global Academic.

Peterson, M. F., Smith, P. B., Akande, A. and Ayestaran, S. (1995) Role conflict, ambiguity, and overload: A 21-nation study, *Academy of Management Journal*, 38(2), pp. 429–452.

Reva, B. B. and Ataalla, M. F. (2002) Cross-cultural perspectives of expatriate managers working in Egypt, *International Journal of Cross-Cultural Management*, 2(1), pp. 83–101.

Rowlinson, S. M. (1988) An analysis of factors affecting project performance in industrial building, unpublished PhD thesis, Brunel University, Middlesex.

Smith, N. (1999) *Managing Risk in Construction Projects*, Oxford: Blackwell Science.

Strategic Forum for Construction (2002) *Rethinking Construction: Accelerating Change Consultation Paper*, London: Strategic Forum for Construction.

Thamhain, H. and Wilemon, D. (1996) Building high performing engineering project teams, in: Katz, R. (Ed.), *The Human Side of Managing Technological Innovation*, New York, NY: Oxford University Press, pp. 122–136.

Thompson, J. D. (1967) *Organisations in Action*, New York, NY: McGraw-Hill.

Tirmizi, S. A. (2008) Effective multicultural teams: Theory and practice, *Advances in Group Decision and Negotiation*, 3, pp. 1–20.

Trompenaars, F. (1993) *Riding the Waves of Culture*, London: Nicholas Brealey.

Trompenaars, F. and Hampden-Turner, C. (1997) *Riding the Waves of Culture*, 2nd ed., London: Nicholas Brealey Publishing.

Turner, J. R. (1998) *The Handbook of Project-Based Management*, 2nd ed., London: McGraw-Hill.

Vidal, L. A. and Marle, F. (2008) Understanding project complexity: Implications on project management, *Keybernetes*, 8(8), pp.1094–1110.

Vonsild, S. (1996) Management of multicultural projects: How does culture influence project management? Paper presented at the "Challenge of the 21st Century: Balancing Teams and Task", Management Association, World Congress on Project Management, International Project, Paris, Vonsild, June.

Weatherley, S. (2006) ECI in partnership with Engineering Construction Industry Training Board (ECITB), ECI UK 2006 masterclass on multi-cultural project team working, London, 6 December [online]. Available at http://www.gdsinternational.com/infocentre/artsum.asp?lang=en&mag=182&iss=149&art=25863 [Accessed December 2006].

Williams, T. M. (1999) The need for new paradigms for complex projects, *International Journal of Project Management*, 17(5), pp. 269–273.

Winter, M., Smith, C., Morris, P. and Cicmil, S. (2006) The main findings of UK government-funded research network, *International Journal of Project Management*, 24(8), pp. 638–649.

Wozniak, T. M. (1993) Significance vs capability: 'Fit for use' project controls. *Proceedings of the American Association of Cost Engineers International (Trans) Conference, Dearborn, MI*, A2, pp. 1–8.

7 Sustainable construction in emerging economies

Andrew Price and Edward Ochieng

1. Introduction

It has been proven that global warming and climate change pose an unparalleled threat to all living beings. Rapid development of developing countries will hasten global warming and exacerbate resource problems (Murota, 1996). This problem is important in the long run as most policy makers recognise (for emerging economies) there are many other critical sustainable development issues that affect human welfare more immediately. The chairman of the Environment Agency has stated that "changing the way we build, produce energy and make technology more efficient must go hand in hand with the changes in behaviour and life style needed if we are not only survive climate change but thrive" (Harman, 2007, p. 9). Therefore, any methods that can overcome climate change are worthwhile and should be considered as part of the growing sustainability agenda (Agyekum-Mensah et al., 2012). In addition, sustainable practice means achieving the right balance between three sustainable principles – social, economic and environment – in implementing construction projects in emerging economies. So, the need to enforce sustainable construction in emerging economies is significant as what people build today will provide the built environment of the future and will influence the ability of future generations to meet their needs (Dickie, 2000, cited in Pitt et al., 2008).

A great deal of research has been done on the issue of sustainable construction practice. For example, Hill and Bowen (1997) proposed a framework for attaining sustainable construction where the major components of the framework included project environmental assessment and environmental policy. Turfil (2006, cited in Tan et al. [2011]) developed a sustainability framework for small and medium contractors to improve their performance against four aspects of sustainability. A number of multinational organisations working in emerging economies are applying sustainability because of the business value in it, given that investigation shows that the tenants of over 2.4 million properties are willing to pay higher base rents for greener buildings (Miller et al., 2008; Presley and Laura, 2010). Thus, governments in emerging economies are increasingly requiring business consultants, contractors and suppliers to adopt sustainable policies in the construction process.

Globally, buildings are responsible for approximately 40% of the total world annual energy consumption (Omer, 2008). Much of the world's energy, however, is currently produced and consumed in ways that cannot be sustained. One way of reducing building energy consumption will be to design buildings that are more economical in their use of energy. For instance, the US Green Building Council (USGBC) developed a green building evaluation system, Leadership in Energy and Environmental Design (LEED), to evaluate the environmental performance of a building; this is an initiative for sustainability in the construction industry (Tsai and Chang, 2011). LEED is a concise framework that will provide building owners in emerging economies with a measurable green building design, construction and maintenance solutions. Construction organisations in emerging economies should strive to obtain this certification, which shows they have addressed the environmental impact of a building. It is worth mentioning effective implementation of project management principles and processes will help the built environment to achieve sustainability (Agyekum-Mensah et al., 2012). As stated by Curwell (1998, p. 15), "the construction industry as a whole has to rapidly come to terms with the broader environmental and social agenda that is presented by the concept of sustainable development because the built environment affects all human activity."

To facilitate an understanding of the nature of sustainability, this chapter examines the application of sustainability principles to construction projects being delivered in emerging economies. After this introductory section, the chapter is organised in ten main sections. The next section looks at the concept of sustainable development. Energy, population growth and pollution issues are then considered, followed by a section that provides an overview of sustainability. Sustainable construction, barriers to sustainability in construction, sustainable innovation, existing smart materials, key economic drivers for construction in emerging economies and policy implementation are then addressed in that order in sections devoted to them. The penultimate section concentrates on developing a framework for managing sustainable construction projects and the chapter is summarised in the last section.

2. Sustainable development

The Brundtland Report (WCED, 1987, p. 5) defined sustainable development as " . . . development that meets the needs of the present without compromising the ability of future generations to meet their needs". This definition shows that the balance of environmental and social issues are as significant as economic issues and suggests that human, natural and economic systems are interdependent. Since the concept gained currency following the publication of the influential Brundtland Report, businesses have been examining ways to assess how sustainable their operations are (Tahir, 2010). Besides, there is a growing concern that in the long run sustainable development may be compromised unless effective measures are taken to achieve a balance between environmental, economic and social outcomes. Although energy is not one of the three components of sustainable development, it is indirectly link to it. Given the close ties between energy and the objective of sustainable development, sustainable energy is considered to be a critical aspect of achieving sustainable development in emerging economies (Oyedepo, 2012). These problems must be taken seriously if humanity is to reach a bright future with minimal environment impacts.

Many other international and national meeting and conferences have shown the concern for conserving the environment for future generations, including the UN "Earth Summit" in 1992 in Rio de Janeiro which declared sustainable development to be overarching policy goal of governments (Parkin, 2000). In 1999, the DEFRA published a new strategy, *A Better Quality of Life: A Strategy for Sustainable Development for the United Kingdom*, which included four aims:

1   "social progress which recognises the needs of everyone;
2   effective protection of the environment;
3   prudent use of natural resources; and
4   maintenance of high and stable levels of economic growth and employment" (DEFRA, 1999).

3. Energy, population growth and pollution

Energy security, economic growth and environmental protection are the national energy drivers of any emerging country. Figure 7.1 demonstrates that as population increases, in many cases faster than the average of 2%, the need for more energy increases dramatically.

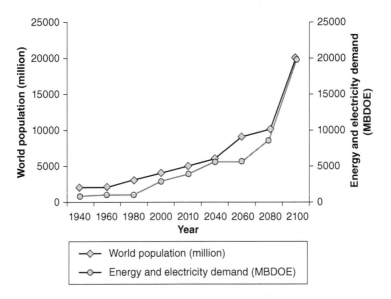

Figure 7.1 Annual estimated world population and energy demand

Source: Omer (2008).

According to WRI (1994), 2% of the world's land surface is covered by cities and the people living inside those cities consume 75% of the resources. If this continues, the rest of the world's population will be requiring the use of all sources of energy when oil and gas are depleted. Improved efficiency in energy use and renewable energy resources will be significant for stabilising populations while providing a standard living style all over the world (WRI, 1994).

At the current rate, taking into account population increases and the higher consumption of energy needed by emerging economies, oil, natural gas and coal will be depleted within a few decades. Consideration of natural resources and energy conservation has become a global issue as a result of climate change, global warming and the inconsistency of natural resources and energy demand due to the energy resources depletion (GhaffarianHoseini, 2013). Recently, environmental concerns have extended to focus on significant pollutions such as carbon dioxide. Buildings are by far the greatest producers of carbon dioxide and this "eco footprint" can only increase with the large population growth predicted to occur by 2050 (CIOB, 2013). Buildings account for more than 330 million tonnes a year of carbon dioxide released in the UK (CIRIA, 2004). The build-up of carbon dioxide (CO_2) and greenhouse gas emissions (GHG) is leading towards global warming, with unpredictable consequences.

4. Sustainability

Lynam (1989, p. 17) defined sustainability as "the capacity of a system to maintain output at a level approximately equal to or greater than its historical average with the approximation determined by the historical level of variability". IUCN (1991, p. 25) defined it as as "development that improves the quality of human life while living with the carrying capacity of the supporting eco systems". There are those researchers who argue, however, that resources should be defined more broadly to include stock of technology and know-how (Russell, 1994).

Sustainability currently forms a cornerstone for most development and socio-economic activities in the built and natural environments (Edum-Fotwe and Price, 2009). With the publication *Our Common Future* (WCED, 1987), sustainability

Table 7.1 Simplified matrix for sustainable practice

Basic views of sustainability	
Natural resources	*Desired outcome*
Energy	Generate clean and efficient energy to meet the needs of humanity
Water	Ensuring water quality and availability
Air	Sustaining clean and healthy air
Materials	Alter industrial patterns to reduce and eliminate unsustainable patterns of production and consumption
Land	Support ecologically sensitive land management and development
Ecosystems	Support biodiversity

became the focus of a major worldwide discussion and reflects predominantly the environmental dimension of sustainability (Lélé, 1991) creating a simplified matrix of the basic tenets of sustainable practice as shown in Table 7.1.

Besides, sustainability is increasingly becoming a common concept in all business, which can also mean the financial sustainability whereby a business entity has a responsibility to remain financially viable over the longer term (Linda, 2011).

5. Sustainable construction

In 1994, sustainable construction was defined as the creation of healthy built environment based on resource-efficient and ecologically based principles (Kibert, 1994 cited by Agyekum-Mensah et al., 2012). Since then, the industry has faced a lot of different interpretations of sustainability and a number of problems or doubts about how sustainability could be achieved. However, the Bruntland Report came out with a definition of "meeting the needs of the present without compromising the needs for the future generation". The definition underlay a variety of efforts to ensure a good quality of life for future generations. Sustainable construction not only refers to the buildings but also includes the processes or activities to build them. The UK's Department of the Environment, Transport and the Regions (2000, p. 20) proposed ten themes for action with the aim of providing individual firms with practical pointers for achieving sustainable construction, as shown in Table 7.2.

As sustainable construction initiatives continue to develop and gain popularity, critics and supporters alike are constantly evaluating progress (Presley and Meade, 2010). The evaluation of sustainability must include more than the

Table 7.2 Ten themes for action (*Building a Better Quality of Life*)

Ten themes for action	
Design for minimum waste	Think about using recycled materials
Re-use existing buildings	Renovate to improve their sustainability if possible
Aim for lean construction	Continuous improvement and high quality work
Minimise energy in construction and usage	Beware of energy consumption during construction processes and building energy efficient solutions
Eliminate pollution	Adopting Environmental Management System (EMS) or International Standard Organisation (ISO)
Preserve and enhance biodiversity	Throughout the construction process
Minimise water usage	Efficiency in use of water in construction and building
Social respect	Responsive to local community and workforce
Setting targets	Benchmarks own performance with others' projects and set targets for continuous improvement

Source: DETR (2000, p. 20).

immediate investors or tenants of the buildings but also consider suppliers, the local community in which the building sits and other stakeholders.

Sustainable construction can be an ingredient of the sustainable development agenda. Thus, sustainable development that balances social, economic and environmental goals (the so-called "triple bottom line") is now on the agenda for the global construction industry (Khalfan, 2002). The literature shows that the construction industry makes a vital contribution to the social and economic development of every country (Xiaoling et al., 2013). Its building sector has major impacts on the environment. It provides the basic living conditions for the sustainability and development of human life on Earth (Xiaoling et al., 2013). The built environment currently has a huge impact on the environment, raw materials and also the physical and economic health of individuals and communities. Sustainable construction can contribute to the achievement of sustainable development themes by embracing the following objectives (DETR, 2000):

- being more profitable and competitive;
- delivering buildings and structures that provide greater satisfaction, well-being and value to customers and users;
- respecting and treating its stakeholders more fairly;
- enhancing and better protecting the natural environment; and
- minimising its consumption of energy (especially carbon-based energy and natural resources).

5.1. Sustainable construction practice

In construction practice, sustainable construction refers to the different methods used in the construction project which bring less harm to the environment, benefit society and increase the profits of the company (Shen et al., 2010). Thus, sustainable construction practice refers to the various methods implemented into the process of construction projects that involve less harm to the environment (that is, minimizing waste) and maximise the use of waste production methods (as opposed to maximise waste produced) for the benefit of society and to add value to the company.

It is a well-known fact that successful delivering of projects in the construction industry depends on effective project management. Whitty (2013) considers project management to be a complex emergent behaviour. He (2013, p. 30) suggests that an integration of science-based techniques and project management tools may lead to an efficient delivery of construction projects. However, the problem of how sustainability could be achieved within the construction industry is compounded by the term "sustainable construction" being used to describe a process that starts well before construction, goes through construction and continues after construction (Hill and Bowen, 1997). Therefore, it is important to the life cycle processes of industrial and commercial building and of the stakeholders who play a role within these processes (Presley and Meade, 2010). Sustainable construction does not simply mean a construction company should continue its business growth but also the need to achieve the principles of sustainable development, which means it may need in some cases to stop growing or to grow in different ways (Du Plessis, 2002). Within the construction business, sustainability is about achieving a win-win outcome by contributing to an improved environment and social advancement, and in the meantime gaining competitive advantages and economic benefits for construction companies (Shen et al., 2010).

Some studies have suggested that embracing sustainable principles in the process of implementing construction projects can contribute to profit making (Tseng, 2009). The three dimensions of the "triple bottom line" concept (Thomas, 2002) are interrelated and they may influence each other in multiple ways. Table 7.3 demonstrate the benefits that a company operating in emerging economies may enjoy by integrating the sustainability principles into practice.

5.1.1 Environmental sustainability

The philosophy of environment sustainability is to preserve the earth for the future generations. It must be able to maintain a stable resource base, avoiding over-exploitation of renewable resource systems or the environmental sink function and depleting non-renewable resources only to the extent that investment is made in adequate substitutes (Harris, 2000). The global building sector needs to

Table 7.3 The benefits of adopting sustainable principles in construction practice

Environmental sustainability
- Sustainable performance can deliver significant business profit.
- Lean construction and reducing pollution can promise quantifiable cost savings.
- Recycling waste can reduce tax and materials cost.
- Improved efficiency through regulatory compliance and preventing fines.
- Improved public image and enhanced company reputation.

Social sustainability
- Emerging evidence of productivity gains for staff involved in environmental and social performance improvement schemes.
- Economic benefits come from good relationships with clients, local communities and other stakeholders.
- Better health and safety practices to improve efficiency and reduce accidents, saving managerial time and money.

Economic sustainability
- Significant opportunities available to enhance or sustain market position.
- Construction and business activities will not be controlled by local opposition, government bodies or client requirements.
- Improve project performance to develop trust with clients, improve profit and investment.
- Long term relationships with clients.

Source: Vitorino (2005).

cut energy consumption 60% by 2050 to meet global climate change, reducing water consumption and protecting its build quality, which are the main objectives in sustainable building. With the increasing population and higher living standards, these requirements will be a big issue and have a big impact on the water supply.

5.1.2 Social sustainability

Social sustainability has been described as a nebulous concept, a "concept in chaos" (Vallance et al., 2011) and there is little consensus on how it might be defined as it is argued that it covers a range of dimensions (Dempsey, 2012). CIRIA (2004) describes social sustainability as commitment by organisations to integrate social principles and the concerns of stakeholders into their organisations in a manner that fulfils and exceeds current legal and commercial expectations. Construction industries in emerging economies should promote a healthy living and socially cohesive communities. In a construction project, social sustainability recognises the needs of everyone related to construction from the initial stages of the project to demolition (including site workers, local communities, supply chain and end user). In emerging economies, societal expectations are changing, especially in relation to lifestyles and workplace practices, and contractors operating in emerging economies need to consider these changes. Perhaps the best way to implementing social responsibility is to examine the principles, which include inclusiveness, integrity, transparency, fairness and diversity (CIRIA, 2004).

5.1.3 Economic sustainability

Economics is important in terms of sustainability because of its broader social sciences meaning that explains production, distribution and consumption of goods and services. Harris (2000) stated that a sustainable economy must be able to produce goods and services on a continuing basis, to maintain manageable levels or government and external debt and to avoid extreme sectorial imbalances which damage agricultural or industrial production. Economically, construction contributes as a major sector for the global economy and a significant employer. The key contribution of the construction industry to economic sustainability is established through the efficient use of resources and materials, constant employment opportunities through formal construction, and through sustained investment and capital formation opportunities for the economy.

6. Barriers to sustainability in construction

Countries in emerging economies are currently facing a number of barriers regarding the implementation of sustainability principles in construction. These barriers have been highlighted by many researchers, for example, Hill and Bowen (1997) Williams and Dair (2006) and Kubba (2012), to include the following.

6.1 Financial resources to address sustainability

One of the main potential barriers for the implementation of sustainable construction is cost. Critically, the increased first cost associated with sustainable buildings is a major barrier for owners to pursue sustainable building objectives (Lapinski et al., 2006). The cost of providing environmental sustainable buildings and other developments is significantly higher than for standard schemes and most are not convinced that there is a potential demand for such buildings (Williams and Dair, 2006). Costs initially increase by an average of 2% to 7% for a sustainable building over an ordinary building and only some projects can recover overall net costs in a short period (Hill and Bowen, 1997). One of the biggest concerns is contractor finance and working capital (Kubba, 2012). Many contractors find it difficult to maintain their cash flow for operations and have sufficient resources for the expansion of working capital. These barriers still can be overcome by changing the view of stakeholders from cost to value and from short term to long term.

6.2 Lack of consideration from client and stakeholders

In the reviewed literature, it has been suggested that in some cases clients lack the information they need to make choices about which development options would be more or less sustainable (Williams and Dair, 2006). Sourani and Sohail (2011) mentioned that a low level of awareness and understanding about sustainability issues not only exists in people working in public client organisations but also among stakeholders, organisations and groups such as contractors, end users and funding organisations. This problem might be attributed to the lack of training on sustainability issues by several institutions and professional bodies, lack of clear structure and guidance, and the nature of relevant codes of practice in terms of being advisory rather than mandatory.

6.3 Insufficient knowledge and skills about sustainability

Although there are assessments tools and indicators already in place, the question of when to use them and who should use them is creating confusion and presents a burden to practitioners (Sourani and Sohail, 2011). There is a need to develop simple but wide-ranging tools and techniques to deal with situations where sustainability needs to be assessed. Addressing these problems will require significant and sustained investment in education and training alongside increased publicity (Marchman and Clarke, 2011).

6.4 Insufficient regulation by governments

Policies, regulations, incentives and commitment by leadership may not be sufficient to move towards the realisation of sustainable development. As Sourani and Sohail (2011) have stated, value added tax (VAT) was enacted on building refurbishments but not on new buildings and as Williams and Dair (2006) have emphasised, refurbishment is the better option from an environmental point of view than a new build. Even though there are many regulations and government policies in place to support sustainability issues, these authors have identified that such regulations and policies may be insufficient. There is, therefore, the need for a more mandatory role in order to better address sustainability in emerging economies.

6.5 Resistance to change

Addressing sustainability requires new ways of thinking, practices and attitudes. Therefore, it requires change. There is resistance to change and to the implementing of new initiatives in companies which have been in the industry for a long time. In client organisations, problems arise due to lack of leadership, restrictions on funding and lack of guidance.

6.6 Fragmentation

Another barrier to the promotion of sustainable construction in emerging economies is the industry's size and fragmentation. In Egan's report *Rethinking Construction* (1998), it was noted that this level of fragmentation has both strengths and weaknesses. On the positive view, it enables the industry to deal with highly variable workloads, whereas the negative view is that the extensive use of subcontracting has brought contractual relations to the front and prevented continuity of contracts to work as a team (Egan, 1998). Egan further suggested that partnering and framework agreements can be used as tools to tackle fragmentation so as to improve performance through agreeing mutual objectives and encouraging sustainable construction.

6.7 Sustainable building

Over the past decades, sustainable building, also known as green building (GB), has emerged as a new building philosophy, encouraging environmentally friendly resources, maximizing recycling and reducing waste production and placing emphasis on indoor environmental quality (Wang et al., 2005). Its approach to the built environment involves a holistic approach to the design of the building (Khalfan, 2002). Khalfan (2002) mentioned that although new technologies are increasingly being developed to cope with the current practice in building greener structures, the basic priorities for sustainable buildings are to reduce the overall impact of the built environment on human health and the natural environment. It is worth noting that the success of a sustainable building in emerging economies will depend on the quality and efficiency of the green systems installed. What surprises many people unfamiliar with this design movement is that good sustainable buildings often cost little or no more to build than conventional designs (Khalfan, 2002).

6.8 Cost–benefit for sustainable building

The perceived extra cost of sustainable building coupled with the perceived low value of its social quality has generally prevented action to date, except by the most committed. However, a significant aspect of the growing interest in green building and sustainable design can be attributed to the recognition by clients that there are direct economic benefits from green buildings (Halliday, 2008). A number of studies from around the world demonstrate a pattern of greener buildings being able to gain tenants and investor interest and commanding higher rents or sale prices (USGBC, 2013). Similarly, RICS (2011) suggested that, contrary to public perception, designing to environmentally efficient specifications does not necessarily result in higher capital cost.

6.9 The effectiveness of the drivers implemented

In considering the effectiveness of the drivers that have been implemented to achieve sustainable construction practices in emerging economies, it would perhaps be wise to consider people's moral views towards the environment and also the process of change. The following have been identified as key drivers:

6.9.1 Ethics and behavioural change

Ethics is the branch of philosophy that investigates morality and the ways of thinking that guide human behaviour (London et al., 2006). The philosophy of ethics requires that people take a step back from experience and reflect critically on it. Steg and Vlek (2009) suggested that when an environmental behaviour has been selected and its casual factors identified, intervention strategies can be targeted on those factors. In addition, Steg and Vlek (2009) note that economic analysis in ethics rests on a serious misunderstanding between "wants" or "preferences" and "beliefs" or "values". However, sustainable economics offers a different way to view economics and the environment, as sustainable economics is concerned with the resource rate flowing through the economy (Dyllick and Hockerts, 2002). Therefore, it makes sense to develop a sustainable economic system that uses resources at a rate that the earth can sustain – if humans realise natural resources originate from the earth.

6.9.2 Legislation and regulation

Cocklin and Blunden (1998) suggested that regulatory analysis and regulation theory provide appropriate foundations for the analysis of environmental problems and encourage firms to invest in green technologies. Tam et al. (2006) and Fraj-Andres (2009) indicated that fines and penalties for non-compliance to regulations will lead to a more cautious attitude to environmental compliance. Recent evidence (Jardins, 2009) provides a different view, that environmental regulations go too far, and mentioning that governments fail to consider economics or whether the overall social benefit provided may be worth the "cost" to the environment. Stricter environment policies are needed as drivers. From the above review, it can be observed that sustainable construction in emerging economies will be based on best practices which emphasise long term affordability, quality and efficiency.

7. Sustainable innovation

According to Edgeman and Eskildsen (2012), innovation has been cited as a key enabler of sustainability. It is also widely accepted that innovation is key to business success (Hansen et al., 2009). Innovation is considered as a key integrative thread of the gains realised by effective environmental policy implementation and effective use of enterprise excellence models. Integrative innovation is innovation which addresses societal, environmental and financial performance and thus integrates sustainability and enterprise excellence and does so regularly, rigorously, comprehensively and profitability (Edgeman and Eskildsen, 2012). These innovations are referred to as socio-ecological innovations and are a result of the strategic integration of sustainable innovation and innovation for sustainability. Edgeman and Eskildsen (2012) identified the following criteria that socio-ecological innovation addresses:

- reduce cost, risks, waste and deliver proof of value;
- focus attention on the redesign of selected products, processes or business functions to optimise performance;
- drive revenue growth by integrating innovative approaches into core strategies; and
- differentiate the enterprise value proposition through new business models that use innovation to enhance enterprise culture, brand leadership and other intangibles to secure durable competitive advantages.

Socio-ecological innovation results from the highly efficient and effective integration of sustainable innovation and innovation for sustainability. Such integration happens neither spontaneously nor accidentally, but rather purposefully (Edgeman and Eskildsen, 2012). Applying sustainability to innovation management is important, both from a moral and a business perspective (Hansen et al., 2009). From a moral perspective, researchers stress the role of organisations in solving global societal and environmental challenges. Multinational construction organisations in emerging economies are both responsible for and capable of engaging in these challenges due to the large amount of resources available to them. From a business perspective, there is a wide agreement that the challenges of sustainability in emerging economies offer significant potential for innovation and related business opportunities. New regulations and laws in emerging economies will increase the pressure for organisations to promote innovation (Hansen et al., 2009).

8. Existing smart materials

Over the years, building materials have changed and led to improved buildings: more structurally sound, better energy performance and less limited in their design. Now, however, more so than before, the advances in technology have led to massive innovations in construction materials. The building industry in emerging economies is known to be slow to advance in adopting new technologies and being innovative in their building design. This may be due to the fact that introducing new materials into a building design is a rather big risk. If the new materials do not perform in the way they were anticipated, then the cost to rectify this will be huge. It is safer to incorporate these innovative materials at a slow and steady pace to monitor how they perform in the long term. The innovation in smart building materials will allow safer, cheaper and quicker construction. Materials are continuously being developed for use in construction, and they are being developed in different fields of applications. Abdellatif (2011) has categorised smart materials into the following fields:

- biomaterials: fire retardant materials and structures;
- thin films: materials for electronics, photonics and magnetics;
- surface and interface characterisation: sensor materials and systems;
- nanoparticles, nano-composites and tissue engineering; and
- high-throughput/combinatorial methods: cementitious and other construction materials.

The uses of smart materials in construction can vary. Smart materials can be considered to be reused or recycled products as they have less or no embodied energy, which makes them more sustainable whilst serving the same function. Smart materials can also be considered to be materials that are modern and can adapt to a building, such as shape memory alloys (SMAs). Sinopoli (2010) suggested that integrating smart materials in a building's technology system will provide a more sustainable building as the aims of the smart materials fundamentally will meet the aims of the components that make up a building technology system. As a result of these combined, carbon emissions will be reduced and there will be increased energy performance.

9. Key economic drivers of construction in emerging economies

According Betts et al. (2011), environmental policy is likely to have an increased influence not only on global economic growth in general but more specifically also on construction activity. It is also difficult to quantify how truly committed governments in emerging economies will be to new environmental policies and whether they will in practice be willing to accept any major negative impact upon economic growth (Betts et al., 2011). Such policies, if aggressively pursued, could have substantial implications for construction activities in emerging economies. Moreover, the impact of construction may not necessarily be negative as climate change policy will probably imply major investment spending in both "smarter" building and improved infrastructure in areas such as water, electricity and other utilities. It is worth noting that better buildings in emerging economies remain more expensive to build. If built to the right standard, savings will come from benefits such as reduced energy costs, reduced maintenance needs and improved health of the users (Du Plessis, 2002). As suggested by Betts et al. (2011) and Du Plessis (2002), this is good for a planet on which the global population has more than doubled in the past 40 years and where more than 50% now live in cities. There has been rapid increase of new buildings in emerging economies. For instance, in China an estimated two billion square metres of new building surface is added to the building stock every year (Betts et al., 2011). In developed nations, the focus is on the refurbishment and modernisation of existing buildings.

10. Policy implementation

In both emerging markets and developed nations, governments are preparing new policies to encourage or force the construction sector to come up with more energy efficient buildings. For instance, the European Union's newly updated Energy Building Performance Directive, the Obama administration's recently launched Better Building Initiative and China's Building Energy Efficiency legislation are just a few examples of this trend (Betts et al., 2011). The need for sustainable construction projects in emerging economies goes beyond energy efficiency and climate change. Betts et al. (2011) found that the construction sector is responsible for more than a third of the global resource consumption, including 12% of all fresh water use, and significantly contributes to the generation of solid waste (estimated at 40%). Thus, there is the need to implement sustainable construction policies at all levels. The rationale for coming up with sustainable construction projects is shared by both developed countries and those with emerging economies. The culture to effect change and to achieve the benefits from sustainable construction needs to be rooted at local levels. The ability and willingness to respond to environmental policies will ultimately decide the shape of future emerging economies. What is clear in this chapter is that practitioners working in emerging economies cannot afford to ignore the environmental demands that are increasingly placed on the performance of construction projects. According to Betts et al. (2011), this will translate to the financial bottom line and be reflected in the economic value of properties. Those who implement the right policies will reap the maximum advantage.

11. Framework for managing sustainable construction projects in emerging economies

Based on the preceding discourse, there is a link between the application of sustainability principles and the success of a sustainable construction project. The integration of sustainability initiatives in emerging economies will fall under the following four themes:

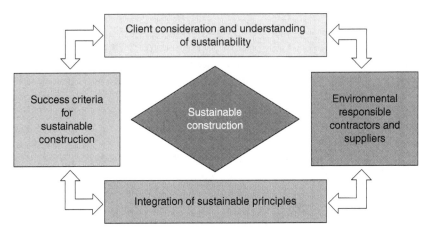

Figure 7.2 Proposed sustainable construction framework

- client consideration and understanding of sustainability;
- environmentally responsible contractors and suppliers;
- integration of sustainable principles;
- success criteria for sustainable construction.

These are also represented diagrammatically in Figure 7.2, whilst the integration of the "triple bottom line" in the construction process is shown in Figure 7.3.

As can be seen in Figure 7.2, consideration and understanding from the clients is essential for the integration of sustainable principles into construction projects being delivered in emerging economies. The right contractors and suppliers must be selected to ensure sustainable project delivery.

As illustrated in Figure 7.3, the triple bottom line should capture the essence of sustainability by measuring the impact of product, planning, process and design. It is worth mentioning that there is no universal standard method for assessing the triple bottom line. This can be viewed as a strength because it will allow clients and contractors operating in emerging economies to adapt general frameworks to the needs of different projects, policies or different geographic boundaries. Additionally, the integration of sustainability principles in construction projects is required for the management of current environmental issues and the attainment of significant improvements in performance in construction projects. What governments in emerging economies must consider is the integration of sustainability principles into construction projects, even though it is worth noting that not all sustainable factors are applicable to every type of project due to the complex nature of the global construction industry. It also appears that progressive training for construction professionals is important for the successful application of sustainability principles in construction projects. However, further research is required to establish whether such training would increase the level of sustainable practice within the construction industry.

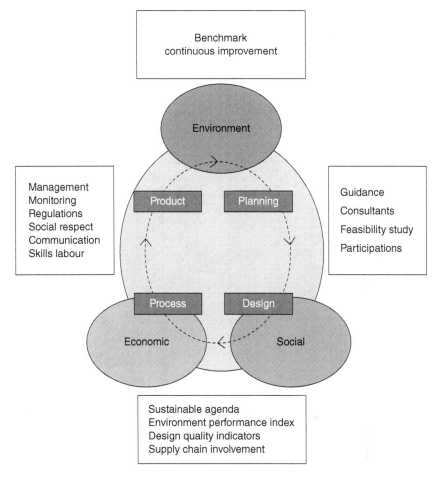

Figure 7.3 Integration of the triple bottom line into construction processes

12. Conclusion

As revealed in this chapter, a number of sustainable factors determine the integration of sustainability principles into construction projects being delivered in emerging economies. The evidence suggests that these factors need to be considered for the successful application of sustainability principles, thereby ensuring sustainable improvements in a construction project. It has been shown that the integration of sustainable principles into construction projects can lead to sustainable construction and improve project delivery. The literature reviewed indicated that LEED is a concise framework that will provide building owners in emerging economies with measureable green building design, construction, and maintenance solutions. Accordingly, these findings suggested that LEED has been designed with the three sustainable principles in mind. The review demonstrated that environmental tools such as LEED could directly affect the overall performance of a construction project. However the core finding of this chapter is the cost of getting environmental certification. It is worth noting that clients tend to aim for the minimum rating to minimise capital cost and expand profit. Therefore, the findings show that contractual obligations, regulations and a favourable cost–benefit ratio are the drivers for practitioners to aim for higher environmental certifications.

Sustainable factors can be incorporated in a construction project, resulting in the successful management and delivery of an environmental construction project. However, not all sustainable factors are applicable to each construction project due to the "variable" nature of the construction industry in emerging economies. Understanding and knowledge of the sustainability factors is necessary to ensure their successful application in construction projects and, most importantly, it is necessary to understand the drivers behind the use of these factors. It appears that practitioners in emerging economies tend to use the strategies implemented by the government only when incentives or benefits are in place. Furthermore, the influence of sustainability is significant in changing people's mind-set about sustainable construction and buildings. The chapter has reaffirmed that drivers are needed to ensure sustainable construction delivery and subsequently improve project performance. Sustainable construction can be successfully implemented in emerging economies when practitioners recognise the benefits and incentives in it.

References

Abdellatif, A. A. (2011) Innovations and sustainability in modern architecture [online]. Available at: http://www.academia.edu/3578261/Innovations_and_sustainability_in_modern_architecture_The_impact_of_Building_materials_selection [Accessed 16 September 2014].

Agyekum-Mensah, G., Knight, A. and Coffey, C. (2012) 4Es and 4 Poles model of sustainability: Redefining sustainability in the built environment, *Structural Survey*, 30(5), pp. 426–442.

Betts, M., Robinson, G., Burton, C., Cooper, A., Godden, D. and Herbert, R. (2011) *Global Construction 2020*, Oxford: Oxford Economics.

CIOB (2013). Sustainability and construction [online]. Available at: http://www.ipbas-sociates.com/uploads/Sustainability_construction_1287073859.pdf [Accessed 21 July 2013].

CIRIA (2004). How to deliver socially responsible construction – a client's guide [online]. Available at: http://www.ciria.org/service/Web_Site/AM/ContentManagerNet/Content Display.aspx?Section=Web_Site&ContentID=8962 [Accessed 19 July 2013].

Cocklin, C. and Blunden, G. (1998) Sustainability, water resources and regulation, *Geoforum*, 29(1), pp. 51–68.

Curwell, S. R. (1998) An introduction to the BEQUEST Project, paper presented at 7th International Workshop in the EU, Organised by SC as part of the BEQUEST project, Manchester, Amsterdam, Turin, Helsinki, Vienna, Milton Keynes, Florence [online]. Available at: http://www.surveying.salford.ac.uk/bq/extra [Accessed 3 July 2014].

Dempsey, N., Brown, C. and Bramley, G. (2012) The key to sustainable urban development in UK cities? The influence of density on social sustainability, *Progress in Planning*, 77, pp. 89–141.

Department for Environment, Food and Rural Affairs (DEFRA) (1999) *A Better Quality of Life: A Strategy for Sustainable Development for the United Kingdom*, Cm 4345, London: HMSO.

Department of the Environment, Transport and the Regions (DETR) (2000) *Building a Better Quality of Life: A Strategy for More Sustainable Construction*, London: DETR.

Du Plessis, C. (2002) *Agenda 21 for Sustainable Construction in Developing Countries*, Pretoria, South Africa: The Council for Scientific and Industrial Research (CSIR) Building and Construction Technology.

Dyllick, T. and Hockerts, K. (2002) Beyond the business case for corporate sustainability, *Business Strategy and the Environment*, 11(2), pp. 130–141.

Edgeman, R. and Eskildsen, J. (2012) *Socio-ecological innovation: Strategic integration of innovation for sustainability and sustainable innovation*, Denmark: Aarhus University.

Edum-Fotwe, F. and Price, A. D. F. (2009) A social ontology for appraising sustainability of construction projects and developments, *International Journal of Project Management*, 27(4), pp. 313–322.

Egan, S. J. (1998) *Rethinking Construction*, London: Department of Trade and Industry.

Fraj-Andres, E. and Martinez-Salinas, E. (2009) Factors affecting corporate environmental strategy in Spanish industrial firms, *Business Strategy and the Environment*, 18(8), pp. 500–514.

Halliday, S. (2008) *Sustainable Construction*, 1st ed., Oxford: Butterworth Heinemann.

GhaffarianHoseini, A-H., Dalilah Dahlan, N., Berardi, U., GhaffarianHoseini, A., Makaremi, N. and GhaffarianHoseini, M. (2013) Sustainable energy performances of green buildings: A review of current theories, implementations and challenges, *Renewable and Sustainable Energy Reviews*, 25, pp. 1–17.

Hansen, E., Grosse-Dunker, F. and Reichwald, R. (2009) Sustainability innovation cube – A framework to evaluate sustainability-orientated innovations, *International Journal of Innovation Management*, 13(4), pp. 683–713.

Harman, J. (2007). Construction industry could be next victim of climate change [online]. Available at: http://www.wwf.org.uk/filelibrary/pdf/allianz_rep_0605.pdf [Accessed 7 January 2013].

Harris, J. M. (2000) Basic principles of sustainable development [online]. Available at: http://ase.tufts.edu/gdae [Accessed 28 June 2013].

Hill, R. and Bowen. P. (1997) Sustainable construction: Principles and a framework for attainment. *Construction Management and Economics*, 15(3), pp. 223–239.

International Union for Conservation of Nature (IUCN), United Nations Environment Programme (UNEP), and World Wildlife Fund (WWF) (1991) *Caring for the Earth: A Strategy for Living*, Gland, Switzerland: IUCN.

Jardins, D. (2001) *Environmental Ethics: An Introduction to Environmental Philosophy*, 3rd ed., Stamford: Wadsworth Group, Thomson Learning Inc.

Khalfan, M. M. A. (2002). Sustainable development and sustainable construction [online]. Available at: http://www.c-sand.org.uk/Documents/WP2001-01-SustainLitRev.pdf [Accessed 8 January 2013].

Kubba, S. (2012) Green design and building economics, in: *Handbook of Green Building Design and Construction*, pp. 493–528.

Lapinski, A. R., Horman, M. J. and Riley, D. R. (2006) Lean processes for sustainable project delivery, *Journal of Construction Engineering and Management,* 132(10), pp. 1083–1091.

Lélé, S. (1991) Sustainable development: A critical review. *World Development,* 19(6), pp. 607–621.

London, K., Oliver, J. and Everingham, P. (2006) Ethical behaviour in the construction procurement process, Research project No: 2002-62 [online]. Available at: http://

www.construction-innovation.info/images/pdfs/Research_library/ResearchLibraryA/
Project_Reports/Ethical_Behaviour_in_the_Construction_Procurement_Process.pdf
[Accessed 12 November 2014].

Lynam, J. K. and Herdt, R. W. (1989) Sense and sustainability: Sustainability as an objective in international agricultural research, *Agricultural Economics*, 2(6), pp. 127–143.

Marchman, M. and Clarke, S. N. (2011) *Overcoming the Barriers to Sustainable Construction and Design through a Cross-Reference of West Coast Practices*, Clemson, SC: Associated Schools of Construction.

Miller, N., Spivey, J. and Florance, A. (2008) Does green pay off? *Journal of Real Estate*, 4, pp. 385–399.

Murota, Y. (1996) Global warming and developing countries: The possibility of a solution by accelerating development. *Energy Policy*, 24(12), pp. 1061–1077.

Omer, A. M. (2008) Energy, environment and sustainable development. *Renewable and Sustainable Energy Reviews*, 12(9), pp. 2265–2300.

Oyedepo, S. O. (2012) On energy for sustainable development in Nigeria, *Renewable and Sustainable Energy Reviews*, 16(5), pp. 2583–2598.

Parkin, S. (2000) Context and drivers for operationalising sustainable development, *Proceedings of ICE*, 138, pp. 9–15.

Pitt, M., Tucker, M., Riley, M. and Longden, J. (2008) Towards sustainable construction: Promotion and best practices, *Construction Innovation: Information, Process, Management*, 9(2), pp. 201–224.

Presley, A. and Meade, L.. (2010) Benchmarking for sustainability: An application to the sustainable construction industry, *Benchmarking: An International Journal*, 17(3), pp. 435–451.

Royal Institution of Chartered Surveyors (RICS) (2011) *Royal Institution of Chartered Surveyors: A Vision For Sustainability*, London: RICS.

Russell, D. (1994) *Theory and Practice in Sustainability and Sustainable Development*, Washington: Research and Reference Services Project.

Shen, L-Y., Tam, V. W. Y., Tam, L. and Ji, Y-B. (2010) Project feasibility study: The key to successful implementation of sustainable and socially responsible construction management practice, *Journal of Cleaner Production*, 18(3), pp. 254–259.

Sinopoli, J. S. (2010) Smart buildings systems for architects, owners and builders [online]. Available at: http://www.sciencedirect.com/science/article/pii/B978185617653800017X [Accessed 16 June 2014].

Sourani, A. and Sohail, K. (2011) Barriers to addressing sustainable *construction in public procurement strategies, Proceedings of the Institution of Civil Engineers: Engineering Sustainability*, 164(4), pp. 229–237.

Steg, L. and. Vlek. C. (2009) Encouraging pro-environmental behaviour: An integrative review and research agenda, *Journal of Environmental Psychology*, 29(3), pp. 309–317.

Tahir, A. C. and Dartin, R. C. (2010) The process analysis method of selecting indicators to quantify the sustainability performance of a business operation, *Journal of Cleaner Production*, 18(16–17), pp. 1598–1607.

Tam, V. W. Y., Tam, C. M., Shen, L. Y., Zeng, S. X. and Ho, C. M. (2006) Environmental performance assessment, perceptions of project managers and the relationship between operational and environmental performance indicators, *Construction Management and Economics*, 24(3), pp. 149–166.

Tan, Y., Shen, L. and Yao, H. (2011) Sustainable construction practice and contractors' competitiveness: A preliminary study, *Habitat International,* 35(2), pp. 225–230.

Thomas, D. K. H. (2002) Beyond the business case for corporate sustainability, *Business Strategy and the Environment*, 11, pp. 130–141.

Tsai, C. Y. and Chang, A. S. (2011) Framework for developing construction sustainability items: The example of highway design, *Journal of Cleaner Production*, 20(1), pp. 127–136.

Tseng, M. L., Lin, Y.-H. and Chiu, A. S. F. (2009) Fuzzy AHP based study of cleaner production implementation in Taiwan PWB manufacturer, *Journal of Cleaner Production*, 17, pp. 1249–1256.

USGBC (2013) United States Green Building Council. http://www.usgbc.org/ [Accessed 4 July 2013].

Vallance, S., Perkins, H. C. and Dixon, J. E. (2011). What is social sustainability? A clarification of concepts, *Geoforum*, 42, pp. 342–348.

Vitorino, M. (2005) *A study investigating how construction companies in the UK interpret and implement sustainable construction practice.* PhD thesis, University of East Anglia.

Wang, W. Zmeureanu, R. and Rivard, H. (2005) Applying multi-objective genetic algorithms in green building design optimization, *Building and Environment*, 40(11), pp. 1512–1525.

Whitty, J. (2013) Thinking in slow motion about project management, in: Drouin, N., Müller, R. and Sankaran, S. (Eds), *Novel Approaches to Organisational Project Management Research: Translational and transformational*, Advances in Organisation Studies 29, Copenhagen: Copenhagen Business School Press, pp. 95–116.

Williams, K. and Dair, C. (2006) *What Is Stopping Sustainable Building in England? Barriers Experienced by Stakeholders in Delivering Sustainable Developments*, Chichester: John Wiley and Sons Ltd.

World Commission on Environment and Development (WCED) (1987) *Our Common Future* (the Brundtland Report) [online]. Available at: http://conspect.nl/pdf/Our_Common_Future-Brundtland_Report_1987.pdf [Accessed 29 November 2012].

WRI (1994) *World Resources: A Guide to the Global Environment,* Washington, USA: People and the Environment.

Xiaoling, Z., Wu, Y., Shen, L. and Skitmore, M. (2013) A prototype system dynamic model for assessing the sustainability of construction projects. *International Journal of Project Management* 32, pp. 66–76.

8 The economic, technological and structural demands of the construction industry in emerging markets south of the Sahara

A case study of Ghana

Divine K. Ahadzie

1. Introduction

That Africa is emerging as the place to invest in the twenty-first century is not in doubt, and the statistics are indeed striking. In 2011, the combined GDP of some of the major players in the region amounted to US$797,797 billion (World Bank, 2014; OECD National Accounts Data Files, 2014). Presently, the combined GDP of these major players, namely Nigeria (US$262,597), South Africa (US$384,313), Kenya (US$40,697), Angola (US$114,147) and Mozambique (US$14,244), amounts to US$815,998 billion (ibid.): that is a 2.3% increase in GDP in just two years. The World Bank (2014) projects that this current GDP will grow at 5.5% annually, which is phenomenal for the growth prospects of the region.

It is worth noting that within the first decade of this century, six of the world's ten fastest economies came from Africa, namely Angola, Nigeria, Ethiopia, Chad, Mozambique and Rwanda (*Africa Watch*, July 2012, p. 48). Also, within the last few years the region is reported to have posted an annual growth rate of 4.8% as against 3.3% in Latin America and 0.5% in the Western economies (IMF Global Monitoring Report, 2013; cited in *Africa Watch*, August 2013, p. 98). This is a region which in the last 40 years or so was largely agrarian and relied very much on traditional exports to offset trade balance. However, many economies have now diversified into processing and manufacturing bases and it is noted that within the last two decades, returns on investment in the region are now the highest in the world (*Africa Watch*, June 2013, p. 80). The net effect is that small and medium sized companies are expanding plus the services industry is in relatively high demand. An example is mobile telephony, which is now at 99% penetration in most countries and accounting for approximately 6% of GDP in 2013 (Delloite LLP GSMA, 2012; *Africa Watch*, June 2013, p. 74; *Africa Watch*, September 2013, p. 50). Another example is the stock exchange, which is reported to have gained 35% and 39% respectively in Nigeria and Uganda in 2012 (*Africa Watch*, August 2013, p. 98). There has also been a marked increase in Foreign Direct Investment (FDI), which currently stands at US$40 billion as against US$6 billion in the year 2000, gaining up to 15% in the last decade (*Africa Watch*, August 2013, p. 98). Following investor confidence and a boost in trade, airline traffic is reported to have achieved an annual growth of 11% in international passenger movement, and expectations are that there is huge potential for further expansion (International Air Transport Association, 2013; cited in *Africa Watch*, November/ December 2013, pp. 76–78). Generally, analysts have commended the continent for the sustained high growth achieved for several years now and indications are that the region will continue to be among the fastest growing economies in the world (World Bank, 2013). This impressive sustained economic growth has inspired analysts to project that the region could indeed be on the verge of being more attractive to investors than Asia or Latin America (IMF Global Monitoring Report, 2013; cited in *Africa Watch*, August 2013, p. 98).

Contextually, the recent economic growth in Africa is significant in how it can be used for the accelerated expansion of "new value fixed capital formation" that can be put towards transforming the ageing and outdated infrastructure in many emerging countries in order to reach international standards. Urbanisation and scarcity of urban land is also pushing for a new paradigm in urban land use, and a number of countries are now pushing for high-rise buildings as against the traditional sprawl development. A number of high profile projects are to be delivered in various parts of Africa, which includes the proposed airport expansions in Burkina Faso, Sierra Leone, Dakar, Nigeria, Ethiopia and Ghana (*Africa Watch*, November/December 2013, pp. 74–77). It is also noted that the unofficial race for the continent's tallest building is in contention, with the Republic of South Africa pushing for the construction of a 110 storey (1,446 feet) building comprising offices and shops; Ethiopia is also looking up to construct a 99 storey complex to reach 1,469 feet for completion by 2017; while Ghana is seeking to build Africa's answer to Silicon Valley, a US$10 billion technology hub of 75 storeys (902 feet) called Hope City (*Africa Watch*, November/December 2013, p. 74). In railway construction, Kenya has formally launched a new US$3.2 billion Chinese-financed railway, which would extend across East Africa to reach South Sudan, DR Congo and Burundi. This is said to be the country's biggest infrastructure project since independence 50 years ago (Kenya Railways, 2013; *Ghanaian Times*, 2014).

Notwithstanding the opportunities offered by the impressive economic fortunes and the prospects they bring for embarking on ambitious infrastructure projects, visible threats also do exist in many ways that need to be seriously addressed. Indeed, while many companies have confidence in investing in Africa, a survey by *Africa Watch* (August 2013, p. 27) suggests that many have a limited understanding of the continent, especially because many microeconomic and demographic factors are lacking. Energy deficiencies cause immense concerns and corruption is also constantly being highlighted. While these concerns are indeed legitimate, it ties in very well with the copious lack of deep industry insights in many emerging sectors in developing countries, including the construction industry. The purpose of this chapter, which focuses on Ghana, is to create a knowledge base for a deeper understanding of the opportunities and threats that exist for using the industry as the engine for engendering growth and job creation in the wider Ghanaian economy. While the focus is on Ghana, there is the potential for analysts to have some basis to reflect on the inherent opportunities and challenges that exist in emerging construction markets in the sub-region.

The rest of the chapter is structured as follows. The first three sections after this introduction provide overviews of Ghana as an emerging economy, the emerging economic outlook of the Ghanaian construction industry, and governance and institutional landscape of the Ghanaian construction sector, respectively. Next, employment opportunities and key market niches are considered. A framework for understanding and isolating economic, technological and structural demands (ETSDs) is then developed, which is used for evaluating the ETSDs in the subsequent section, whilst the last section deals with conclusions.

2. Ghana as an emerging economy

Ghana was the first "black" country south of the Sahara to achieve independence from British rule, which it did in 1957. A prosperous country at independence, Ghana began to experience socio-economic hardships soon after independence and this has, among other reasons, largely been attributed to political instability and economic misrule (Breisinger et al., 2008). The tip of the iceberg was in the mid-1970s and early 1980s when growth stagnated into negative values (Breisinger et al., 2008). It was not until 1984, following prescriptions by the International Monetary Fund (IMF) underpinned by a relatively long period of political stability, that the country was put back on a recovery track under a long period of Economic Recovery Programme (ERP) work (World Bank, 2010). By all standards, the ERP was declared a success, helping to achieve a consistent positive growth rate of 4% to 5% in the ensuing years (ISSER, 2008). This notwithstanding, Ghana still faced many developmental challenges such as poverty, high indebtedness and a dependency on limited primary products to the extent that livelihood remained heavily skewed toward agricultural and rural activities. However, the economy was to receive a boost following governance reforms culminating in multi-party elections in 1992, which have since generated investor confidence.

Ghana is indeed now starting to reap the full benefits of its democratic credentials and is now touted as one of the fastest growing developing countries in sub-Saharan Africa with an impressive GDP growth rate of 8% in recent times (ISSER, 2008, 2013). Following rebasing of the growth indicators in 2011, Ghana now has the one of highest levels of per capita income (that is, GDP/capita over US$1,000) in West Africa and is classified as a low–middle income emerging country (ISSER, 2013). The trend in the growth of GDP and GDP per capita is shown in Figures 8.1 and 8.2. Figure 8.1 shows that within the last decade, the GDP of Ghana has increased substantially from US$5 billion in 2000 to US$40 billion in 2011. Similarly Figure 8.2 indicates that GDP per capita increased from US$200 in 2000 to US$1,600 in 2011, representing a 700% increase over the last decade. Admittedly, the momentum slowed down in 2012 but only marginally as Ghana still performed better in the sub-region, growing over 2% higher than the sub-regional average (ISSER, 2013). Indeed Ghana's development trajectory in recent times has been so striking and some analysts have suggested this growth performance does not only contrast with the country's turbulent experience in the

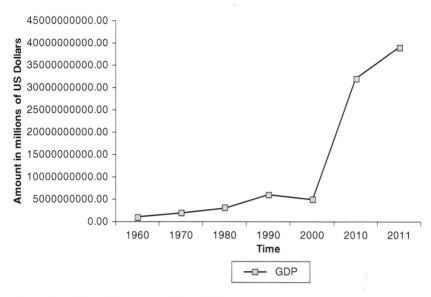

Figure 8.1 GDP of Ghana from 1960 to 2011

Source: World Bank (2014) and OECD National Accounts Data Files (2014).

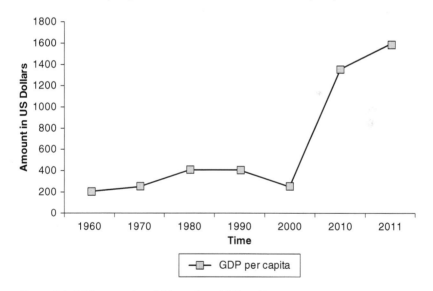

Figure 8.2 GDP per capita of Ghana from 1960 to 2011

Source: World Bank (2014) and OECD National Accounts Data Files (2014).

late 1970s or early 1980s but it is also globally unusual (Breisinger et al., 2008). Among many economies, both developed and developing, only a few are noted to have maintained such a consistent growth rate for such a considerably long time (Breisinger et al., 2008).

According to *Africa Watch* (August 2013, p. 26), analysts believe that the elevated confidence that the global business and capital markets have in Ghana was demonstrated recently in its second Eurobond, which was oversubscribed to the tune of US$2.2 billion when the country needed just US$1 billion to help finance infrastructural and other projects. Indeed, the indicators are good for Ghana as the most attractive emerging African market with political stability; an oil production economy; the second largest gold-mining industry; the largest cocoa production in Africa; a boom in commodity prices spurred by private sector spending; plus an increasing household consumption and service industry spurred by fast growing middle and upper classes. Infrastructure such as business and shopping malls are also catching up, to the extent that market observers are predicting that soon Ghana will be the preferred shopping destination of the upper class of nearby countries, instead of South Africa and Britain. It is refreshing to note that in a recently conducted survey, a panel of 62 experts rated Ghana as the preferred investment destination for business in the sub-region, which can be attributed to its oil emerging status, natural resource wealth, robust economic growth, and commitment to democratic principles (*Africa Watch*, August 2013, pp. 27–38).

The signs are indeed positive for Ghana, yet analysts are also cautioning that Ghana needs to be extra vigilant, especially as the last couple of months of 2014 turned rather turbulent with the free fall of the cedi and huge external debts. Added to this is the thorny issue of corruption, which keeps cropping up every now and then and appears to be gaining infamous momentum (see Transparency International Index 2013 on Corruption, cited by *Africa Watch*, February 2014, p. 24). These could indeed threaten Ghana's promising future if care is not taken. Naturally, the government of Ghana has had to assure the business and investor community that the economy is resilient and that it is doing everything possible to stamp out the shady deals that could derail the good fortunes that the country is enjoining. The Bank of Ghana also instituted some tough measures to help stem the free-fall of the cedi (Bank of Ghana, 2014). However, in the latter part of 2014, the Cedi began to stabilise while the government started engaging with the IMF to help sanitise and bring the economy back on stream. The engagement with the IMF has recently been concluded and Ghana is now on a bailout package of US$ 918 million dollars over a three year period (IMF, 2015).

3. Economic outlook of the Ghanaian construction industry

In a typical modern society, approximately half of all physical assets (that is, the gross fixed capital formation) are generated by the construction industry, accounting for up to about 10% of the national wealth (GDP) (Winch, 2002). In 1980, overall construction output in Ghana was estimated at US$180 million (Edmonds and Miles, 1984) and this has more than doubled to about US$600 million[1] within the last two decades. Following the rebasing of the economy in 2011, it is projected that construction could be accounting for up to US$4 billion of GDP, with labour earnings totaling up to US$1 billion. The industry currently also employs[2] approximately 400,000 people, many of whom operate in the informal sector (Ghana Statistical Service, 2013), which has an annual projected growth rate of 7–9%; this could help put construction employment pass the one million mark in the next decade.[3] The growth rates attained by the construction sector over the last decade are shown in Figures 8.3, 8.4, 8.5 and 8.6. As shown in Figure 8.3, in 2006 the sector emerged as the fastest growing sub-sector in the economy, with a growth rate of 8.2% as against a national average of 6.2%; while in 2007 it attained a growth rate of 11%, the highest for a decade, the sector's share being 8.9% to GDP (see also Anaman and Osei-Amponsah, 2007). The trend also shows that in 2009 and 2010 there was indeed a drop in the general economy (Figure 8.4) and the construction sector also failed to maintain the high growth rate of 17% achieved the year before.

However, it still fared better in overall industry performance, bouncing back to maintain a GDP contribution of 8.9% in 2011 (see Figures 8.5 and 8.6). Indeed, the sector continued to rank as the most important industrial sector posting 10.5% contribution to GDP in 2012 (see Figure 8.6). This impressive GDP rate could be much higher if construction activities in the informal sector (for example, self-build housing and the construction of small retail shops), which are now very

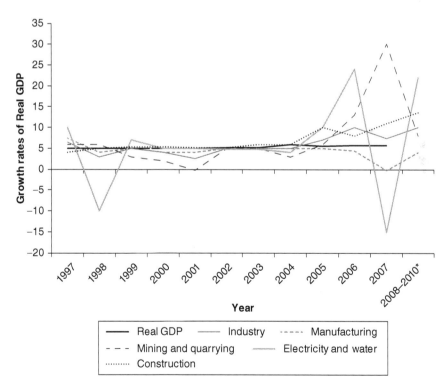

Figure 8.3 Growth rates of real GDP, industry and its sub-sectors

Source: Adapted from ISSER (2008).

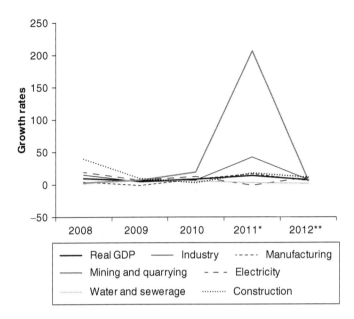

Figure 8.4 Sectoral growth rates

Source: Adopted from ISSER (2013).

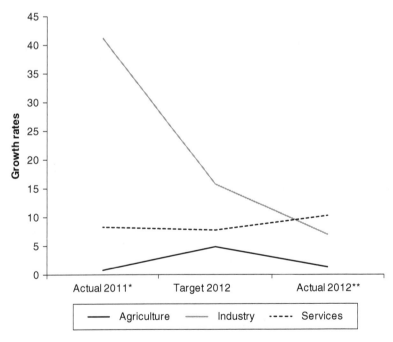

Figure 8.5 Annual rates of real GDP and industrial sub-sectors

Source: Adapted from ISSER (2013).

dominant in the Ghanaian market, are properly accounted for. It was anticipated that with the economy showing signs of recovery from the 2009 and 2010 growth performance, both domestic and foreign investors would be attracted to boost infrastructure development in the ensuing years (ISSER, 2013). However, with the recent set-back for which Ghana is now seeking engagement with the IMF for assistance, it is not exactly clear how things are going to turn out. Naturally, the government keeps assuring the business community that, notwithstanding the engagement with the IMF, the economy is still resilient and will bounce back to the enviable status quo very soon. What is at least refreshing here is that the government recognises the need for robust fiscal measures to turn the challenging situation around.[4]

Indeed with the stability achieved over the past two decades, there has been increased international presence in the industry, which heralds well for technology transfer in the industry. Some of the main international groups are Bilfinger Berger (Germany), Sogea-Satom (France), Sade (France), China International Construction Co. (China) and Odebrecht International, the global engineering and construction firm selected to construct a portion of Ghana's Eastern Corridor road system (Anaman and Osei-Amponsah, 2007; *Africa Watch*, August 2013, p. 64). These companies are actively present in major infrastructure projects in the markets. Presently, the Institute of Social Statistical and Economic Research (ISSER), Ghana's most recognised think tank, projects that the industrial sector will be the driver of the economy in the medium to long term (ISSER, 2013). With the construction sector playing the leading role, there is no doubt that it holds much potential in leading the industrial sector in engendering significant growth and job creation in the wider economy.

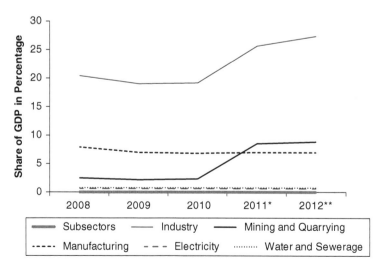

Figure 8.6 Sectorial contribution to GDP

Source: Adapted from ISSER (2013).

4. An overview of the governance and institutional landscape of the Ghanaian construction sector

Figure 8.7 provides an overview of the governance and institutional landscape existing in the Ghanaian construction industry. As in many Commonwealth countries, a key feature of the Ghanaian construction environment is the tripartite nature of clients, consultants and contractors as independent actors. However, while the consultants and contractors have a common front through their professional and association affiliates, clients are yet to come to terms to the fact that they also need to group and help exert their influence on the development of the industry. The stronghold on traditional procurement practices, whereby the responsibility for the design of the project is separated from construction, is also copiously manifested. The underlying legislative framework is enshrined in the Public Procurement Act 2003 (Act 663); LI 1630 of the National Building Regulations; and the bylaws of the various decentralised government areas, which are set out in the Local Government Act 1993 (Act 462). Under Ghana's Public Procurement Act, the standard form of contract for construction work is based on the International Federation of Consulting Engineers (FIDIC) version. Prior to the enactment of the Public Procurement Act in 2003, the conditions of contract then in use (popularly called the pink form[5]) were based on the Joint Contract Tribunal (JCT) version of the UK. A number of policy initiatives such as the National Housing Policy, National Water Policy, National Transport Policy and the National Urban Policy Framework (launched in May 2012) are in operation. An engineering bill has also been in Parliament for quite some time, waiting to be passed into law.

The Ministry of Water Resources, Works and Housing (MWRWH) has oversight responsibility for policy implementation for designated areas and also oversees the activities of building contractors, especially on public projects. Two main classifications exist for contractors: category "D" for general building works; and category "K" for civil works. The classifications are based on financial standing, equipment holding, personnel employed and past experience. D1/K1[6] contractors can tender for jobs over US$500,000, D2/K2 between US$200,000–500,000, D3/K3 between US$75,000–200,000, and D4/K4[7] up to US$75,000. Electrical contractors are classified as E1 (which can execute works above US$200,000), E2 (US$75,000–200, 000) and E3 (up to US$75,000). Plumbing contractors are classified as G1 (for works over US$75,000) and G2 (up to US$50,000). The Ministry of Roads and Highways is also responsible for contractors that work in the road sector and oversees the Ghana Highway Authority (GHA), Department of Urban Roads (DUR) and Department of Feeder Roads (DFR).

Class A– contractors are qualified to carry out roadworks, airports and related works, Class B– contractors are qualified to undertake bridge construction, culverts and other drainage structures, while Class C– contractors are qualified to carry out labour based works. Class S– contractors are qualified to construct structures and Class M– contractors are qualified for miscellaneous (for example, minor specialist) road related works. The Ministry of Environment, Science and Technology (MEST)[8] oversees the Environmental Protection Agency (EPA) and Town and Country Planning (TOCP) to implement policies towards enforcement of environmental and zoning laws and building regulations, while the Occupational Health Unit of the Ministry of Health (MOH) is responsible for occupational and public health issues relating to construction works. The Ministry of Lands, Forestry and Mines (MLFM) has many departments and agencies such as the Lands Commission, Survey Department, Office of the Administrator of Stool Lands, Land Valuation Board and Land Title Registry, and these agencies and bodies have roles that could impact on the success of construction activities.

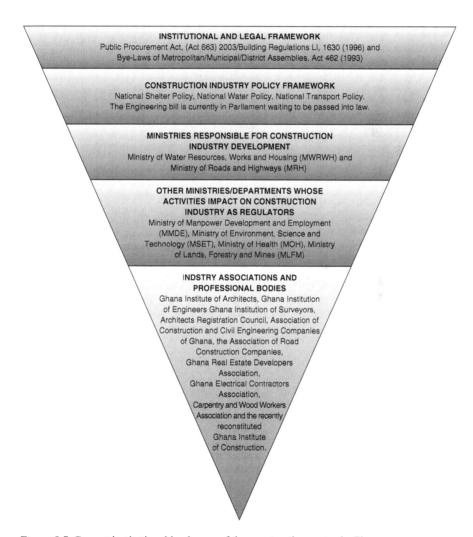

Figure 8.7 Current institutional landscape of the construction sector in Ghana

There is a lack of firm data on exactly how many contractors are operating in the Ghanaian construction sector. Some estimates[9] suggest that there could be over 50,000 firms operating[10] in various sub-sectors of the industry. These companies constitute several intermediary players and key amongst them are the Association of Construction and Civil Engineering Companies of Ghana (ABCCG), the Association of Road Construction Companies (ASROC), the Ghana Real Estate Developers Association (GREDA) and many more who are in the informal sector. Professional bodies are also affiliated, key amongst them the Ghana Institution of Engineers (GhIE), Ghana Institution of Surveyors (GhIS), Ghana Institute of Architects (GIA) and Ghana Institute of Construction (GIOC).[11] While these organisations are playing their respective roles towards the development of the Ghanaian construction industry, what is missing is the institution of an apex body such as the Construction Industry Development Board (CIDB) that could provide decisive and requisite leadership in coordinating the sector towards engendering a systematic construction industry development agenda. It is also worth noting that while professional project management practice as a discipline/profession is gaining some grounds and acceptance within the construction sector, the project management environment is still characterised by inefficiencies in construction procurement practices, excessive bureaucratic conditions, poor communications and administrative practices, lengthy payment procedures and excessive delay in projects. Hitherto, the pink form made specific reference to the architect and/or engineer (especially for civil engineering works) as team leader and responsible for the overall administration of public projects. However, the title Project Manager is now preferred as per the spirit of the manuals of the Public Procurement Act 2003 (Act 663). However, empirical evidence indicates that on projects where the title has been assumed, the PM generally plays the traditional administrative roles reminiscent of the traditional procurement practices rather than assuming the authority and responsibility for the management of the design and construction of the project from inception to completion. Thus, it can be argued that although there is room for improvement, Ghana is still a long way from integrating professional project management practices as a core managerial discipline in the construction sector (Ahadzie et al., 2012, 2014).

The lack of pressure groups from client organisations in the Ghanaian industry is also a major weakness which needs to be addressed urgently. Given that the industry is now largely private sector led, it is high time that major client bodies with large portfolios in construction formed a recognised pressure group to demand improved performance from the sector. Here, lessons could be learnt from the British Property Federation, which was formed in the 1980s to influence performance in quality and project delivery in the UK industry (Masterman, 1992). Within this context of professional influence, it is refreshing to note that the GIOC has now been reconstituted and it is hoped that it will finally come to make its impact felt following a number of earlier failed attempts. Hopefully, this should help weed out unprofessionalism in the industry, as currently many of the registered contractors who go under the umbrella body of the Association of Building and Civil Engineering Contractors cannot "technically" be classified as professional builders but instead are what Fellows et al. (2001), for instance, describe as "hopeful entrepreneurs" who find themselves in construction because they have the funds and entry to the industry is easy.

5. Employment opportunities and key market niches

Overall construction employment prior to 1966 was 18% of total recorded employment. Between 1966 and 1990, it dipped to 10% of total recorded employment and estimates from Ghana's national population census in 2010 have put the figure at around 3% of total recorded employment from the 1990s to date. Notwithstanding the significant percentage drop over the decades, in terms of real numbers, the current estimated rate of 3% represents approximately 400,000 employees as against approximately 70,000 achieved prior to 1966 (Ghana Statistical Service, 2013). It is also important to note that prior to 1966 construction activities were largely publicly funded, accounting for as much as 80% employment in the public sector; between 1966 and 1990, employment trends stood at 67% and 34% for the public and private sectors respectively (Edmonds and Miles, 1984; Ghana Statistical Service, 1995–2008). However, following privatisation of the then State Construction Company (SSC) and reforms that took place in many public institutions that operated construction departments, employment in the public sector dipped from the high of 67% to a mere 5% by the late 1990s. Presently, the government of Ghana remains the major investor in the heavy engineering and road construction sectors while private financing dominates particularly in the housing and building sub-sectors. Typical of the construction industry in many developing countries, close to 70–80% of the employment generated in the private sector, especially at the semi-skilled and non-skilled levels, is found in the informal[12] sector (Ganesan, 1983; ILO, 2001; Wells, 2007; Ghana Statistical Service, 2013). Within this context, employment at the professional level could be accounting for 8% of the total and the skilled level could be accounting for 25%, while semi-skilled and non-skilled levels could be accounting for 67% (Ghana Statistical Service, 2008). With increasing sophistication and emerging heavy technological use in construction, there is set to be a decline in the high use of semi-skilled/ unskilled labour and this will be one of the future socio-economic challenges that the sector has to counter by way of appropriate training. Importantly, it is high time that the government of Ghana draws on its urbanisation agenda to support the private sector to bring employment[13] levels to higher percentage levels than those previously achieved in the 1960s. This among others is the way forward if the government is to reap the full potential of the industry going into the future.

In Ghana, housing provision is largely of two forms: large scale houses built by commercial speculative residential building companies, such as those belonging to the Ghana Real Estate Developers Association (GREDA), and house building financed by individual owner-occupiers (hereinafter referred to as "self-build"[14]). In the self-build sub-sub-sector, clients mostly engage artisans in the form of "labour only contracts"[15] and this is highly informal. This self-build housing provision model does not promote the development of mass housing schemes, yet it remains the housing realisation culture in Ghana and currently accounts for about 90% of the housing stock.

All in all, the residential sub-sector represents the greatest opportunity for growth and job creation, especially in the private sector. Drawing from Amoa-Mensah (2003), the sector could be accounting for over 50% of the total market share with annual construction output in the region of US$900 million. A key opportunity in Ghana's residential sector is the vast market size of about one million houses needed in the short term and about four million house-units[16] needed in the long term[17]. Currently the trend (especially with private sector housing) is to build single-family houses; however, with the increasing population and the looming scarcity of land, there is a gradual paradigm shift towards apartments, multi-storeys and condominiums and this has particular technological challenges for many of the artisans who operate in the informal sector. That is, the suburban framework that enjoins families to normally enjoy their own spatial environment in single-family detached houses is gradually being subject to new thinking towards greater density and far less personal space for both individual and collective use. This trend is going to determine the skills and technology that will be needed in the future, particularly relating to high-rise construction. The challenge for analysts is to determine the precise economic potential of these emerging technological demands so that appropriate policy guidelines can be injected for the accelerated growth of the sector (technological demands are considered in more detail in section 7.2 below). In the long term, the expansive growth of the GREDA will pave the way to determine the future housing model and that will entrench the construction of high-rise residential structures.

While in terms of market size the building sector could also be accounting for 25% of the total market share, in terms of sheer size of projects and finance, the sector is also huge. Until the late early 2000, this sector was heavily publicly funded but it is now largely private sector dominated. Current market niches comprise retail stores, fuel filling stations, shopping malls and warehouses; these are all now private sector led and expanding rapidly. Indeed, presently, large-scale fuel filling stations (approximately two acres of plot size per station) appear very profitable and average 10 facilities per square kilometre. Similarly, with the expansion of the educational sector and particularly the proliferation of tertiary institutions, many large campuses and their associated dorms are in huge demand and being constructed. Thus, even though the government continues to invest heavily in educational buildings using a special tax that is paid into a fund (called GETFUND), private sector investment is also huge, with many opportunities. With a growing population and Ghana being the preferred destination to shop in the sub-region, shopping mall, recreation centres and theatres will be in demand in the medium to long terms. Currently, evidence of formal classifications are only required for contractors who execute public buildings but with the trend shifting to private sector dominance, there will be the need for a strong private sector led registration and licensing system to vouch for the quality of contractors who are to be engaged on these huge infrastructure opportunities. That is why it is now necessary for client organisations to foster a common front and voice that will vouch for and push for excellence and quality service in the rapidly expanding industry. Here, emphasis on the requisite and continued training of the pool of artisans in the informal sector is crucial to help keep the skills base continuously solid and attractive.

The total road network in Ghana is in the region of 70,000 kilometres and less than 10% are in acceptable condition as per international benchmarks (ISSER, 2013). Installed energy[18] generating capacity is about 2,000 MW, far below levels for a country with population of 24 million and expanding rapidly (ISSER, 2013). Given the growth potential of Ghana, if the huge energy deficit is not addressed sooner rather than later, it will inevitably lead to increased production costs and eventually collapse many industries including those in construction that depend on the energy sector. The government is committed to increasing total energy capacity to 5,000 MW in 2015 and power sector reforms are currently ongoing with the government seeking partnerships to expand hydro, thermal and gas power plants, which are in high demand as investment opportunities. Indeed Ghana recently signed a 1,000 MW contract with American multinational conglomerate General Electric (GE) (*The Finder*, 2014, cited in *Joy Online*, 2014) and the government is requesting that many more investors come on board.

Ghana's long term objective is to make its ports the preferred hubs in West Africa, having recently grown from 200,000 containers per year to over one million (*Africa Watch*, August 2013, p. 56). Ghana's two major ports in Tema and Takoradi are strategically placed in the sub-region, serving countries such as Burkina Faso, Mali, Niger, Cote d'Ivoire, Togo, Benin and Nigeria. The future direction of the heavy engineering sector including ports and harbours is summarised by quoting an interview that *Africa Watch* (August 2013, p. 61) had with the Director of the Ghana Ports and Harbours Authority about shaping the future of Ghana's ports. The Director said:

> The long term, or the medium-to-long term, is to expand facilities, to expand the number of beds, to expand the operational areas. Business at the port has increased significantly. Five or ten years ago, we were talking of 100,000 TEUs [twenty-foot equivalent units]; 200,000 TEU; 300,000 TEU. Now we are in excess of 800,000 and we are hitting one million TEU . . . so we have come up with schemes to expand the ports – both Tema and Takoradi.

Although the industrial sub-sector is just emerging, there is hope for a very promising future. In the short term, local capacity for steel frame construction has to be developed to capture jobs in the sector. Ghana was supposed to have an US$850 million Gas Infrastructure Project on stream in 2013. While there has been some real logistical problems causing delay in this project, there is still hope that this when finally completed it will help boost Ghana's industrial sector for the future. Currently, over 50% of construction equipment passes through the port, which is indeed a boost for the growth of the industrial sector. While the current opportunities with this sector largely are offshore, the spin-on effect in the maintenance and repairs of oil rigs can be huge for the local market. According to the Director of the Ghana Ports and Harbours Authority: "There are about 35 rigs between Senegal and Benin and Ghana is strategically located to win a significant slice as the only suitable facilities are in South Africa and Las Palmas" (*Africa Watch*, 2013, pp. 56–63). Building local capacity to take advantage of this competitive edge is paramount and would require a decisive action plan for both training and creating the necessary entrepreneurial environment.

6. Framework for understanding and isolating the ESTDs

The framework for evaluating the ETSDs (economic, technological and structural demands) is adapted from Barrie and Paulson (1992). They argued that the construction industry can be classified into four main sub-sectors: residential, building, heavy engineering and industrial. They went further to describe what could be the potential mini-sectors under each sub-section and these have been adapted and diagrammatically represented in Figure 8.7. Albeit proposed in 1992, Barrie and Paulson's conceptualisation of the industry is still very much relevant in modern day infrastructure development, especially in the context of emerging countries such as Ghana. Besides the conceptualisation of the sub-sectors, Barrie and Paulson also argued for the potential challenges that the industry has to face going into the future, namely rising cost, quality, coordination and control. These are indeed also very relevant to modern construction management and very much reflected in construction business performance measurements. However, Barrie and Paulson failed to address issues of sustainability, health and safety, and ecological and environmental degradation, which have become critical construction performance measures in recent times. Drawing on these emerging concerns, Barrie and Paulson's conceptualisation of the classification of the sub-sectors and challenges is, thus, as per the modified framework shown in Figure 8.8. This modification is potentially representative of current trends in the global construction industry and, therefore, is useful for evaluating the ETSDs in the section that follows.

7. ETSDs

7.1 Economic demands

Presently, global infrastructure demand is estimated at approximately US$4 trillion with the estimated shortfall put at US$1 trillion annually (World Economic Forum and Boston Consulting Group, 2013). Contextually, most sustainable economies including some emerging ones (such as South Africa, China, Indonesia, Poland, Brazil and India) have core infrastructural stock value of approximately 70% of GDP as compared to Ghana whose stock could be put around 10% of GDP (see, for example, Olorunfemi, 2014). That is, among many other emerging economies whose core infrastructure is substantially untapped, Ghana will need to invest very heavily for the next 30 years to bridge its share of the US$1 trillion shortfall. Drawing on a simplistic proportional representation of African countries with

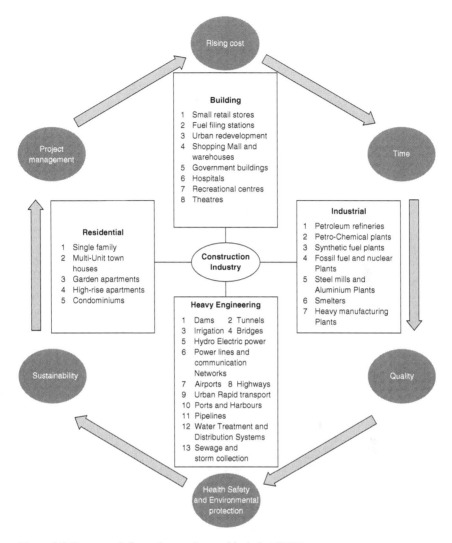

Figure 8.8 Framework for understanding and isolating ETSDs

Source: Adapted from Barrie and Paulson (1992).

inadequate infrastructure, Ghana could be requiring up to US$20 billion annually towards bridging the infrastructure gap to acceptable international benchmarks. The World Bank is reported to have noted in 2012 that Ghana would need to spend US$26 billion on infrastructure to support economic growth, including basic infrastructure upgrades in roads, housing and utilities (Oxford Business Group, 2014). However, with the huge financial constraints that many emerging economies face regarding competing socio-economic demands, raising the needed investment alone is particularly challenging and almost impossible. While governments will continue to play a leading role in financing the vast majority of strategic infrastructure projects, the reality is that modern economic demands now render them incapable of closing the gap alone, and they will have to seek strong private sector participation. Without innovative financing and delivery models as well as private companies that are suited to carrying out the much-needed infra-structure projects, it will not be possible to meet the demand (World Economic Forum and Boston Consulting Group, 2013; Olorunfemi, 2014).

Thus, if Ghana is to remain competitive and sustain the impressive growth rate it has achieved over the years, a more strategic initiative towards acceler-ated economic development is required. Given that the fiscal situation of many of these countries is increasingly constrained, many analysts including the World Economic Forum and Boston Consulting Group (2013) have strongly advocated for Public–Private Partnership (PPP) as the plausible way forward. Currently, Ghana is in the process of initiating some milestone PPP[19] projects involving a number of capital-intensive and strategic public infrastructure projects to attract the private sector for speedy execution. These projects include the dualisation of the Accra–Takoradi highway, the Accra–Tema motorway overlay, the Asutsuare water project and a health diagnostic centre at the Korle Bu Teaching Hospital; all these projects are at various stages of preparation for PPP implementation (*Ghanaweb Business News*, 2014). Other examples are the Accra plains irrigation project and the Ghana Port project (InfraPPP Infrastructure Knowledge, 2014). This is indeed a positive sign; however, Ghana is still far from integrating PPP as a core strategic investment approach in the Ghanaian construction industry as it has been particularly slow in putting in place an appropriate legislative instrument (LI). It is high time that Ghana lived by the old adage that "time is money" by placing urgency and timely delivery into policy initiatives affecting the construc-tion sector.

Several studies (see, for example, Edmonds and Miles, 1984; Ahadzie, 1993; World Bank, 2003; Fugar and Agyarkwa-Baah, 2010) have documented the poor image of the industry by virtue of poor construction performance, excessive delayed payments, long suspended projects, and poor procurement and communi-cation practices, and for several years authorities in Ghana have been too slow at taking decisive action to address these perennial structural and management prob-lems. Within this context, recent experiences from emerging countries show that the effective exploitation of construction resources coupled with effective agenda and project management practices can provide a net gain for socio-economic improvement, especially with regard to job creation and poverty alleviation. For instance, in Singapore, it has been established that the construction industry played a significant role in the development process of the country. The success story of Hong Kong from 1985 to 1995 is also strongly linked to an increase in construction activities during that period. There are also success stories in many other emerging countries such as Malaysia, Chile, China and Brazil where the construction industry is being used as a powerful engine for economic growth, thereby creating many jobs to lift millions of people from poverty (ILO, 2001; World Economic Forum and Boston Consulting Group, 2013). In Africa, Ghana can also draw some useful lessons from South Africa and Mauritius, where the construction industry is faring relatively better.

7.2 Technological demands

The boom days of the early independence era witnessed a growing trend of apartment-size family[20] houses, normally of two storeys, in the central business district of Kumasi, Accra and Takoradi, the three biggest cities in Ghana. Analysts believe[21] that these structures were built to high quality standards and indications are that the craftsmen available then were also well trained.[22] This was at a time when the government of Ghana was the biggest employer of construction workers and had a well set-out apprenticeship scheme for manpower training in the industry. It is worth noting that even though many of these structures built over 50 years ago are going through adaptations to meet emerging and modern functional needs, they are still in very good structural conditions compared to some recent buildings. Following the 1980s and the emerging focus on core family values, the trend in family apartments gradually dwindled, largely giving way to detached single-family houses, which have remained the case.

However, spurred by the challenges of urban sprawl and the imminent shortage of land, in recent times the trend in housing development has been gradually shifting from detached single-family houses to apartments once again, and possibly to high-rise residential buildings. This will demand that the artisans who currently operate informally are conversant with emerging technologies such as the reinforced concrete works and scaffolding technology often associated with such projects. The increasing sophistication of electrical and mechanical and fire safety installations also demand a very thorough knowledge of building services, especially regarding electrical, mechanical and fire systems management in the home. Similarly, there is a knowledge gap regarding relevant training on emerging technologies and specifications for the correct use of the many modern imported building materials that have found their way into the market. While there is a huge taste for these materials, there is no systematic approach to testing the quality of the materials and training artisans in the correct choice and application of these materials.

In the last decade, glass panelling and façade has become synonymous with both residential and commercial buildings and there is pressure on, particularly, small and medium scale steel fabricators, who dominate this market to be very scientific in the selection and use of steel products including improving their management skills. Improved mobile communication has also led vast construction activities in the installation and maintenance of mobile phone mast, increasingly making the steel and metal works sector quite significant in the Ghanaian economy.

Expansion of the market and the rising demand for multi-storey and high-rise buildings will require very high knowledge and skills in advanced concrete production and reinforced concrete practices. Ghana has in recent times witnessed too many collapsed structures, which has raised concern about the quality of skilled personnel in the system, especially the artisans who serve as the point of first contact for many of these structures in the informal but vibrant sector. The increasing sophistication of projects demands that tradesmen are properly trained in modern concrete technology, advanced blockwork/brickwork, advanced glass technology, roof technology, scaffolding technology, electric and mechanical systems technology, fire systems technology and management, plant, machine and crane technology and handling, advanced steel work and welding technology, quality awareness, materials and waste control, and architectural sheet metal construction. There is also the need to develop the analytical and critical thinking skills of tradesmen as well as their business and entrepreneurship skills.

Ghana, like most emerging countries, is now a global village characterised by urbanisation, international migration and increasing communication and information technological advancement. Embedded in these characterisations will be how the construction industry is used to shape the connectivity between urban housing and infrastructure development in cities. However, ageing infrastructure and old stocks of properties are also in need of repairs, refurbishment and in some cases, complete regeneration is urgently needed. Specialist skills in demolition, which are currently at a rudimentary stage, and especially the demolition of multistorey structures with close proximity to other buildings, will therefore be in huge demand going into the future.

7.3 Structural demands

One principal way in which modern societies generate new value is through the exploitation of construction activities. In a typical modern society, half of all physical assets are created by the industry, generating approximately 5–10% of GDP (Winch, 2002). Thus, the emerging growth of the Ghanaian industry, if well exploited, could help engender the accelerated growth that will help champion growth in the wider economy and take millions out of poverty. Here, the major structural weakness of the Ghanaian industry is the lack of a well thought-out construction industry development agenda to spearhead the development of the industry. Ofori et al. (2012) outline some interesting construction industry development initiatives and describes how, for instance, countries such as the UK, Hong Kong, Malaysia and Singapore have initiated well considered programmes towards making their industry successful. In Africa, too, recent construction industry development programmes can be found in Tanzania, South Africa, Malawi and Zambia, and quite recently in Rwanda and Uganda. Evidence suggests that the role of the industry reaches its peak when a country attains middle income status; as Ghana has just entered this "club" (albeit at the lower level), it has become necessary that it takes these initiative quite seriously (Bon, 1992, cited in Ng et al., 2009).

In the recent past, one of the major policy initiatives in Ghana and, for that matter, the industry, is the introduction of the Public Procurement Act 2003 (Act 663) to help ensure value for money in the procurement for goods and services, including construction. Whilst this is definitely good, it is yet to help change the poor image of the industry with respect to poor quality service, excessive delays and unprofessionalism, in many instances. With all of the potential of the industry as shown by the indicators, it is high time that a construction industry development agenda was robustly and deliberately embedded into the national development agenda for the industry, so that construction can be used as an engine of growth that engenders the accelerated development of the economy. There are issues relating to the competitive participation of local contractors, technical and managerial skills improvement, growth and expansion of domestic construction enterprise, employment generation capacity, delivery of quality work, efficiency/ timely delivery of work, and sustainability that need to be rigorously addressed.

Other issues to be considered are private sector involvement in infrastructure, local response to the formation of regional blocs and common markets such as the ECOWAS, local response to the need to use the industry to fight poverty and local response to issues of sustainable development, especially environmental and ecological degradation. When the State Construction Company (SCC) was in existence in the early days of independence and for some years later, it was able to internationalise itself by executing international projects in Lome (Togo) and Angola (Edmonds and Miles, 1984). Ghana also had a speciality Bank for Housing and Construction devoted solely to providing finance to the construction industry, which has been defunct since the 1990s due to mismanagement. It is contended that Ghana missed the opportunity to use these competitive advantages in the early days of the industry to build a strong construction culture founded on efficiency and professionalism and on which the now dominant private sector would have ridden.

Currently, Ghana urgently needs the embodiment of a deliberate attempt to improve the capacity and effectiveness of the construction industry in order that it may meet the increasing demands of the emerging economy. However, indications are that the Ministry of Water Resources Works and Housing (MWRWH) and the Ministry of Roads and Highways (MRH), two major ministries in charge of the development of the industry, are yet to provide the decisive holistic policy direction required to move the industry forward. It appears that the ministries are more concerned with policy issues that affect public infrastructure rather than a holistic development of the industry including private sector practices. It is interesting to note that while the MWRWH operates a registration regime for building contractors, they have often failed to provide an up to date list of the total number of contractors actually registered and operating in the country.

Given that the Ghanaian construction industry is now predominantly private sector led, it is high time that a strong private sector initiative is encouraged by the government so that the development of the industry can be viewed from a holistic perspective. It is for this reason that client organisations need to come together and demand action from all other stakeholders in the industry. Together with other professionals and association affiliates, the private sector led industry has many areas of interest including business development; technology development and transfer; and human resources development. Fox and Skitmore's (2007) framework for developing the construction industry – which identifies industry-led better practices, the need for financial resources and skills training, government policies and strategies for supporting the construction business, research and development for construction and exploring alternative markets – could be adopted beneficially (Ng et al., 2009). Here, it is proposed that in addition to the mandate of those ministries with oversight responsibility over the industry, the various trade and professional associations, contractor associations and the proposed client association should all together adopt a decisive framework for the creation of a development agenda for industry. Among others, key issues to be addressed would be:

- the lack of robust leadership in creating and driving a national philosophy for the construction industry through a construction industry development agenda;
- the lack of coordination of the numerous interest groups, professional bodies and industry associations that exists in the industry;
- the lack of comprehensive database for all sub-sectors of the industry including the informal sector;
- poor productivity and lack of professionalism in the industry;
- institution of a strong and vibrant apprenticeship and certification programme for artisans and tradesmen;
- quality assurance; and
- health and safety and environmental concerns.

There are many countries where it is generally agreed that significant progress has been made in the construction industry through a systematic development initiative that Ghana can benefit from. Figure 8.9 is a proposed framework of a typical Construction Industry Development Board that could be established to help provide effective leadership for the numerous interest groups and industry associations and also provide a more focused direction of the industry on performance measurement, training and development. The framework proposes strong stakeholder collaboration between the major players in the industry, focussing on developing workable industry led best practices, financial resources and skills training as well as support for construction business, research, and development. While government support is obviously needed, it is proposed that the Board should be largely private sector led and it is important that the Ghana Association of Industries should play a leading role in setting this important agenda. Also, it is important for clients' organisations with major portfolios in the construction industry to now associate and exert their influence on the other major stakeholders, in order that there be decisive action on improvement in the industry. As noted earlier, there are various critical issues such as the competitive participation of local contractors, technical and managerial skills improvement, strong artisanal apprenticeship and certification programmes, and quality assurance which require specific action, going into the future, for the growth of the industry. It is to be noted that Ghana as an emerging country needs a functional city that offers appropriate infrastructure for sustainable patterns of urbanisation. A well thought-out agenda on the sub-themes captured in Figure 8.9 (namely industry led best practices, financial resources and skills training, and government policies supporting construction business research and development) should be helpful in allowing Ghana to develop the local industry's capacity for wealth creation.

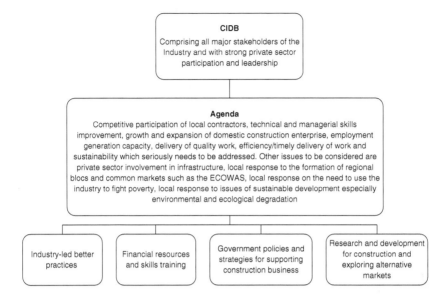

Figure 8.9 Proposed structure for providing leadership for improvement of the Ghanaian construction industry

8. Conclusion

Substantial theoretical and empirical evidence exists between growth in the construction industry and growth in the wider economy in many countries. The Ghanaian construction industry has demonstrated over the last two decades or more that it has the potential to be the engine of economic growth. Analysts believe that the construction industry, being the most important industrial sector, will be driving the economy in the medium to long term. This obviously demands prudent exploitation of its activities/outcomes, which depends on aligning robust decision making to the existence of a well thought-out construction industry development agenda. This is what Ghana has failed to do to date despite all the promising indicators that the industry has shown. Given the impressive confidence that the construction industry inspires as an emerging influential economic sector, it is important to clearly understand the economic, technological and structural demands (ETSDs) required for its growth and competitiveness in an increasingly challenging global business environment. This chapter highlights these ESTDs to help generate an understanding of the potential way forward in addressing them. Indications are that the Ghanaian construction industry progresses consistently and contributes substantially to GDP. The residential and building sub-sectors, which are now heavily private sector financed, offer the greatest opportunity for growth and job creation in the short to medium terms. At present, the heavy engineering sector is heavily government funded but there is every indication that, given the global fiscal regime, the government cannot go alone; hence the strong advocacy from many quarters to speed up policy guidelines and the drawing up of legislative instruments necessary for the adoption of PPP schemes. The importance of the industrial sector, with a promising emerging oil sector plus spin-offs into steelworks, is also recognised.

As it is now, over 70% of employment in the industry is in the informal sector responsible for many large scale projects, especially in the residential and building sub-sectors. Ghana lacks a robust apprenticeship training regime and skills acquisition and development by artisans is largely informal, in that the labour pool learns its trade from the gangmasters. Given that the trend in these structures is now turning towards multi-storeys, the challenge is how these artisans will cope with the increasing sophistication of the industry, including the use of modern technology and materials and heavy equipment. The informal training has limitations as the master craftsmen who do the training might themselves have limited skills. Furthermore, the informal method of technological knowledge acquisition and skill formation does not comply with the actual demands for construction output required of an emerging economy.

As noted by Ng et al. (2009), as economic activities expand and more facilities are demanded, the volume of construction activities increases. Typically, this seems to mature to its peak when the economy is emerging and/or expanding rapidly, as in the case of Ghana. Thus, the Ghanaian construction industry is at the threshold where its activities and resources can be tapped to its full potential for the benefit of the wider economy and strategies that need to be put in place have been suggested. The indications are that Ghana can benefit from the considerable experience of many countries, both developed and developing, that have had to initiate systematic and decisive construction development agendas. The pressing need for the industry to respond to competitive global standards is apparent. Concomitantly, achieving these global standard practices remains highly challenging as the dynamics of the industry are directed by organisational and country specific needs. The list of strategies provided in this chapter would be helpful to all the stakeholders and decision makers forging ahead for excellence and wealth creation in the Ghanaian construction industry.

Acknowledgements

The author would like to thank Messrs Salifu Abdul-Hafiz and Kingsley Atta Nyamekye, Research Assistants (on National Service) at the Centre for Settlements Studies, Kwame Nkrumah University of Science and Technology (KNUST), Kumasi, Ghana, for conducting the desktop study and also producing the graphs. The author would also like to acknowledge the assistance of Mr. William K. Ahadzie, formerly Technical Instructor of the Department of Building Technology, KNUST, for proofreading the manuscript.

Notes

1 This projected estimate for overall construction output was derived from the average annual metric tonnage of cement production as published in the Bank of Ghana's *Quarterly Bulletin.*

2   Employment status would be much higher if the informal sector is properly accounted for.

3   The growth rate is derived from the Kenyan experience as published in Mitullah and Wachira (2003).

4   See for instance <www.citifmonline.com/Ghana> dated 13th October 2014, captioned "Ghana opens discussion with IMF for bailout programme".

5   So called because of the pink cover of the articles of agreement.

6   D1/K1 and D2/K2 contractors are generically referred to as "large scale contractors" and constitute 5% of total registered building contractors.

7   D4/K4 and D3/K3 are generically referred to as "small" and "medium scale contractors" respectively and constitute 95% of the total registered building contractors.

8   The MEST has since be re-designated as MESTI: Ministry of Environment, Science, Technology and Innovation.

9   See for instance Paul (1989). Paul reported about 23,000 self employed contractors in the informal sector and about 2,000 formally registered. A decade on, a 1998 database of registered contractors at the MWRWH suggested approximately 6,000 (Ayisi, 2000), indicating a 300% increase. Allowing for exiting through insolvency and for new entries, it could be argued based on proportional representation over that years contractors operating in Ghana, both in the formal and informal sectors, may be more than 50,000 in number.

10   While the keyword here is operating, it is difficult to vouch for how many contractors are actually operating regularly as against those who, albeit registered, might not be fully engaged as contractors.

11   An association of professional builders which was reconstituted in 2014 following long years of dormancy to help bring sanity and professionalism into contractor organisations and development in construction practice. The GIOC is to function on similar lines to the CIOB of the UK and the Nigerian Institute of Building (NIB).

12   Informality means that many of these artisan are not registered formally by any government institution and also do not belong to any union or association.

13   Potential strategies that the government can draw upon to use the industry more efficiently for job creation are discussed in section 7.3.

14   Self-build houses refer to the gradual piecemeal acquisition of housing inputs such as plots of land, sand, and various building materials in the housing construction process in tandem with the owners' resources flow and executed through the engagement of professional builders and artisans. Typically, such self-build houses could take between 5 and 10 years to complete (Ahadzie and Amoa-Mensah, 2010).

15   A labour-only contract means that the client takes responsibility for providing the materials required while the artisans provide only labour services.

16   Associated water and sanitation facilities, which are hugely inadequate, are also in huge demand. Over 80% of homeowners now rely on self-build bore-holes/hand-dug wells for household water supply.

17   Figures are derived from secondary data provided in the Bank of Ghana's *Quarterly Bulletin*, the *Ghana Living Standard Surveys Report* published by the Ghana Statistical Service, the Bank of Ghana's *Report on Housing Trends in Ghana* (2007) and the works of Amoa-Mensah (2003).

18   Recent developments in Ghana, particularly, in the year 2014, suggest that the energy problem is becoming a crises and this has the potential of derailing the impressive growth potential of many private sector businesses, including construction.

19   Ghana even has a Minister of State with sole responsibility for PPP.

20   The focus on the family was then on the extended family system typical of many African countries.

21 Although it might be difficult to provide hard evidence, it is believed that many structures that were built in the early years of independence and up to the late 1970s still look more structurally durable, at least by "visual inspection", than many modern day structures. Indeed there is enough evidence to suggest that the main block building material used in modern times, sandcrete blocks, does not meet quality standards in many production sites (Baiden and Asante, 2004; Baiden and Tuuli, 2004).

22 There used be a formal apprenticeship programme by the Public Works Department (PWD) and the then State Construction Corporation (SCC), which ensured a system and well defined training regime for artisans. Dependency on informality has now created a situation where one cannot vouch for any credible and legitimate training regime for artisans. Most now acquire their training from their gangmasters who may have limited knowledge in many aspects of modern construction technology and practice.

References

Africa Watch (2012) Donors not keeping pace with African Progress, *Africa Watch*, July, pp. 48–49.

Africa Watch (2013) Architecture: Africa race to the top, *Africa Watch*, November/December, pp. 75–76.

Africa Watch (2013) Africa's progress: When will investors catch up, *Africa Watch*, June, pp. 80–88.

Africa Watch (2013) Airports: upgrading to first-class, *Africa Watch,* November/December, pp. 76–78.

Africa Watch (2013) Commentary: This time Africa economic growth is real, *Africa Watch*, August, p. 98.

Africa Watch (2013) Finance Minister Seth Terkper talks Eurobond – and beyond, *Africa Watch*, August, pp. 38–41.

Africa Watch (2013) Ghana: Economic growth that inspires confidence, *Africa Watch*, August, pp. 25–37.

Africa Watch (2013) Two sides to Africa's mobile revolution, *Africa Watch*, September, p. 58.

Africa Watch (2014) The free fall of the cedi, *Africa Watch*, February, pp. 74–75.

Africa Watch (2014) The republic of corruption, *Africa Watch*, February, pp. 18–24.

Africa Watch (2014) Uneasy times: nation hits economic turbulence, *Africa Watch*, March/April, pp. 24–29.

Ahadzie D. K (1993) A strategic guide for improving productivity in the construction industry in Ghana, unpublished MSc dissertation, University of Manchester Institute of Science and Technology, Manchester, UK.

Ahadzie D. K. and Amoa-Mensah, K. (2010) Management practices in the Ghanaian house building industry, *Journal of Science and Technology*, 30(2), pp. 62–75.

Ahadzie, D. K., Kissi, E. and Kumi, T. (2012) The status of project management practices in the Ghanaian construction industry, in: Badu, E., Dinye, R., Ahadzie, D. and Owusu-Manu, D. (Eds), *Proceedings of the 1st International Conference on Infrastructure Development in Africa (ICIDA)*, March 22–24, Kumasi, Ghana.

Ahadzie. D. K., Proverbs, D. and Sarkodie-Poku, I. (2014) Competencies required of project managers at the design phase of mass house building projects, *International Journal of Project Management*, 32(6), pp. 958–969.

Amoa-Mensah K. (2003) Housing in Ghana: A search for sustainable options as the way forward for enhanced output – years 2003 and beyond, paper presented at the International Building Exhibition, 27–29 August, Accra International Conference Centre, Accra, Ghana.

Anaman K. A. and Ose-Amponbsah, C. (2007) Analysis of the causality links between the growth of the construction industry and growth of the macro-economic in Ghana, *Construction Management and Economics*, 25, pp. 951–961.

Ayisi, P. (2000) Cost control practices in the Ghanaian construction industry, MSc dissertation, Department of Building Technology, Kwame Nkrumah University of Science and Technology, Kumasi, Ghana.

Baiden, B. K. and Asante, C. K. O. (2004) Effects of orientation and compaction methods of manufacture of strength properties of sandcrete blocks, *Construction and Building Materials*, 18(10), pp. 717–725.

Baiden, B. and Tuuli, M. (2004) Impact of quality control practices in sandcrete blocks production, *Journal of Architectural Engineering*, 10(2), pp. 53–60.

Bank of Ghana (2001–2008) *Bank of Ghana Quarterly Bulletin*, Accra: Bank of Ghana.

Barrie, D. S and Paulson, B. C. (1992) *Professional Construction Management*, New York: McGraw Hill.

Breisinger, C., Diao, X., Thurlow, J. and Hassanel, R. A. (2008) International Food Policy Research Institute and University of Ghana paper presented at the PEGNet Conference Assessing Development Impact – Learning from Experience, Agriculture for Development in Ghana, 11–12 September, Accra, Ghana.

Deloitte LLP GSMA (2012) What is the impact of mobile telephony on economic growth? [online]. Available at: http://www.gsma.com/publicpolicy/wp-content/uploads/2012/11/gsma-deloitte-impact-mobile-telephony-economic-growth.pdf [Accessed 20 February 2014].

Edmonds G. A. and Miles, D. W. J. (1984) *Foundations for Change Aspects of the Construction Industry in Developing Countries*, London, UK: ITG Publications.

Fellows, R., Langford, D. and Newcombe, R. (2001) *Construction Management in Practice*, London, UK: Wiley-Blackwell.

Fox, P and Skitmore, R. M. (2007) Factors facilitating construction development, *Building Research and Information*, 35(2), pp. 178–188.

Fugar, F. D. K. and Agyarkwa-Baah, A. (2010) Delays in building construction in Ghana, *Australian Journal of Construction Economics and Building*, 10 (1/2), pp. 103–106.

Ganesan, S. (1983) Housing and construction: Major constraints and development measures, *Habitat International*, 7(5/6), pp. 173–194.

Ghana Procurement Authority (2003) *Procurement Act 663 Tender Documents for the Procurement of Works: Medium/Large Contracts*, Accra: Public Procurement Authority.

Ghana Statistical Service (1995–2008) *Ghana Living Standard Surveys: Report on the 1st, 2nd, 3rd, 4th and 5th Round*, Accra: Ghana Statistical Service.

Ghana Statistical Service (2013) *National Population and Housing Census*, Accra: Ghana Statistical Service.

Ghanaian Times (2014) Kenya high speed rail project to proceed, 30 January [online]. Available at: http://www.ghanaiantimes.com.gh/ [Accessed 25 February 2014].

GhanaWeb Business News (2014) Projects gather steam, Tuesday, 18 February [online]. Available at: http://www.ghanaweb.com/GhanaHomePage/business/artikel.php?ID=301072 [Accessed May 2014].

IMF (2015) Ghana and the IMF [online]. Available at: www.Imf.org/external/country/Ghana [Accessed 18 August 2015].

InfraPPP Infrastructure Knowledge (2014) Projects in Ghana [online]. Available at: http://infrapppworld.com/pipeline-html/projects-in-ghana [Accessed 8 May 2014].

Institute of Statistical, Social and Economic Research (ISSER) (2008) *The State of the Ghanaian Economy,* Accra: ISSER: University of Ghana, Legon.

Institute of Statistical, Social and Economic Research (ISSER) (2013) *The State of the Ghanaian Economy 2012*, Accra: ISSER, University of Ghana, Legon.

International Labour Organization (2001) *The Construction Industry in the Twenty-First Century: Its Image, Employment, Prospects and Skills Requirement*, Geneva, Switzerland: ILO.

Joy Online (2014) General Electric to produce 1000 MW power [online]. Available at: http://www.myjoyonline.com/business/2014/May-5th/general-electric-to-produce-1000mw-power.php [Accessed 15 June 2015].

Kenya Railways (2013) Standard Gauge Railway [online]. Available at: www.google.com/Kenya Railways [Accessed 25 February 2014].

Lopes, G. (1998) The construction industry and macro economy in sub-Saharan African country, *Construction Management and Economics*, 16, pp. 637–649.

Masterman, J. W. E. (1992) *An Introduction to Building Procurement Systems*, London: E. & F. N. Spon Ltd.

Ministry of Finance (2009) *Budget Statement of Ghana*, Accra: Ministry of Finance.

Mitullah, W. V. and Wachira, I. S. (2003) *Informal Labour in the Construction Industry in Kenya: A Case Study of Nairobi*, a Sectorial Activities Programme working paper, International Labour Office, Geneva, Switzerland.

Ng, T. S., Fan, R. Y. C., Wong, W. M. J., Chan C. A., Chang H. Y., Lam I. T. P. and Kumaraswamy M. (2008) Coping with structural change in construction: Experience gained from advanced economies, *Construction Management and Economics*, 27, pp. 165–180.

OECD (Organisation for Economic Co-Operation and Development) (2014) National Accounts Data Files [online]. Available at: www://data.worldbank.org/indicators/NY/GDP.MKTP.CD [Accessed 19th February 2014].

Ofori, G., Ai Lin, E. T. and Tjandra, I. K. (2012) Construction industry development initiatives: Lessons for Ghana from some overseas examples, in: Badu, E., Dinye, R., Ahadzie, D.K. and Owusu-Manu, D. (Eds), *1st International Conference on Infrastructure Development in Africa*, 22–24 March, Kumasi, Ghana, pp. 1–19.

Olorunfemi, A. I. (2014) The future of sustainable infrastructure in Nigeria in the wake of privatization of the power sector, in: Ejohwuomu, O. and Oshodi, O. (Eds), *3rd International Conference on Infrastructure Development in Africa*, 17–19 March, Abeokuta, Nigeria, pp. 2–6.

Oxford Business Group (2014) Laying the cornerstones: Government expenditure is supporting increased activity, in: The Report: Ghana 2013: Construction and Real Estate [online]. Available at: http://www.oxfordbusinessgroup.com/news/laying-cornerstones-government-expenditure-supporting-increased-activity [Accessed, 17 February 2014].

Paul, S. (1989) *Private Sector Assessment: A Pilot Exercise in Ghana*, Washington, DC: Country Economics Department, World Bank [online]. Available at: http://www-wds.worldbank.org/servlet/WDSContentServer/IW3P/IB/2000/08/18/000009265_3960927224727/Rendered/PDF/multi_page.pdf [Accessed October 2014].

UK Trade and Investment (2004) *Building, Construction and Property Market in Ghana* [online]. Available at: http://www.uktradeinvest.gov.uk/building/ghana/profile/overview.shtml [Accessed 25 February 2014].

Wells, J. (2007) Informality in the construction sector in developing countries, *Construction Management and Economics*, 25, pp. 87–93.

Winch, M. G. (2002) *Managing Construction Projects*, London, UK: Blackwell.

World Bank (2003) *Ghana 2003 Country Procurement Assessment Report*, Washington DC, USA.

World Bank (2010) *Ghana Skills and Technology Development Project (GSTDP)*, Accra: World Bank.

World Bank (2013) African Competitiveness Report [online]. Available at: http://www.worldbank.org/content/dam/Worldbank/document/Africa/Report/africa-competitiveness-report-2013-main-report-web.pdf [Accessed 19 February 2014].

World Bank (2014) *World Development Indicators: GDP* [online]. Available at: http://data.worldbank.org/indicator/NY.GDP.MKTP.CD [Accessed 19 February, 2014].

World Economic Forum in collaboration in conjunction with the Boston Consulting Group (2013) *Strategic Infrastructure Steps to Prepare and Accelerate Public–Private Partnerships*, The World Economic Forum, 10th May, Cape Town, South Africa.

Legislation

Local Government Act 1993 (Act 462), Accra, Ghana.
Public Procurement Act 2003 (Act 663), Accra, Ghana.
National Building Regulations 1996 (LI 1630), Accra, Ghana.

9 The challenges of infrastructure procurement in emerging economies and implications for economic development

A case study of Ghana

Nii Ankrah, Joseph Mante and Issaka Ndekugri

1. Introduction

Infrastructure comprises the physical facilities, institutions and organisational structures or the social and economic foundations for the operation of a society (UNCTAD, 2008). The World Bank (1994) defines infrastructure in physical and economic terms as public utilities (power, telecommunications, piped water supply, sanitation and sewerage, solid waste collection and disposal, and piped gas), public works (roads and major dam and canal works for irrigation and drainage) and transport facilities (urban and inter-urban railways, urban transport, ports and waterways, and airports). Whilst the "public" tag may no longer be relevant in the light of widespread private sector participation and ownership, one can agree with the examples of infrastructure projects cited in the definition above. UNCTAD (2008) provides a similar list with a caveat that this is changing with the growth of information communication technology (ICT). Many authors such as Prud'homme (2004) and Kessides (1993) also define infrastructure projects in similar terms as the World Bank report.

In Ghana, infrastructure has been defined to include "immovable capital such as, roads, power plants, water delivery systems, sewerage treatment plants, telecommunication and transport facilities, electrification, hospitals and schools" (World Bank, 1997, p. 2). All these facilities share common characteristics (UNCTAD, 2008). Firstly, they are capital-intensive. They are "formidable undertakings" (p. 88) involving huge financial outlay. Secondly, they often involve physical networks of strategic importance. Thirdly, they are also major determinants of the competitiveness of an economy; good infrastructure can play a major role in the decision of an investor to set up in a particular economy. Fourthly, in many societies, services associated with infrastructure are thorny social and political issues and, thus, subject to public interventions. Finally, infrastructure projects are relevant to economic development and global integration.

Prud'homme (2004) adds that infrastructure projects are capital goods in themselves; they are often "lumpy" and not "incremental" (p. 4); they are long-lasting and space-specific. According to Prud'homme (2004), they often benefit both enterprises and households. Odams and Higgins (1996) identify five additional characteristics of major infrastructure projects that are particularly relevant in the context of emerging economies. Firstly, there is often an external funder who plays an active role in determining the project structure; secondly, the client is often the State or a State-owned entity; thirdly, the presence of a foreign element either in the form of an investor or a contractor; fourthly, the contractor often plays a more active role in what is traditionally the role of the client, as in the case of a concession agreement; and finally, the contractor tends to assume much more significant risks.

A number of significant conclusions can be drawn from the literature on the features of infrastructure projects; they are complex projects with significant impact on the public. Hence, their procurement often involves the State and its agencies on one hand and major contractors on the other – in the case of Ghana and other emerging economies, these would often be foreign contractors. Similarly, funding for such projects, particularly in the case of developing countries, comes from external sources. Additionally, these projects are often laden with complex administrative and legal arrangements which make them high risk ventures, especially in developing countries.

The importance of infrastructure projects, however, makes them indispensable in spite of their associated challenges. The past two decades, especially, have witnessed increased research into the relationship between infrastructure development and economic development with growing evidence of positive association between them (see for example, Estache, 2004; ADB et al., 2005; Andrés et al., 2008; Calderón and Servén, 2010a; Calderón and Servén, 2010b; Ncube, 2010; Foster and Briceño-Garmendia, 2010; PEI, 2011). This has spurred key players (states and international institutions, but increasingly also the private sector) to invest more in infrastructure delivery.

Using Ghana as a case study, the aim of this chapter is twofold: (a) to critically review the procurement regime in emerging economies and current trends towards private participation in infrastructure procurement; and (b) to examine the likely impact of such issues on economic development. Regarding how the rest of the chapter is structured, section 2 examines infrastructure and economic development whilst section 3 provides an overview of the concept of procurement. Sections 4, 5, 6 and 7 consider the framework for procurement, procurement practices, contract formation and inefficiencies in the procurement regime respectively, all in Ghana. The last section concludes the chapter.

2. Infrastructure and economic development

The issue of the impact of infrastructure development on economic growth has engaged the attention of many authors for years (Estache, 2004). Research on Latin America (Andrés et al., 2008; Calderón and Servén, 2010b), sub-Saharan Africa (Calderón and Servén, 2010a; Ncube, 2010; Foster and Briceño-Garmendia, 2010; PEI, 2011) and East Asia (ADB et al., 2005) have all shown positive linkages between infrastructure development and economic growth and productivity. These reports have also indicated regression in growth where there have

Table 9.1 Distribution of findings on the impact of infrastructure investment on productivity or growth

Area studied	Number of studies	Percentage showing positive effect	Percentage showing no significant effect	Percentage showing a negative effect
Multiple countries	30	40	50	10
United States	41	41	54	5
Spain	19	74	26	0
Developing countries	12	100	0	0
Total/average	102	53	42	5

Source: de la Fuente and Estache (2004, cited in Briceno-Garmendia et al., 2004).

been cuts in infrastructure development. In their attempt to evaluate the trend of research on this subject, Briceno-Garmendia et al. (2004) reproduced findings of a study conducted by de la Fuente and Estache (2004, cited in Briceno-Garmendia et al., 2004) to illustrate the impact of infrastructure development on growth. The report critically reviewed 102 published studies on infrastructure development in a range of countries and the impact of such infrastructure on productivity, economic growth and other development goals in those countries. The findings are shown in Table 9.1.

Although the study revealed varied impact of infrastructure development on economic growth and productivity in multiple countries, United States and Spain, the verdict on the developing countries captured was unequivocally positive. Since the pioneering work of Aschauer (1989) on the subject, many studies have confirmed that infrastructure development is crucial to economic development (see for example Kessides, 1993; World Bank, 1994; Sanchez-Robles, 1998; Canning and Pedroni, 1999; Kirkpatrick et al., 2006; Harris, 2003; Briceno-Garmendia et al., 2004; UNCTAD, 2008; Calderón and Serven, 2010c; Giang and Sui Pheng, 2011). Briceno-Garmendia et al. (2004) indicated that reliable and affordable infrastructure can reduce poverty and, thus, help achieve the Millennium Development Goals. This finding has been independently corroborated through empirical research conducted on sub-Saharan Africa (see Agénor et al., 2005).

Again, Sanchez-Robles (1998) found a positive impact on economic growth after a study of road length and electricity generating capacity, similar to findings arrived at by Canning and Pedroni (1999). A study which examined the impact of investment in telecommunication infrastructure in Nigeria on economic growth found a positive correlation (Osotimehin et al., 2010). Giang and Sui Pheng (2011) suggested that infrastructure has the potential to raise the productivity of other factors of production. After an assessment of empirical data from sub-Saharan Africa and comparative data from over 100 countries, Calderón et al. (2008) also concluded that infrastructure development impacts economic growth and equity. It is, therefore, not surprising that studies such as those by Estache and

Vagliasindi (2007) and Foster and Pushak (2011) have submitted that the infrastructure deficit has limited growth in Ghana.

In more specific terms, it has been argued that lack of adequate infrastructure in sub-Saharan Africa is holding back GDP growth by 2.2% (Foster and Briceño-Garmendia, 2010; PEI, 2011). Prud'homme (2004) also asserts that infrastructure development can result in lower cost and enlarged markets. These factors, in turn, can lead to growth, which will eventually culminate in improved welfare. It has also been found in support of Prud'homme's (2004) argument that the provision of infrastructure for potable water, electricity, health and sanitation will directly and dramatically benefit and improve the welfare of households and thereby impact poverty reduction (Briceno-Garmendia et al., 2004). This position is further confirmed by research pointing out that apart from the impact on productivity and growth, infrastructure development also complements the private sector and other sectors of the economy, and affects the durability of private capital and investment adjustment cost, health and education (Agénor and Moreno-Dodson, 2006; Estache and Fay, 2007). In contributing to a country's international competitiveness, infrastructure also enhances the participation of the developing world in the global economy. Access to infrastructure services has been and remains an important determinant of the standard of living of inhabitants of countries all over the world.

The general consensus in the rapidly increasing literature on the subject is that infrastructure matters to growth and development, though its impact may differ on the basis of levels of income. It is predicted that under the right conditions, infrastructure development can play a major role in both productivity and equity and thereby help reduce poverty. It is, therefore, not surprising that both states and international institutions focusing on development have placed a lot of premium on infrastructure development across the globe. Indeed these institutions, including the International Bank for Reconstruction and Development (IBRD) and the various regional development banks, namely the African Development Bank (AfDB), Asian Development Bank (AsDB), European Bank for Reconstruction and Development (EBRD) and Inter-American Development Bank (IDB), have identified infrastructure development as an essential part of any effective strategy for alleviating poverty in the developing world.

In Ghana, for instance, typical annual infrastructure spending is circa US$1.2 billion, equivalent to 11% of its 2006 GDP (Foster and Pushak, 2011). This expenditure is from four main sources: overseas development assistance (ODA) represents 35%; public investment constitutes 28%; private investment 24%; and the remainder from non-OECD sources (Foster and Pushak, 2011). With the strong and rising involvement of China and other non-OECD members such as Brazil and India in infrastructure provision in Ghana, the percentage contribution of the non-OECD members is likely to rise (Foster et al., 2009). As is the case with many emerging economies, Ghana's infrastructure development challenges are many (see Table 9.2). To satisfy the huge demand for infrastructure, the Government will need to spend more on infrastructure procurement. Public procurement thus constitutes a substantial portion of government spending and can be considered an important policy tool. However, it is also a business, management

Table 9.2 Achievements and challenges in Ghana's infrastructure sectors

	Achievements	Challenges
Air transport	Emerging role of Accra in serving sub-region, renewal of aircraft fleet.	Improving air safety and security.
ICT	Very competitive market with high levels of mobile penetration at relatively low cost.	Improving the quality of mobile services. Harnessing market to complete universal access agenda (Internet and mobile).
Ports	Advanced institutional reform and private sector participation.	Alleviating capacity constraints that are currently holding back performance.
Power	Well-endowed with generation capacity. Good electrification rate.	Improving resilience to hydrological shocks by developing gas fired power and upgrading aging transmission network. Tackling huge hidden costs due to underpricing.
Railways		Funding the rehabilitation of the network. Improving performance of GRC to recapture mining traffic.
Roads	Good performer on road network, both in terms of financing and road network quality.	Preserving the real value of the fuel levy. Improving rural connectivity.
Water resources	Substantial volume of water storage available. Organisation by Africa standards.	Strengthening capacity of new River Basin. Developing irrigation potential.
Water and sanitation	Reached MDG for water. Significant improvements in utility finances.	Improving reliability of water supply. Reducing non-revenue water. Reaching the MDG for sanitation.

Source: Foster and Pushak (2011).

and administrative activity requiring efficiency in its execution to optimise economic development outcomes. This activity, therefore, requires close scrutiny.

3. Procurement

At the heart of infrastructure development is procurement. The concept of procurement in the context of construction is broad and covers virtually the entire process of acquisition, procurement planning, the process of contractor selection, negotiation of contract terms, contract formation and contract administration 2010). Based on the CIB W92 definition of procurement as the framework through which construction is brought about, acquired or obtained (see McDermott, 1999), Akintoye et al. (2003) have opined that procurement entails the acquisition of land, design, construction, management and commissioning of a project. Contract strategy and formation that define the allocation of risk on infrastructure projects are at the core of the process. Love et al. (1998) consider procurement as an organisational system that identifies relationships and assigns responsibilities among key players in the construction process. This definition, like the others, presents the contract formation process as integral to procurement. Throughout this chapter, therefore, contract formation and all issues relating to the construction contract are treated under procurement.

A myriad of standardised procurement methods exist to provide a systematic approach to navigating all the issues that must be dealt with under the ambit of procurement as identified above. Procurement methods used in building and civil engineering works include the traditional, integrated, management-centred and collaborative methods. These methods define inter alia the organisational structures and roles of key participants and government agencies, processes to be followed and timeframes for action as well as applicable contract strategies. For a critical, in-depth and practical exploration of infrastructure procurement processes in emerging economies, this chapter examines the procurement regime in Ghana. Apart from its typicality as an emerging economy, the focus on Ghana will also provide an understanding as to how context-specific issues can influence the efficiency of the procurement process.

In Ghana, procurement of major projects is mainly carried out by the State, represented by ministries, departments and agencies of government (MDAs) who are the procurement entities. These organisations use a number of procurement methods including the traditional procurement method (the commonly used method); design and build; and engineer, procure and construct (EPC) (the integrated methods). These different approaches offer several different advantages such as the assurance of competition; fairness and minimised tender cost due to the availability of bills of quantities; the potential to achieve low project cost, quality and functionality (in the case of traditional procurement); single point responsibility, early contractor involvement; and buildability (in the case of the integrated methods).

As will be revealed in the next section, dysfunctional processes and institutional structures in Ghana largely confound the realisation of such benefits and amplify the well-known disadvantages of these approaches which include excessive cost overruns as a result of incomplete designs; fragmentation; excessive variations; disruption of work; increased completion time and overall cost of project; and incidence of disputes (Latham, 1994; Franks, 1998; National Audit Office, 2001; Morledge et al., 2006).

4. Framework for procurement in Ghana

4.1 A historical perspective: Pre-Public Procurement Act 2003 (Act 663)

Public procurement during the pre-independence era was the function of the colonial administration performed by Crown agents and the Public Works Department (PWD). The former was responsible for the procurement of goods, and the latter, works. After independence in 1957, a number of MDAs were established in the 1960s and entrusted with responsibilities including carrying out infrastructure projects and providing consultancy for such acquisitions. These included the Ghana National Construction Corporation (GNCC), Electricity Corporation of Ghana, Ghana Water and Sewerage Corporation (GWSC) and Architectural Engineering Services Corporation (AESC). Section 2 of the Ghana Water and Sewerage Corporation Act 1965 (Act 310), for instance, gave the GWSC a mandate, inter alia, to make engineering survey plans and construct and operate works relating to water and sewerage. Similarly, the objects of AESC under Section 3 of the Architectural Engineering Services Corporation Act 1973 (NRCD 193) included carrying out technical studies in planning, design and supervision of infrastructural works. Central, Regional and District Tender Boards were set up to advise on the procurement of works.

By the mid-1990s, the public entities set up as conduits for procurement had become overwhelmed by the growing demands from the MDA and had become inefficient (World Bank, 2003b). In 1993, the Statutory Corporations (Conversion to Companies) Act 1993 (Act 461) was enacted to enable existing corporations to be converted into companies. The AESC, ECG and GWLC were all transformed into limited liability companies. Public entities were no longer obliged to use State institutions to carry out works on their behalf. State entities increasingly relied on private consultants and contractors to execute projects.

The literature points to the traditional method of procurement, with design split from construction both in time and space, as the dominant procurement method used during the Pre-Act 663 era (Anvuur et al., 2006; Kheni, 2008). The World Bank (2003b) identified selective tendering and sole-sourcing as the most widely used tendering methods prior to the enactment of Act 663. The two methods were used in about two-thirds of all projects within the public sector. The other tendering method used was competitive tendering. Tender Boards set up in the 1960s and subsequently regulated under the District Tender Board Regulations 1995 (LI 1606) continued to perform their roles until the coming into force of Act 663.

Procurement during the pre-Act 663 period was plagued with several deficiencies (World Bank, 2003b). These included lack of a comprehensive legal framework with clear procedures on procurement, weak capacity of procurement staff and an unclear institutional and organisational framework for procurement. There were delays in contract closure, preparation of technical specifications and drawings, evaluation, approvals and payments (World Bank, 2003a). These had a snowballing effect on contract delivery, performance issues, avoidable claims and disputes. Challenges associated with procurement of works in Ghana during the period before the enactment of Act 663 are well documented in Westring (1997), Eyiah and Cook (2003), World Bank (2003b) and Anvuur et al. (2006).

4.2 Current procurement regime: Post-Public Procurement Act 2003 (Act 663)

For the first time in Ghana, a new unified law on procurement was enacted in 2003. Act 663 has nine parts, which cover issues such as the establishment of a procurement authority and structures (see Parts One and Two), general rules on procurement (Part Three), methods of procurement (Part Four) and tendering procedures (Part Five). There are separate rules on engaging services of consultants (Part Six). The law applies to all procurement of goods, works and services financed in whole or in part from public funds, loans obtained or guaranteed by the State and foreign aid, and activities incidental thereto such as description of requirements, invitation of sources, preparation, selection, award of contract and contract administration. It, therefore, clearly covers all aspects of procurement of infrastructure projects.

Under Act 663, competitive tendering (national and international) is the main method for contractor selection except in cases where a justification exists for the use of other tendering methods such as two-staged tendering, restricted tendering and sole-sourcing. Conditions and procedures for the use of these tendering methods are outlined in the Act. All procurement entities are required to use the appropriate tender documents as provided in the Fourth Schedule to the Act. Section 50 of Act 663 requires that these documents be used with minimum modifications to be introduced through the contract data sheet and the Special Conditions of Contract. No changes are to be made in the standard tender documents. Bids are to be opened at the time and place stipulated in the invitation documents and in the presence of all bidders. Bid evaluation criteria are to be predetermined as per the invitation documents and must be objective and quantifiable.

Evaluation is not to be based solely on the lowest tender price but also other weighted criteria provided in the bid document. In arriving at the lowest evaluated tender, the committee must consider the tender price in the light of any margin of preference applied, the cost of operating or maintaining the works, the functional characteristics of the works, payment or guarantee terms and national security. Section 59 of the Act additionally requires that the effect of the acceptance of the tender on the national economy be considered in terms of the balance of payment position and foreign exchange reserves of the country, counter trade arrangements offered by suppliers and contractors, extent of local content, and the overall economic development potential offered by tenders.

4.3 Other rules on infrastructure procurement

There are two instances where the provisions of Act 663 do not apply. Firstly, an applicable loan agreement, guarantee contract or foreign agreement may provide a different procedure for the utilisation of such funds. Secondly, the Minister of State responsible for a particular procurement can decide that it is in the national interest to use a different procedure. The choice of procurement method for major projects in Ghana is therefore not only guided by the requirements of Act 663, but it is also tied to considerations of national interest and donor funding requirements. As a result, there exist two streams of procurement rules, namely those under Act 663 and those contained in agreements with donors or creditors (these two constituting what may be regarded as the default strategies).

5. Current procurement practices

Infrastructure procurement strategies currently in use in Ghana can be classified into three categories, namely the Default Strategies (used typically on Government of Ghana and donor-funded projects); the Exigency-Driven Strategy; and increasingly a Public–Private Partnership (PPP) strategy. This classification is based on their fit with the requirements of Act 663. These strategies are explained in detail next.

5.1 Default strategies

On annual basis, budgetary allocations are made for the procurement of major projects in the road, water, energy, health, education and other sectors of the economy by the Government, the main provider of public physical infrastructure. Once budgetary allocations are made in line with the procurement plans of the various ministries, the Government, through the various implementing agencies under the ministries, implements such projects either in-house or through contracting, using mainly traditional procurement methods. The various tendering processes approved by Act 663 are used, consultants and contractors are selected, contracts are negotiated and signed, designs are produced, construction works are carried out and projects delivered. These processes are regulated under the Act and managed by the implementing agencies. This is one of the default strategies.

Under the traditional procurement method commonly used under the default strategy, design is split from construction both in time and in space (Anvuur et al., 2006; Kheni, 2008). Survey, design and estimation are often treated as a package distinct from the construction phase. Different funding arrangements would usually be made for the feasibility studies and design phase on one hand and the construction phase on the other. In many instances, there is considerable time lag between the period when such studies and designs are conducted and when the construction takes place. Thus, updating technical reports and designs prior to construction is a common occurrence. Sometimes dated designs are used to procure works. In some cases, only quick reviews of old designs are done immediately before execution of projects. There are also occasions where designs are actually only identified to be inadequate in the course of construction. These scenarios often lead to issues of buildability, change of designs and sometimes extensive changes in the scope of works. The effects of such variations on cost and delay are enormous. In most cases, designers for the initial detailed designs do not supervise the work, leading to the appointment of new consultants who invariably review the earlier design.

Another type of default procurement strategy is the donor-driven approach. Sections 14(1)(d) and 96 of Act 663 permit funding agencies and donors to apply rules distinct from those under Act 663. Section 96 specifically provides as follows: "Notwithstanding the extent of the application of this Act to procurement, procurement with international obligations arising from a grant or concessionary loan to the government shall be in accordance with the terms of the grant or loan."

Thus, the procurement under this strategy is dictated by the procurement guidelines or rules of the funding agency concerned. Notwithstanding the diversity of rules and practices of funding organisations, some common trends are observable. Usually, initial needs assessment and selection of projects are reviewed and streamlined through the funding agency's project appraisal document (PAD). The PAD does not only examine the viability and feasibility of the project but also outlines how the project is to be executed with tentative timetables. Once the funding agency's requirements are met, an elaborate process of due diligence and financial or loan negotiations are carried out between the parties, with the Ministry of Finance and Economic Planning (MOFEP) leading the Government team.

Other Government institutions such as the Cabinet, Parliament, Attorney-General's department and the sector ministry may all play their respective roles in securing and approving funding for the project. Once funding is secured, procurement of the project is undertaken in accordance with the procurement strategy, methods and rules of the funding organisation. Depending on the funding organisation's threshold, every major step in the procurement is to be approved by the funding agency concerned. Tendering under this strategy is by international competitive bidding. In some cases, particularly under bilateral funding arrangements, consultants and contractors are nominated by the home State of the funding agency. Under this strategy, the procurement method used will often depend on the rules of the funding agency and the complexity of the project. In practice, the traditional and the integrated methods (design and build; and engineer, procure and construct) have all been employed. The donor-driven strategy is widespread in the road and energy sectors. There are indications that the World Bank procurement procedures are the most dominant procurement procedures in Ghana.

The use of traditional procurement is particularly rife in the road sector, as Table 9.3 shows. For all the projects indicated, the choice of method was attributed partly to funding challenges. As a result, seamless execution of projects through both the design and construction phases is rare.

The fragmentation of project components results in serious lack of synchrony between the design phase and the construction phase. Insufficient designs are often detected during the construction phase, leading to disruption of progress of work, excessive variations and cost and time overruns. An earlier study in 2002 of procurement of 132 works indicated that up to 84% of works contracts studied incurred cost overruns of up to 30% of the initial cost (World Bank, 2003a and 2003b). The reports attributed the cost overruns to excessive variations and unrealistic time extensions by consultants. It was also discovered that contractors will often lower their bid prices in order to become competitive. During the execution of contracts, contractors found all possible means to recoup whatever they lost.

The natural consequence of this widespread practice was increased claims, some of which eventually resulted in disputes. These views on the traditional procurement system re-echo the position of the UK National Audit Office report (National Audit Office, 2001, p. 6) on the negative effects of the traditional procurement method, such as cost and time overrun and lack of innovation. The difficulties with the use of the traditional method in Ghana are exacerbated by

Table 9.3 Major road infrastructure projects in Ghana as of 2010 procured through traditional methods

Projects	Length	Funding source	Status as of 2010
1. Achimota-Apedwa	64.8 km	Yet to be determined	All designs completed
2. Apedwa-Bunso	22.0 km	ADB/OPEC	Works in progress
3. Bunso-Anyinam	11.5 km	BADEA/OPEC	Ditto
4. Anyinam-Kumasi	136 km	ADB/OPEC	Ditto
5. Accra-Yamoransa	115 km	JBIC	Bidding process in progress
6. Three (3) Roads Study (feasibility study and design):			Consultancy services signed
a. Sawla-Fufulso	167 km	ADB	
b. Wa-Han	77 km	ADB	
c. Lawra-Han-Tumu	108 km	ADB	
7. Navrongo-Tumu-Kupulima (feasibility study and design)	128 km	DANIDA	Detailed engineering design in progress
8. Akatsi-Noepe (feasibility study and design)	30 km	DANIDA	Detailed engineering design completed
9. a. Kumasi-Techiman Lot 1: Feasibility and detailed design and supervision	113.3 km	EU	Procurement of works starting October 2002
b. Bibiani-Abuakwa	73.6 km	EU	Ditto
c. Tarkwa-Agona Junction.	58.8 km	EU	Ditto
10. Tetteh Quarshie-Mamfe Road (design review and supervision)			Procurement stage. General Procurement Notice published, short-listing of consultants done for lots 1 & 3.
a. Lot 1: Tetteh Quarshie Inter-change	10 km	ADB	
b. Lot 2: Tetteh Quarshie-Pantang	30 km	IDA proposed	
c. Lot 3: Pantang-Mamfe Road		ADB	

11.	Reconstruction. of Tema-Sogakope Road (design review and supervision)	88 km	KfW	Procurement of consultant for design review and for works in progress
12.	Selected roads in Ashanti and Brong-Ahafo regions	–	KfW	Detailed engineering design completed; procurement of civil works in progress
13.	Proposed road sector development programme (RSDP)		IDA	
Lot 1:	a. Brewaniase-Oti Damanko	50 km		Procurement of works in progress
	b. Oti Damanko-Yendi	50 km		
Lot 2:	a. Sogakope-Ho	15.6 km		
	b. Ho-Fume	50 km		Procurement of works in progress
Lot 3:	a. Techiman-Kintampo	30 km		Procurement of works in progress
	b. Berekum-Sampa	14 km		
Lot 4:	a. Accra-Tema Motorway II and III	33 km		
	b. Jasikan-Brewaniase	78 km		
Lot 5:	a. Wa-Han	35 km		
	b. Bamboi-Bole	48 km		

Source: Ministry of Road Transport (2010).

administrative lapses, which result in inadequate preparation for projects and use of incomplete designs. Essentially, the appeal of the traditional procurement method in the Ghanaian context appears to be its ability to allow projects to be fragmented and funded separately. Again, it seems the overwhelming use of the traditional procurement method is not due to any logical assessment of its ability to deliver a project effectively and efficiently, but its capacity to allow the client to stagger delivery.

The traditional procurement method is not the only method of procuring major externally funded infrastructure projects in Ghana. Design and build as well as engineer, procure and construct procurement methods are increasingly being used (Ameyaw, 2011). Some of the notable design and build and engineer, procure and construct projects in Ghana include the Accra and Kumasi sports stadia; two new stadia, one at Essipong in Sekondi and the other in Tamale; the Accra Waste Project; the 400 MW capacity hydro-electric dam at Bui in the Brong-Ahafo region of Ghana (Baah and Jauch, 2009; Hensengerth, 2011); and the €45 million Tamale Water Supply Extension Project completed in 2008 (Ghana Water Company Limited, 2011). The rationale is that the design and build method is suitable for funding agencies who require home consultants and contractors to execute the projects they fund. The use of design and build, and engineer, procure and construct are relatively prevalent within the water and the energy sectors. The road sector has witnessed the use of engineer, procure and construct only in relation to highly specialised projects such as the rehabilitation of the Adomi Bridge at Atimpoku in the eastern region of Ghana. Once again, it appears that the choice of procurement method is motivated by funding requirements rather than considerations relating to project success. Contractor selection under the donor-driven strategy is mainly by international competitive tendering.

5.2 Exigency-driven strategy

It is a common practice in Ghana to see major projects procured by strategies other than the default approaches due to exigencies. In such cases, projects do not go through the normal default rules on procurement. For example, in 2007, at the peak of an energy crisis in Ghana, the Government entered into a power purchase agreement with Balkan Energy Ghana. The agreement was on a lease-build-operate basis. There is no evidence that there was any open tender during the award of the contract.

Another example of the exigency-driven procurement strategy is what is referred to as "unsolicited proposal" or single-source procurement (see Section 40 of Act 663). With this procurement strategy, Government agencies conduct a needs assessment and identify major projects within their sector which they hope to implement but are unable to due to budgetary constraints and donor fatigue.

The agencies discuss their intention to have such projects undertaken with prospective investors and contractors at various fora. Interested contractors then make contact with the agencies either directly or through their respective sector ministries. The contractors or investors are allowed to do their own feasibility studies and select which projects they would want to implement. On the basis of the preference of a particular contractor and a promise to secure funding for the chosen project, a memorandum of understanding is signed between the sector ministries and the prospective contractor. The latter then takes steps to secure funding or identify funding sources for the selected project.

A prospective contractor who is willing to self-finance a project is offered supplier credit arrangements. If it secures funding from other sources, it would be required to provide details of the funding institutions and financial term sheets to MOFEP, which then undertakes due diligence on the funding organisations involved. Once MOFEP is satisfied that the facility meets its requirements, negotiation of the terms of the facility is undertaken. A financial agreement is prepared or studied by MOFEP and further reviewed by the Attorney-General's department, which has to give "no objection" to the transaction. Once Cabinet and Parliamentary approvals are secured for the facility, the sector ministry and the agency concerned are notified to proceed with the construction agreement with the contractor.

In most cases, the memorandum of understanding signed at the initial stages of the transaction will endorse the use of sole-sourcing as the tendering process. Act 663 has rules on when sole-sourcing should be used. Where the transaction is to be procured by sole-sourcing, a value for money audit is carried out by MOFEP. The essence of this exercise is to compensate for the absence of the use of open tender procedures, which are widely believed to have inherent mechanisms to achieve accountability, fairness and value for money. Projects granted through the unsolicited proposals strategy are often awarded as design and build or engineer, procure and construct projects. The successful contractor is, therefore, responsible for the delivery of the entire project at the agreed cost.

This strategy appears to have become dominant in the water sector and an example is an agreement signed in September 2009 regarding the Kpong Water Expansion Supply system valued at US$273 million. Reliance on unsolicited proposals for infrastructure procurement is not limited to the water sector. This process has also gained grounds in the energy sector. The proliferation of the unsolicited proposals strategy has, however, been viewed by some as an attempt to side-step the provisions of Act 663. This has led to the issuance of a Cabinet directive in 2012 asking for all such practices to be halted. It is, therefore, not surprising that the Government is seeking to channel the soaring private interests in the development of public infrastructure into the PPP strategy.

5.3 PPP strategy

Since the late 1980s, there has been an increase in private sector participation in infrastructure development across the globe (World Bank, 1994; UNCTAD, 2008). These developments have occurred through the establishment, operation and maintenance of infrastructure facilities by private sector entities (UNCTAD, 2008). Many reasons have been assigned for these developments. Financial constraints faced by many developing countries seem to be the key reason (UNCTAD,

2008, p. 85). In the 1990s, many governments in the developing world saw the global trend towards private sector involvement as an opportunity to deal with the increased pressures of fiscal adjustment (Annez, 2006; Calderón and Servén, 2010b). Kessides (2004) argues that the main challenge of the government controlled infrastructure development model was underpricing, which consequently led to under-investment. Under-investment invariably led to huge deficits, which needed to be catered for by governments. Governments in turn needed more funds, which were not available. One of the options was to turn to the private sector. In effect, the State retrenched from infrastructure development as a result of high cost, poor performance, inefficiency and inability to expand to meet rapidly growing demands (Harris, 2003; Kessides, 2004). The hope was that the private sector would bear the cost of infrastructure and services whilst government plays the role of a regulator (Estache and Fay, 2007). Consequently, many developing countries have opened up what was once the preserve of the State to the private sector.

From 21 projects involving an amount of about US$11,787 million between 1984 and 1989, private sector investment in infrastructure (PPI) projects in East Asia and the Pacific region rose to 871 projects with a total investment amounting to U$135.5 billion between 2000 and 2009 (World Bank/PPIAF, 2010; Park, 2010). These projects in addition to investments in existing projects in the region brought the total for the region to U$181 billion, constituting 36% of the total investment in infrastructure with private participation for the period 2000–2009 in developing countries (Park, 2010). According to the World Bank, 43 out of 48 sub-Saharan African countries implemented 238 infrastructure projects between 2000 and 2009 with private participation and a total investment commitment of US$47.6 billion (Izaguirre and Perard, 2010). Added to existing investment, the total for this region was US$79 billion, accounting for about 10% of total investment in infrastructure in developing countries for the period (Izaguirre and Perard, 2010), as against about 3% during the period between 1990 and 2000. Eight developing countries in the South Asian region implemented 361 infrastructure projects with PPI during the last decade. This constituted 15% of the total PPI investment in developing countries with a total investment of US$174.4 billion (Jett, 2010). Compared to the relatively negligible investment in the 1980s, PPI has seen astronomical increase over the past two decades.

Overall, data captured by the World Bank's PPIAF shows that between 1990 and 2009, Latin America and the Caribbean saw the largest number of PPI projects, totalling 1,425, and attracted a total investment of US$578,760 million; East Asia and the Pacific followed closely with 1,403 projects which attracted a total investment of US$308,306 whilst sub-Saharan Africa attracted 382 projects involving investment totalling US$95,040 during the period (World Bank/PPIAF, 2010b). In terms of the countries which attracted the PPI projects and investment during this period, China, India and Brazil received many projects and more investment commitments. China received the highest project numbers (931) while Brazil and India had 463 and 380 projects respectively (World Bank/PPIAF, 2010b). According to the World Bank/PPIAF's (2010b) records, Brazil received the highest investment commitments totalling US$270,346 during the period 1990–2009 as against US$158,397 and US$111,806 for India and China respectively. In terms of sectors, although the energy sector ranked highest in respect of the number of projects (1,852) between 1990 and 2009, the telecommunication sector attracted more investment (US$719,645) (World Bank/PPIAF, 2010b). In spite of the effect of the recent global financial crises and the fact that private capital is becoming much more selective (Izaguirre, 2010a, 2010b), PPI across the developing world continues to remain steady, with the upward trajectory set to continue.

In June 2011, the Cabinet of Ghana approved a national policy on PPP. The rationale was to address the funding constraints facing the Government in respect of infrastructure development. Inadequate internal resources and donor fatigue led the Government to turn to private investor participation (in PPP models) for the delivery of major infrastructure projects. In the PPP policy document, the Government states that Ghana's infrastructure deficit demands that the State makes a sustained spending of US$1.5 billion per annum over the next 10 years. The policy document further indicates that the Government intends to use PPP as a strategy to assist in bridging the infrastructure spending gap, reinforcing the point that PPP is on an upward trajectory. PPP is yet to receive legislative backing but the prospects of this are very good.

6. Contract formation and review process

6.1 Contract formation

At the end of a successful tender evaluation process, a consultant or a contractor is selected. The selection of such an entity or entities paves the way for contract formation where parties order their relationships. Sections 65(2) and (3) of Act 663 provides that where the tender documents require the signing of a written contract, such document shall be signed within thirty days after a notice of acceptance of a bid has been dispatched. The contract begins on the commencement date stipulated on the contract document. However, the commencement is subject to parliamentary approval of the relevant contract documents in compliance with the constitutional requirement under Article 181(5) of the 1992 Constitution of Ghana. Section 65(2) of Act 663 stresses the need for the written contract to conform to the tender. There is overwhelming evidence in support of the practice of contract negotiation prior to the signing of the written agreement.

For most major construction transactions, there are two sets of agreements: the financial agreement and the construction agreement. Both require Parliamentary approval in order to be operational [see the 1992 Constitution of Ghana, Article 181(5)]. In relation to the construction contract, it is common for standard form contracts to be used. It is also common knowledge among construction practitioners that the construction contract is not made up of a single document. There is the agreement and other documents deemed to be part of the agreement. These include the letter of acceptance, the bid and appendix to the bid, the conditions of contract (general and special), the designs (specifications and drawings) and the priced bill. These documents are hierarchically arranged in order of importance.

190   *Nii Ankrah et al.*

6.2 Conditions of contract

The Fédération Internationale des Ingénieurs-Conseils (FIDIC) suite of contracts constitutes the main standard forms or conditions of contract in use in the procurement of major infrastructure projects in Ghana (especially the Red book, the Yellow book and the Silver book). In fact, the use of the FIDIC suite of contracts have become so entrenched that even major projects funded by the Government are awarded using the FIDIC conditions of contract. The conditions of contract developed by the Public Procurement Authority are also based on FIDIC but are used for small and medium sized projects.

The dominance of the FIDIC range of conditions of contract for major projects is attributed to several reasons. Firstly, key funding agencies such as the World Bank and some of the other multilateral development banks as well as bilateral funding institutions require that the FIDIC conditions of contract be used for projects they sponsor. Secondly, the FIDIC conditions are widely accepted standard forms, tried and tested internationally. These features endeared it to foreign contractors. Thirdly, parties to major construction contracts generally agree that the provisions in the FIDIC conditions are fairly balanced and address concerns of both employers and contractors. The fourth reason is the idea that the FIDIC range of contracts is developed by international experts. Finally, the FIDIC suite of forms has benefitted from years of research and thus addresses old as well as current challenges with the relationship between employers and contractors (Mante, 2014). In addition to the FIDIC suite of contracts, there is also evidence of the use of EU conditions of contract, mainly for EU sponsored projects. Chinese contractors prefer using their own versions of contracts. On some highly technical projects, bespoke contracts are used.

Regardless of which contract forms are used, it is common to have two sections: the general conditions, which are made of standard clauses of general application, and the special conditions or Conditions of Particular Application (COPA). The special conditions are tailored to suit the specific understandings between the parties involved in the particular transaction. The terms of the special conditions are, therefore, subject to negotiation between the parties and typically address payment terms, labour, protection of utility services, performance security, termination of the contract for the employer's convenience, and dispute resolution. According to Mante (2014), contract documents (including the COPA) are often prepared and negotiated by contract specialists within their institutions with little or no involvement of the Attorney-General's department. This state of affairs, as will be argued later, can be problematic.

6.3 Contract review

The legal framework for major infrastructure procurement in Ghana mandates the Attorney-General's department and Parliament to review major transactions involving the State (see Figure 9.1). The review process aims, inter alia, at ensuring that what escapes the attention of government representatives responsible for agreeing contract terms with foreign contractors can be identified and remedied.

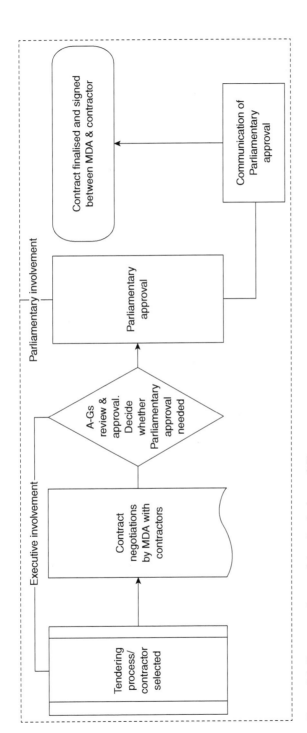

Figure 9.1 Contract review model under the laws of Ghana

Source: Public Procurement Act 2003 (Act 663).

However, these institutions are resource-constrained and, in the case of Parliament, totally inactive when it comes to reviewing and approving terms of construction contracts. By Article 88 of the Constitution of Ghana 1992 and Sections 20–25 of the State Property and Contract Act 1960 (CA 6), the Attorney-General's department has a mandate to examine transactions negotiated by procurement entities, propose changes, proffer legal advice, and/or vet and approve contract documents covering such transactions. Section 22(2) of the State Property and Contracts Act 1960 (CA 6) expressly gives the Attorney-General the right to "examine and approve" contracts involving the State. In the case of Parliament, Article 181(5) of the Constitution makes it mandatory for all "international business and economic transactions to which the government is a party" to be submitted to Parliament for its scrutiny and approval. The Constitutional provision expressly indicates that an international economic transaction which does not satisfy this criterion cannot become operational.

In practice, however, Parliament has not carried out this Constitutional function for the nearly twenty years of its existence as a Constitutional requirement and, thus, remains in no position to remedy any oversights from implementing agencies (see Figure 9.2). Even if Parliament was to play its required role, without a deliberate policy to guide its review and approval system, thorough scrutiny may not be achieved. Moreover the Attorney-General's department, which has borne the review responsibility, is plagued with human resource challenges and lack of expertise in the face of volumes of work. This calls into question the extent of rigour that goes into review of issues like pricing and value for money. Similar to the Parliamentary situation, there is an absence of formal guidelines on what Attorneys should look out for during the review process.

6.4 Contract administration and performance

Procurement also covers the execution and contract administration stage. In Ghana, the emphasis is often on the planning and the contract placement stage with very little in terms of statutory provisions existing on contract performance issues. Although Section 14 of Act 663 brings functions related to the "phases of contract administration" under the scope of the Act, neither these "phases of contract administration" nor the functions related to them are elaborated on. This lacuna is attributable to the fact that the procurement process culminates in the signing of a contract and the ensuing relationship remains contractual. The contract determines performance indicators and what transpires when there is a breach. The contract defines the scope of work to be performed, the rights and obligations of the parties, the functions and the responsibilities of the engineer, architect, construction or project manager regarding supervision, and general administration of the contract . The contract also provides a dispute governance system which the parties are to rely on in case of a dispute. In other words, the efficiency of the contract administration process derives from the robustness of the underlying contract and the commitment of the parties to uphold its terms and

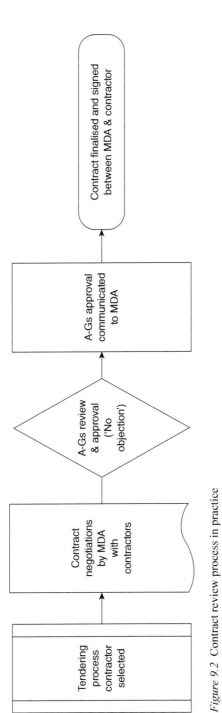

Figure 9.2 Contract review process in practice

Source: Mante (2014).

conditions. The procurement manual of Act 663 provides some general guidance on the role of the project manager or supervisor and issues to be addressed as specified in the contract, but lacks necessary detail in terms of how such functions should be monitored.

Whilst the above discussions appear to affirm the acclaimed nature of public procurement as a formalised, standardised, bureaucratic, and rigid process, it is also evident that there are elements of arbitrariness and lack of compliance with procedures, as well as other issues that undermine the realisation of efficiency and what the World Bank (2010) terms the development effectiveness of public expenditure. The next section distils these emerging issues and considers their wider implications in the context of infrastructure delivery across the developing world.

7. Inefficiencies in the procurement regime

The procurement reforms in Ghana which culminated in the enactment of Act 663 aimed to achieve value for money, and ensure efficiency, transparency, equity and fairness, and accountability in public procurement. However, these ideals have not been achieved in practice due to inefficiencies associated with the procurement process. These include lack of cooperation and coordination among the employer's sub-units, constraints associated with the legal framework, lack of compliance with the law on procurement, political interference and corruption, human resource constraints, and challenges with funding requirements. Some of the listed issues are examined briefly.

There are several organisations involved in the public procurement process in Ghana. For instance, the development of a policy framework for infrastructure acquisition is the responsibility of Cabinet, the sector ministry and the National Development Planning Commission [see the 1992 Constitution at Article 86 and Civil Service Law 1993 (PNDCL 327) at Section 13]. Technical preparations for projects are the responsibility of the implementing agencies [see Ghana Highway Authority Act 1997 (Act 540), Section 3]. The tender committees and review boards of the various procurement entities are responsible for the contractor selection process. Financial arrangements for infrastructure procurement including payment for works is also a multi-organisational activity involving Cabinet, Parliament, the sector ministries and the agencies of MOFEP at various stages. Consequently, the efficiency of the process depends very much on the ability of the institutions to cooperate with each other and to coordinate their activities. In reality, this quality is lacking. For instance, the MDAs renege on the obligation to refer infrastructure-related contracts to the Attorney-General's department for review because of lack of capacity and the inability of that department to provide timely feedback. Lengthy chains of inter-organisational consultations and approvals delay contract formation, execution and payment for work completed. Generally, individual entities are eager to protect their "turfs" and are unwilling to embrace changes which are likely to impact their roles in procurement. These challenges have had significant implications for the effectiveness of the current institutional framework for public procurement of infrastructure.

Closely related to the institutional weaknesses discussed above are problems associated with the existing legal framework for procurement. The State operates an elaborate legal system which determines the functions of each sub-unit and the consultations and approvals necessary. Consequently, the employer's performance under construction contracts naturally suffers delays due to the complex nature of its decision-making processes. For instance, Section 65 of Act 663 and Clause 8 of the 1999 FIDIC Conditions of Contract for Building and Engineering Works (the Red Book) provide timetables for the execution of written contracts and commencement of works, respectively. For major construction projects, Parliamentary approval under Article 181(5) of the Constitution is required. Under the said Constitutional provision, international business or economic transaction to which the Government of Ghana is a party requires Parliamentary approval in order to be operational. A contract signed in compliance with the timetable under section 65 of Act 663 remains unenforceable until such approval is obtained. Similarly, any commencement of work pursuant to the default position under Clause 8 of the 1999 FIDIC Redbook prior to Parliament's approval of such transactions is void.

The validity of some major contracts which have not received Parliamentary approval have been challenged in a number of judicial decisions by the Supreme Court of Ghana, for example: *A-G v Faroe Atlantic Company Limited* (the *Faroe Atlantic* case) [2005–2006] SCGLR 271; *A-G v. Balkan Energy (Ghana) Limited and Ors* (the *Balkan Energy* case) [2012] 2 SCGLR 998; *Martin Amidu v A-G and 2 Ors.* (the *Waterville* case) suit number J1/15/2012, judgment of 14th June 2013; *Amidu v Attorney-General and 2 others* (the *Isofoton* case), 21st June 2013, Supreme Court (unreported); and *Klomega v Attorney-General and 3 Others*, 19th July 2013, Supreme Court (unreported).

In the *Faroe Atlantic* case, for instance, the Supreme Court held that a power purchase agreement between the Government and a UK-registered company for the refurbishment of a power barge and the generation of power was an international business and economic transaction within the meaning of the provision of Article 181(5) of the Constitution of Ghana. The agreement, the Court held, ought to have satisfied two requirements under Article 181: firstly, it should have been laid before Parliament; and secondly, it was not to come into operation unless approved by a Resolution of Parliament. The contract on which performance was based was, therefore, declared unconstitutional. Thus, the demands of the legal system of the employer necessarily prolong timeframes for decision-making by the employer and this has implications for delays in contract formation and execution. This calls into question the effectiveness of the current legal framework for procurement.

Another of the inefficiencies plaguing the extant procurement regime is the failure on the part of some of the entities entrusted with a role in the process to comply with legal requirements for procurement. A study conducted after Act 663 came into force revealed very low compliance levels and a continuation of old practices and challenges (Osei-Tutu et al., 2008). Although single-source procurement under Act 663 is allowed only in exceptional circumstances (see Section 40 of Act 663), it is widely used and generally encouraged. The Public Procurement Authority, the entity with responsibility for approving the use of single-source procurement, appears to be very liberal in permitting such exceptions. Practice associated with this procurement strategy generally defeats the accountability, competition and transparency expected to be engendered by Act 663.

Politicians often interfere in the procurement process. These interferences take several forms including influencing the contractor selection process. There were instances in Ghana where interference by politicians allegedly went beyond influencing the contract award to taking technical decisions and abrogating contractual obligations (see the *Isofoton* case). Contractors were sometimes instructed to commence projects prior to the execution of the construction contract (see the *Waterville* case). Whilst sometimes these interferences may be attributed to a genuine desire to protect national interest, in many other instances the motive may be personal interest. The consequences of these interferences are that procurement procedures are ignored or side-stepped leading to poor contract administration and, ultimately, poor project delivery. In instances where contractual requirements are ignored, these have led to claims and even commencement of dispute resolution processes by aggrieved parties. Political interference and corruption stifle proper implementation of procurement rules and best practices.

Human resource constraints still remain a key problem with the procurement regime in Ghana. Many procurement staff lack the requisite training to administer the process effectively. Similarly, inadequacies in budgetary allocations and fund disbursement still remain a critical problem with the Ghanaian system over a decade after a World Bank study identified this issue as a critical problem causing payment delays (World Bank, 2003b). Reviews by the World Bank in 2007 and 2010 of the procurement regime in Ghana found that though substantial progress has been made since a 2003 assessment, there were still issues with, inter alia, the legal framework, institutional structures, capacity building and procurement practices (World Bank, 2008a, 2008b, 2010a).

The inefficiencies identified with the current procurement regime in Ghana are not peculiar to the Ghanaian situation. Examination of country procurement assessment reports on many emerging economies reveal striking similarities in the issues facing procurement in such countries. For instance, various assessments, reviews and updates on Nigeria, Zambia, Mozambique, South Africa and Cote d'Ivoire prepared between 2000 and 2010 by the World Bank disclose the persistence of challenges akin to those identified above (World Bank, 2014). Common problems cutting across most of the reports include weak legal and institutional framework, weak procurement oversight, lack of procurement capacity, problems with procurement documentation and inadequacies with budget allocation and the manner in which project funds are released.

8. Conclusion

The chapter has examined public procurement and its impact on economic development. The nature of infrastructure and the methods by which they are generally procured have also been discussed. Using Ghana as a case study, the framework for public procurement, particularly those related to infrastructure acquisition in emerging economies, has been explored. A number of conclusions can be reached from the discussions above. Firstly, there is increasing evidence in support of the position that infrastructure development has a significant impact on economic development. Secondly, in most emerging economies, the State still remains the major client for infrastructure procurement. Thirdly, infrastructure is acquired through public procurement, which constitutes a substantial part of the GDP of emerging economies. In effect, public procurement has a direct impact on the performance of the economies of many developing countries. As the process which ultimately delivers the needed infrastructure capital in support of economic activity, public procurement also has significant indirect effect on economic development through projects which are successfully procured. Consequently, efficient procurement remains a necessity for emerging economies.

The evidence from Ghana reveals inefficiencies in the procurement process common to other emerging economies. Weak legal and institutional frameworks, non-compliance with procurement rules and procedure, lack of training for staff involved in procurement, political interference, inadequate budgetary allocations and erratic release of funds are but a few of the inadequacies associated with many procurement regimes in emerging economies. These have an effect on contract execution, project completion and the delivery of projects according to specifications.

For many developing countries, some efforts are already being made to reverse the trend of inefficient public procurement. Many of the country procurement assessment reports sponsored by organisations such as the World Bank and OECD also provide lengthy lists of recommendations for improvement. These recommendations include reforming procurement policy, strengthening legal and institutional frameworks, compliance with laid-down procurement procedures and fighting corruption. In effect, there is an urgent need for emerging economies to fully implement such recommendations to ensure improvement of their procurement systems, if they are to benefit from the impact of public procurement on economic development.

References

ADB, JBIC and World Bank (2005) *Connecting East Asia: A New Framework for Infrastructure*, Washington, DC: World Bank/ADB/JBIC.

Agénor, P.-R., Bayraktar, N., Aynaoui, K. and Moreira, E.P. (2005) *Achieving the Millennium Development Goals in Sub-Saharan Africa: A Macroeconomic Monitoring Framework* [online]. Available at: http://elibrary.worldbank.org/doi/pdf/10.1596/1813-9450-3750 [Accessed 8 June, 2011].

Agénor, P.-R. and Moreno-Dodson, B. (2006) *Public Infrastructure and Growth: New Channels and Policy Implications* [online]. Available at: http://elibrary.worldbank.org/content/workingpaper/10.1596/1813-9450-4064 [Accessed 19 December 2014].

Akintoye, A., Beck, M. and Hardcastle, C. (2003) *Public–Private Partnerships* [Wiley Online Library]. Available at: http://onlinelibrary.wiley.com/doi/10.1002/9780470690703.fmatter/summary [Accessed 19 December 2014].

Ameyaw, C. (2011) *Comparative Performance Evaluation of the Traditional Design-Bid-Build (DBB) and Design-Build (DB) Procurement Methods in Ghana*, MSc thesis (unpublished), Kwame Nkrumah University of Science and Technology, Kumasi, Ghana.

Andrés, L. A., Guasch, J. L., Haven, T. and Foster, V. (2008) *The Impact of Private Sector Participation in Infrastructure: Lights, Shadows, and the Road Ahead*, Washington DC: The World Bank/ PPIAF.

Annez, P. C. (2006) *Urban Infrastructure Finance from Private Operators: What Have We Learned From Recent Experience* [online]. Available at: http://elibrary.worldbank.org/doi/pdf/10.1596/1813-9450-4045 [Accessed 19 December 2014].

Anvuur, A., Kumaraswamy, M. and Male, S. (2006) Taking Forward Public Procurement Reforms in Ghana, paper presented at CIB W107 Construction in Developing Economies International Symposium on Construction in Developing Economies: New Issues and Challenges, 18–20 January, Santiago, Chile.

Arrowsmith, S. (2005) *The Law of Public and Utilities Procurement* (2nd ed.), London: Sweet and Maxwell.

Arrowsmith, S. (Ed.) (2010) *Public Procurement Regulation: An Introduction*, EU Asia Inter-University Network for Teaching and Research in Public Procurement Regulation Consortium [online]. Available at: http://www.nottingham.ac.uk/pprg/documentsarchive/asialinkmaterials/publicprocurementregulationintroduction.pdf [Accessed 20 August, 2011].

Aschauer, D. A. (1989) Is Public Expenditure Productive? *Journal of Monetary Economics*, 23(2), pp. 177–200.

Baah, A. Y. and Jauch, H. (2009) Chinese Investments in Africa, in: Baah, A. Y. and Jauch, H. (Eds), *Chinese Investments in Africa: A Labour Perspective* [online]. Available at: http://www.cebri.org/midia/documentos/315.pdf [Accessed 19 December 2014].

Bower, D. (2003) *Management of Procurement*, London: Thomas Telford Services Ltd.

Briceno-Garmendia, C., Estache, A. and Shafik, N. (2004) *Infrastructure Services in Developing Countries: Access, Quality, Costs and Policy Reform*, Washington DC: World Bank.

Calderón, C. and Servén, L. (2010a) Infrastructure and Economic Development in Sub-Saharan Africa, *Journal of African Economies*, 19(Suppl 1), pp. i13–i87.

Calderón, C. and Servén, L. (2010b) Infrastructure in Latin America, in: Ocampo, J. A. and Ros, J. (Eds), *Handbook of Latin American Economies*, Oxford: Oxford University Press.

Calderón, C. and Serven, L. (2010c) *Infrastructure in Latin America*, Washington DC: World Bank.

Calderón, C., Servén, L. and Bank, W. (2008) *Infrastructure and Economic Development in Sub-Saharan Africa*, Washington DC: World Bank.

Canning D. and Pedroni P. (1999) *Infrastructure and Long Run Economic Growth, Consulting Assistance on Economic Reform II Discussion Paper No. 57*, Harvard Institute for International Development.

Estache, A. (2004) *A Selected Survey of Recent Economic Literature on Emerging Infrastructure Policy Issues in Developing Countries*, Washington DC: World Bank.

Estache, A. and Fay, M. (2007) *Current Debates on Infrastructure Policy* [online]. Available at: http://elibrary.worldbank.org/content/workingpaper/10.1596/1813-9450-4410 [Accessed 19 December 2014], pp. 1–43.

Estache, A. and Vagliasindi, M. (2007) *Infrastructure for Accelerated Growth in Ghana: Needs and Challenges* [online]. Available at: http://siteresources.worldbank.org/INTGHANA/Resources/CEM_infrastructure_presentation.pdf [Accessed 19 December 2014].

Eyiah, A. K. and Cook, P. (2003) Financing Small and Medium-Scale Contractors in Developing Countries: A Ghana case study, *Construction Management and Economics*, 21(4), pp. 357–367.

Foster, V. and Briceño-Garmendia, C. (2010) *Africa's Infrastructure: A Time for Transformation*, Washington, DC: The World Bank/Agence Française de Développement.

Foster, V. and Pushak, N. (2011) *Ghana's Infrastructure: A Continental Perspective*, Washington, DC: World Bank.

Foster, V., Butterfield, W., Chen, C. and Pushak, N. (2009) *Building Bridges: China's Growing Role as Infrastructure Financier for Sub-Saharan Africa*, Washington DC: The World Bank/PPIAF, p. xi.

Franks, J. (1998) *Building Procurement Systems: A Client's Guide* (3rd ed.), Ascot: Chartered Institute of Building/Englemere Limited.

Giang, D. T. H. and Sui Pheng, L. (2011) Role of Construction in Economic Development: Review of Key Concepts in the Past 40 years, *Habitat International*, 35(1), pp. 118–125.

Ghana Water Company Limited (2011) Projects: Tamale Water Supply Expansion [online]. Available at: http://www.gwcl.com.gh/projects.php [Accessed 22 November 2011].

Harris, C. (2003) *Private Participation in Infrastructure in Developing Countries: Trends, Impacts, and Policy Lessons*, Washington DC: World Bank.

Hensengerth, O. (2011) *Interaction of Chinese Institutions with Host Governments in Dam Construction–The Bui Dam in Ghana*, Bonn: Deutsches Institut für Entwicklungspolitik.

Izaguirre, A. K. (2010a) *PPI Data Update Note 38*, Washington DC: The World Bank/ PPIAF.

Izaguirre, A. K. (2010b) *PPI Data Update Note 42*, Washington DC: The World Bank/PPIAF.

Izaguirre, A. K. and Perard, E. (2010) *Private Participation in Infrastructure in Sub-Saharan Africa in the Last Decade (Data Update Notes)*, Washington DC: The World Bank/PPIAF, p. 1.

Jett, A., Nicholas (2010) *Private Participation in Infrastructure in Latin America and the Caribbean in the Last Decade (Data Update Notes)*, Washington DC: The World Bank/ PPIAF.

Kessides, C. (1993) *The Contributions of Infrastructure to Economic Development: A Review of Experience and Policy Implications*, Washington DC: World Bank.

Kessides, I. N. (2004) *Reforming Infrastructure: Privatization, Regulation, and Competition*, Washington DC: World Bank.

Kheni, N. A. (2008) *Impact of Health and Safety Management on Safety Performance of Small and Medium-Sized Construction Businesses in Ghana* [online], PhD thesis, University of Loughborough, Loughborough, UK, Available at: https://dspace.lboro. ac.uk/ [Accessed 24 October, 2011].

Kirkpatrick, C., Parker, D.S. and Zhang, Y.-F. (2006) Foreign Direct Investment in Infrastructure in Developing Countries: Does Regulation Make a Difference? *Transnational Corporation*, 15(1), pp. 143–171.

Latham, M. (1994) *Constructing the Team – Joint Review of Procurement and Contractual Arrangements in the United Kingdom Construction Industry: Final Report*, London: Department of the Environment.

Love, P. E. D., Skitmore, M. and Earl, G. (1998) Selecting a Suitable Procurement Method for a Building Project, *Construction Management & Economics*, 16(2), pp. 221–233.

Mante, J. (2014) *Resolution of Construction Disputes Arising from Major Infrastructure Projects in Developing Countries – Case Study of Ghana*, PhD thesis, University of Wolverhampton, UK [online]. Available at: http://hdl.handle.net/2436/333130 [Accessed 19 December, 2014].

McDermott, P. (1999) Strategic and emergent issues in construction procurement, in: Rowlinson, S. and McDermott, P. (Eds), *Procurement Systems: A Guide to Best Practice in Construction*, London: E. & F. N. Spon.

Morledge, R., Smith, A. and Kashiwagi, D. T. (2006) *Building Procurement*, Oxford and London: Blackwell Science.

National Audit Office (2001) *Modernizing Construction (HC87 Session 2000–2001)*. London: National Audit Office.

Ncube, M. (2010) Financing and Managing Infrastructure in Africa, *Journal of African Economies*, 19 (Suppl 1), pp. i114–i164.

Odams, A. M. and Higgins, J. (Eds) (1996) *Commercial Dispute Resolution*, London: Construction Law Press.

Osei-Tutu, E., Sarfo, M. and Collins, A. (2011) The Level of Compliance with the Public Procurement Act (Act 663) in Ghana, in: *Proceedings of the International Conference on Management and Innovation for a Sustainable Built Environment*, Delft University of Technology, 20–23 June, Amsterdam, The Netherlands.

Osotimehin, K. O., Akinkoye, E. Y. and Olasanmi, O. O. (2010) The Effects of Investment in Telecommunication Infrastructure on Economic Growth in Nigeria (1992–2007), paper presented at the Oxford Business and Economics Conference, St Hugh's College, Oxford University 28–29 June, Oxford, UK.

Park, J. (2010) *Private Participation in Infrastructure in East Asia and Pacific in the Last Decade (Data Update Notes)*, Washington DC: The World Bank/PPIAF.

PEI (2011) *Infrastructure Investor Africa: An Intelligence Report*, London: PEI Media.

Prud'Homme, R. (2004) *Infrastructure and Development*, Washington DC: World Bank [online]. Available at: http://www.rprudhomme.com/resources/Prud$27homme+2005a. pdf [Accessed 20 December 2014].

Sanchez-Robles, B. (1998) Infrastructure Investment and Growth: Some Empirical Evidence, *Contemporary Economic Policy*, 16(1), pp. 98–108.

UNCTAD (2008) *World Investment Report*, New York and Geneva: United Nations [online]. Available at: http://unctad.org/en/Docs/wir2008_en.pdf [Accessed 20 December 2014].

Westring, G (1997) *Ghana Public Procurement Reform*. An Audit Report prepared for the World Bank, Stockholm: Advokatfirman Cederquist KB.

World Bank (1994) *World Development Report 1994: Infrastructure for Development*, New York: World Bank/Oxford University Press.

World Bank (1997) *Public Expenditure Review: The Delivery of Economic Infrastructure*, Washington, DC: World Bank. Available at: http://www-wds.worldbank.org/external/default/WDSContentServer/WDSP/IB/2011/02/10/000356161_20110210234807/Rendered/PDF/595640ESW0P0431enditure0review01996.pdf [Accessed 20 December 2014].

World Bank (2003a) *Country Procurement Assessment Report: Ghana 1*, Washington DC: Ghana Country Department, World Bank.

World Bank (2003b) *Country Procurement Assessment Report: Ghana 2*, Washington DC: Ghana Country Department, World Bank.

World Bank (2008a) *Main Report: Ghana: 2007 External Review of Public Financial Management, Public Expenditure Review (PER) 1* [online]. Available at: http://documents.worldbank.org/curated/en/2008/06/9727338/ghana-2007-external-review-public-financial-management-vol-1-2-main-report [Accessed 19 December 2014], Washington, DC: World Bank.

World Bank (2008b) *Public Procurement Assessment Report: Ghana: 2007 External Review of Public Financial Management, Public Expenditure Review (PER) 2* [online].

Available at: http://documents.worldbank.org/curated/en/2008/06/9727354/ghana-2007-external-review-public-financial-management-vol-2-2-public-procurement-assessment-report [Accessed 19 December 2014], Washington, DC: World Bank.

World Bank (2010) *Ghana: Assessment of Stage 1 – World Bank – Use of Country Systems in Bank Supported Operations*, World Bank Report, Washington, DC: World Bank.

World Bank/PPIAF (2010a) *Investment Summary – East Asia and the Pacific (1984–1989)* [online]. Available at: http://ppi.worldbank.org/explore/Report.aspx?mode=1 [Acessed 18 June 2011].

World Bank/PPIAF (2010b) *Featured Rankings, 1990–2009* [online]. Available at: http://ppi.worldbank.org/explore/ppi_exploreRankings.aspx [Accessed 9 June 2011].

World Bank (2011) *Guidelines: Procurement of Goods, Works, and Non-Consulting Services under IBRD Loans and IDA Credits and Grants by World Bank Borrowers*, Washington, DC: International Bank for Reconstruction and Development/World Bank, p. 21.

World Bank (2014) *Assessment of Country's Public Procurement System* [online]. Available at: http://web.worldbank.org/WBSITE/EXTERNAL/PROJECTS/PROCUR EMENT/0,,contentMDK:20108359~menuPK:84285~pagePK:84269~piPK:60001558 ~theSitePK:84266,00.html [Accessed 13 June 2014].

Legislation

Architectural Engineering Services Corporation Act 1973 (NRCD 193).
Civil Service Law 1993 (PNDCL 327).
Constitution of Republic of Ghana 1992.
District Tender Board Regulations 1995 (LI 1606).
Ghana Highway Authority Act 1997 (Act 540).
Public Procurement Act 2003 (Act 663).
Statutory Corporations (Conversion to Companies) Act 1993 (Act 461).

Cases

A-G v Faroe Atlantic Company Limited (the *Faroe Atlantic* case) [2005–2006] SCGLR 271.
A-G v Balkan Energy (Ghana) Limited and Ors (the *Balkan Energy* case) [2012] 2 SCGLR 998.
Martin Amidu v A-G and 2 Ors. (The *Waterville* case) suit number J1/15/2012.
Amidu v Attorney-General and 2 Others (The Isofoton Case), 21st June 2013, Supreme Court (Unreported).
Klomega v Attorney-General and 3 Others, 19th July 2013, Supreme Court (Unreported).
A-G v Faroe Atlantic Company Limited (The *Faroe Atlantic* Case) [2005–2006] SCGLR 271.

Part III

Real estate and economic development

Part III explores how topics in real estate can contribute to wealth creation, poverty reduction and economic development.

10 Management of real estate market information and economic development in developing countries

Raymond T. Abdulai and Anthony Owusu-Ansah

1. Introduction

Real Estate (RE)[1] plays a very critical role in the economies of nations. Economic historians like Rosenberg and Birdzell (1986), Torstensson (1994) and Goldsmith (1995) have documented how RE has immensely contributed to the economic development of the advanced world. In many developing countries land, for example, accounts for 50% to 75% of national wealth (Bell, 2006). Indeed, land is the most basic and vital aspect of subsistence for many people around the world and thus a strategic socio-economic asset, especially in poor societies where wealth and survival are measured by control of and access to land (Deininger, 2003; USAID, 2005; Lund, 2008). Thus, RE is far too important a subject to be left out of consideration in any serious macroeconomic deliberation and in the collective quest for poverty alleviation and economic development.

Due to the important role that RE plays, RE ownership insecurity is a major concern to both developed and developing countries as well as the international donor community. The concern is based on the negative ramifications of RE ownership insecurity. Firstly, ownership insecurity in the form of land disputes negatively affects infrastructure and RE development projects as well as other economic activities like agriculture. For example, when a dispute arises over a plot of land where a RE development project is to be carried out, the development cannot proceed until the land dispute is effectively settled. This constitutes a source of major risks to investors. In terms of agricultural activities, a study in Uganda, for instance, established that land disputes reduced the output on a plot of land by at least 30% (Deininger and Castagnini, 2002). The negative impact is even more pronounced where there are delays in settling the land dispute in the State-sponsored courts. Such protracted litigation often stifles RE-based economic activities as court injunctions are issued against any use of the RE until the cases are decided by the courts.

Secondly, in a climate of insecure RE ownership, entrepreneurs are often compelled to spend valuable resources defending their RE ownership, thereby diverting effort and those resources meant for other productive purposes (Deininger, 2003). Thirdly, there is a direct nexus between land disputes and civil strife. In explaining the causes of civil strife, traditional models do not consider the role of land disputes and rather focus on rent-seeking arguments (Collier and Hoeffler, 1998).

However, although not all wars are caused by land disputes, since the beginning of recorded history, disputes over land have often led to some of the most damaging forms of civil strife, including large-scale wars. Indeed, population growth, globalisation and environmental stresses have tended to exacerbate many people's perception of land as an essential but dwindling resource thereby strengthening and tightening the connection between land disputes and violent conflict (USAID, 2005).

Civil strife arising from land disputes with devastating human and economic consequences has been reported in many developing countries, such as Uganda, Angola, Tajikistan, Kyrgyzstan, Uzbekistan, Kazakhstan, Namibia, Papua New Guinea, Peru, Brazil, East Timor, Kosovo, Mozambique, Mexico, Iraq, Nigeria, Ethiopia, Nepal and Venezuela (USAID, 2005); Zimbabwe, Guatemala, Columbia and El Salvador (Deininger, 2003); Democratic Republic of Congo (Huggins et al., 2005); India (Conroy et al., 1998); Nicaragua (Powelson and Stock, 1990); Kenya (Lumumba, 2004); Somalia (Farah, 2004); Sierra Leone (Richard, 2003); South Africa (Bullard and Waters, 1996); and Ghana (Abdulai and Ndekugri, 2007; Abdulai, 2010).

Land disputes also provide fodder for conflict entrepreneurs who use them to manipulate the emotional, cultural and symbolic dimensions of land for personal, political or material gain thereby fomenting civil strife as in the cases of Rwanda and Burundi (Andre and Platteau, 1998; USAID, 2005). It is self-evident in war-torn countries in the developing world that civil strife normally reverses the clock of progress or economic development as many people are displaced and impoverished, human resources are lost via deaths, children are orphaned, a country's infrastructural base is destroyed and assets worth billions of US$ destroyed.

Thus, RE ownership security is considered to be critical in establishing a structure of economic incentives for investing in RE-related activities (Deininger, 2003). Potential investors are unlikely to invest in RE-related activities unless they are certain that their ownership will be protected. It is therefore not surprising that the World Bank has identified RE ownership insecurity together with poor governance as a major factor inhibiting economic development in the developing world (World Bank, 2007) and the Millennium Development Goals give prominence to the role of secure RE ownership in helping to reduce poverty and to achieve economic development (Payne et al., 2007).

Admittedly, there is equally an argument that RE ownership insecurity provides an incentive for investment in RE-related activities. Sjaastad and Bromley (1997) assert that investments in trees, irrigation furrows, buildings or other fixed structures may provide a litigant in a land dispute with an unassailable case and that even though insecurity is a disincentive to invest, it is paradoxically often also an incentive to invest for security. A critical examination of the basis of this assertion, however, shows that it cannot be sustained. Obviously, such an argument implicitly assumes that the legal framework guarantees the protection of investors in RE-based activities whether or not they truly own the RE.

As Abdulai and Domeher (2012) and Abdulai and Owusu-Ansah (2014) aptly note, undertaking visible investment on land may demonstrate an individual's presence or occupation of the land that associates the investor with the land. However, this association with the land cannot be equated to recognition as the rightful owner by members of the community and by the legal system, which is what RE ownership security is about, as will be seen later. According to these authors, investing in a plot of land that one does not rightfully own will in itself trigger disputes or insecurity rather than reduce or eliminate the disputes. They therefore conclude that investing in land-based activities per se cannot provide the investor with indefeasible/guaranteed land rights; the mere fact that a disputant has constructed buildings or any other permanent structures on a disputed plot of land cannot in any way provide him with an unassailable case.

The importance of RE ownership security and the negative impacts of ownership insecurity have precipitated a search for systems that would establish security in the developing world. Management of RE market information via land registration (LR) has been embraced as the panacea to the problem of RE ownership insecurity. The argument is that LR guarantees RE ownership security (Bloch, 2003; MacGee, 2006; Wannasai and Shrestha, 2007) and based on this argument, it is further asserted that LR provides a secure form of collateral for mortgage purposes and therefore guarantees access to formal capital for investment, wealth creation and economic development (Derban et al., 2002; de Soto, 2000). Thus, efforts at securing RE ownership security and supposedly guaranteeing access to formal capital for investment, wealth creation and economic development have often concentrated on the formulation and implementation of LR policies and programmes respectively.

Although LR has been equated to ownership security, empirical evidence from an overwhelming number of studies conducted in various countries such as Afghanistan (World Bank, 2006), Cambodia and Rwanda (Durand-Lasserve and Payne, 2006), Philippines and Honduras (World Bank, 2005), Egypt (Sims, 2002), Ghana (Abdulai and Ndekugri, 2007; Abdulai, 2010; Abdulai and Owusu-Ansah, 2014), Kenya (Migot-Adholla et al., 1994; McAuslan, 2000), Uganda (McAuslan, 2000; West, 2000), Ivory Coast (Stamm, 2000), and India (Banerjee, 2002) has shown that RE ownership security cannot be assured via LR. The same evidence has been established by the studies of Barrows and Roth (1990), de Janvry et al. (2001), Fitzpatrick (2005), Abdulai (2006), Toulmin (2008), Bromley (2008) and Payne et al. (2009).

It is therefore not surprising that Payne et al. (2007) have done a critical analysis of various studies that have supposedly linked LR to ownership security in the developing world and conclude that the residents in those studies already enjoyed ownership security before the introduction of LR programmes. In unauthorised settlements, empirical evidence from studies conducted in Mexico (Angel et al., 2006), Tanzania (Kironde, 2006), South Africa (Allanic, 2003) and Peru (Kagawa and Turkstra, 2002; Ramirez et al., 2005) has shown that residents in such settlements already enjoyed de facto ownership security before the introduction of LR programmes and therefore ownership security did not emanate from LR. Abdulai (2010) concludes that studies that have supposedly linked LR to ownership security either have research methodological problems or what constitutes ownership security appears to have been misinterpreted.

Indeed, some studies have established that LR can be a source of insecurity in some circumstances. According to Durand-Lasserve and Payne (2006), LR can disadvantage poor people who lose the security provided by the traditional systems of RE ownership whilst being unable to complete the bureaucratic process of LR; in worst cases, it has created opportunities for the powerful in society to override traditional systems of ownership, thereby displacing vulnerable owners. Echoing this point, Deininger and Feder (2001), de Janvry et al. (2001) and Deininger (2003) note that LR is not always necessary or sufficient for a high level of ownership security and that in most cases it creates new sources of conflicts if formal landed property rights are assigned without due recognition of traditional arrangements. In Rwanda, Zimbabwe and Kenya, LR created more land disputes and uncertainty as well as denied land access to the poor and other marginalised groups like women whose rights were well protected under the traditional systems of land ownership (Migot-Adholla et al., 1994; Plateau, 1996).

The preceding discourse sets an appropriate academic context for this chapter. It seeks to: (a) critically examine the underpinning theoretical principles of LR systems in order to explain why LR per se cannot guarantee RE ownership security and consequently cannot "unlock" formal capital for investment purposes; and (b) define the actual role of LR and how such a role can contribute to economic development. The next section explains RE ownership security followed by a section that critically examines the bases of equating LR to ownership security. The penultimate section concentrates on defining the actual role of LR and its likely links to economic development, whilst the last section concludes the chapter.

2. Meaning of RE ownership security

According to Barrows and Roth (1990) ownership security is the perception of the likelihood of losing a specific right to cultivate, graze, fallow, transfer or mortgage. As Sjaastad and Bromley (1997) rightly observe, this is actually a definition of insecurity rather than security. It is also a limited definition of insecurity since insecurity can be associated with any of the 12 rights of ownership identified by Honoré (1961) as: (1) right to possess (exclusive control); (2) right to use (usufruct); (3) right to manage; (4) right to income; (5) right to capital; (6) right to security; (7) right to transmissibility; (8) right to divisibility; (9) prohibition of harmful use; (10) absence of term (duration); (11) liability to execution; and (12) residual character (reversionary right).

Ownership security is defined by authors such as Atwood (1990), Place et al. (1994), Bruce and Migot-Adholla (1994), Roth and Haase (1998), Bruce (1998), Brasselle et al. (2001), Kvitashvili (2004) and Place (2009) in terms of breadth, duration and assurance of rights. Regarding duration, Bruce (1998), for instance, explains security as the degree of confidence with which landholders expect to reap the fruits of their investments in land – it is thus about how long one can exercise one's rights on land. Due to the fact that it may take some time before investments begin to generate enough returns, it is argued that land rights are only secure when landholders can exercise their rights for a period of time sufficient to allow them to enjoy the fruits of their labour. Any bundle of rights that can only be exercised over a short duration is therefore regarded as insecure.

Breadth (scope or size) of rights refers to the bundle of rights that one possesses and it is asserted that if a person does not possess certain landed property rights which are considered key, ownership is said to be insecure because that person can only use the RE in a limited way. Thus Place (2009) and Kvitashvili (2004), for example, are of the view that ownership becomes insecure whenever any of the following exists: (a) lack or perceived lack of some key landed property rights; (b) lack of the right duration; and (c) lack of certainty in continuous exercise of one's landed property rights. Also, Place et al. (1994) have observed that maximum security is achieved when an individual has rights to land on a continuous basis, free from imposition of interference from outside sources and the ability to reap the benefits of labour and capital invested in the land whether in use or on transfer to another person. Sjaastad and Bromley (2000) have explained assurance of landed property rights as the risk of losing such rights but that is rather a definition of insecurity. Assurance of landed property rights should be appropriately defined as the degree of certainty that an owner will not lose his landed property rights.

As aptly noted by Sjaastad and Bromley (2000, p. 370), defining ownership security in terms of size, duration and assurance of landed property rights makes the concept of ownership security intractable, and thus for security to survive as a coherent concept, "breadth of rights and duration of rights must be jettisoned, leaving only the idea of security-assurance to do the necessary work". This is because duration cannot be a measure of ownership security. Where a person is entitled to the exercise of, for example, rights over land for any given period of time, security is supposed to be measured in terms of that particular duration. Thus, if James is entitled to exercise any set of rights over, for example, a parcel of land in just one day, security would have to be measured in terms of that one day and not beyond that. Duration may only determine the type of investment that can be undertaken on the land. Potential investors will acquire land for the duration that will serve the purpose for which the land is acquired. An investor who acquires land for a short duration is therefore likely to go for an investment venture that would mature within that short period and security should be measured in terms of that short period.

In terms of size, although the possession of a wide range of landed property rights may be described as desirable, the possession of a single right cannot amount to insecurity. Security in this instance has to be measured in terms of the right that an individual is entitled to exercise over the landed property; it cannot be measured in relation to what the person is not entitled to do. Consequently, the fact that one is entitled to the exercise of a single landed property right cannot amount to insecurity.

Thus it is only assurance of landed property rights (the degree of certainty that an owner will not lose his landed property rights), which adequately and aptly describes ownership security. It is premised on this assurance concept of ownership security that the Food and Agriculture Organisation (2005), Abdulai and Ndekugri. (2007) and Abdulai (2010) have defined ownership security as the degree of certainty that a person's landed property rights will be recognized by law and especially by members of the society and protected when there are challenges to such rights. Ownership security thus involves two forms of validation, which are State validation by legal recognition and validation at the local level through recognition of one's landed property rights by one's neighbours and other persons (Toulmin and Longbottom, 2001). According to Abdulai (2010), if a person's ownership is secure, that person should be able to exercise his landed property rights peacefully or devoid of contestation, and where disputes do occur, the ownership should be protected. Based on this definition, social and legal recognition of landed property rights, the absence of disputes over landed property as well as enforceability and clarity of landed property rights are all parameters of ownership security. Premised on the same conception of security, Honoré (1961), Schlager and Ostrom (1992), Besley (1995), Li et al. (1998) and Antwi (2000) have explained ownership security to be the enforceability of landed property rights against whom they are supposed to be enforced and immunity from expropriation.

Roth and Haase (1998) therefore appear to be right by describing ownership security as a kind of perception held by individuals regarding their ability to exercise landed property rights, both now and in the future, in a manner that is devoid of interferences from others and at the same time that allows them to benefit from any investment made in the landed property. De Souza (2001) describes it as the perception of the probability of eviction. The only way an individual can exercise his landed property rights without interference is when his entitlement to these rights are recognised by law and especially by members of society. Perception is a qualitative or subjective variable that may be difficult to measure.

Finally, there is another legal angle of ownership security which is explained by authors such as Bruce (1998). The assertion is that lawyers are more interested in how to argue cases in the law courts; legal arguments are based on facts and the ability to provide hard evidence to prove one's case on the balance of probabilities and in the case of RE ownership, such evidence is achieved via LR. Thus, according to this school of thought, the attainment of security is synonymous with the ability of RE owners to provide documentary evidence of their ownership. Premised on this, Bloch (2003) argues that security of ownership is an objective measurable variable in that it can be quantified and proven via documentary evidence. Wannasai and Shrestha (2007), in their classification of ownership security, also note that ownership is secure if RE holders possess registered titles or land certificates whilst the ownership is insecure where there is no documentary evidence of ownership. Furthermore, according to Brasselle et al. (2001) it is only when land is registered and protected by legal title that maximum ownership security is afforded. It is based on this concept of security that in the developing world the traditional landownership systems, which are not based on any form of documentation, are often described as pure insecurity. Legality as noted above is a parameter of ownership security but whether it emanates from only documentation or LR will be subjected to analysis later.

This chapter has adopted the concept of ownership security as explained by Honoré (1961), Schlager and Ostrom (1992), Besley (1995), Li et al. (1998), Sjaastad and Bromley (2000), Antwi (2000), Food and Agriculture Organisation (2005), Abdulai et al. (2007) and Abdulai (2010), since such an explanation adequately captures the right parameters on which RE ownership security should be based. Thus, ownership security is simply the degree of certainty that a person's ownership will be recognized by law and by members of the community, especially adjoining owners, and will be protected when there are challenges to it; in other words, it is about legal and societal recognition and enforceability of landed property rights.

3. Reasons for equating LR to RE ownership security

Before considering the reasons for equating LR to security, it is expedient to first of all explain the two types of LR systems that are used globally, which are deed and title registration systems. In deed registration (DR), legally recognized and protected landed property rights arise upon conclusion of an agreement or contract between the RE grantor and grantee; the entry of the agreement or contract and its key contents into the public registry is to provide public notice of the existence of the landed property rights, and challenges to such rights will be handled through civil litigation (Deininger, 2003). It is thus a record of RE ownership instruments or transactions and does not provide any guarantees regarding the actual legal ownership status of the RE (Abdulai, 2010; Abdulai and Domeher, 2012; Awuah and Hammond, 2013). The history of DR is often traced to the Romans who introduced it in, for example, England and Wales in 397 AD when Britain became part of the Roman Empire where ownership and productivity of land was recorded – it formed the basis of a land tax called "tributum soli" in England and Wales (Pemberton, 1992; Dark, 2000). It is commonly used in South America, parts of Asia and Africa, most parts of the United States of America and in Latin cultures in Europe, for example, France, Spain and Italy (Enemark, 2005).

However, in title registration (TR), it is the entry into the registry that gives RE ownership legal validity, guaranteed by the State – all entries in the register are prima facie evidence of the actual legal status of the RE (Deininger, 2003). The State guarantees the accuracy of the data entered in the title register and in some jurisdictions the State indemnifies or pays compensation (from an indemnity fund set up) to owners who suffer any loss due to negligence, mistakes, errors and omissions from TR as well as fraud, unless the owners contributed substantially to the occurrence of these events. The TR system (also called Torrens system) is based on the work of Sir Robert Richard Torrens (1814–1884), which has been described in Abbott (2005) and by Clay County (2014). Sir Robert Richard Torrens was an Irish emigrant to Australia and a land law reformer who devised a system of LR for Australia in 1858 based on the British ship registration system. This explains why TR is often referred to as the Torrens system. The purpose of the system, as Sir Robert himself aptly put it, was to simplify RE transfer. He was looking for five qualities when he introduced TR: reliability, simplicity, low cost, speed and suitability. TR is commonly used in Central European countries such as Germany, Austria and Switzerland (Enemark, 2005).

Hogg (1920) – in an attempt to differentiate between the DR and TR systems – notes their close similarity and the difficulty in distinguishing them but argues that the presence of a statutory provision of title warranty in the title registration system is the single most important distinguishing feature between the two systems. It is important to note that Hogg has completely downplayed one other equally critical distinguishing characteristic between the two systems, which is source of legal validity: whilst TR is a source of legal validity of ownership, DR is not. There are therefore two important factors that differentiate one registration system from the other and that is even from a theoretical perspective. According to Zevenbergen (2002), under TR in countries like Germany, Sweden and Denmark, there are no State guarantees as protection for registrants is only derived from "public faith". Consequently, from a practical perspective, the main difference between the two registration systems is the source of ownership legality. Thus, in practical terms, DR can be appropriately described as the recording of mere transactions that have taken place between RE grantors and grantees whereas TR is about the recording of legal consequences of transactions.

Having explicated the two LR systems, the three main dimensions to the argument that LR is the answer to insecurity of landed property rights are subjected to critical analyses in the following sections.

3.1 Title warranty provided by the State

One basis of equating LR to ownership security relates to the title warranty provided by the State under the Torrens system. Commentators such as Simpson (1984), Jacoby and Minten (2005) and MacGee (2006) equate registration to security based on such warranty. Simpson, for example, argues that because of such guarantee, TR makes RE ownership indefeasible or unimpeachable. Indeed, TR laws of some countries specifically state that TR is conclusive evidence of ownership and that it makes title indefeasible. In Ghana, for instance, section 18 of the Land Title Registration Law 1986 (PNDCL 152) provides that the title register is conclusive evidence of title to any RE and interest in it and therefore indefeasible, which is reinforced in section 43 of the same statute.

Although it is argued that the Torrens system guarantees ownership security because of the title warranty provided by the State, it is obvious that the guarantee is provided by the State and not the registration system per se. Thus, the State could decide not to provide that guarantee, as in the case of the DR system. Indeed, it is possible for the State to guarantee title without any form of registration. Therefore, granted that the State warranty can potently protect ownership, it is clear that LR by itself cannot provide that protection since it actually emanates from the State; if the registration system could provide that warranty there will be no need for the State to provide it. It is also significant to note that in some jurisdictions like Germany, Sweden and Denmark, as earlier noted, no State warranties are provided under the Torrens system and so protection for registrants is only derived from "public faith", just like the DR system.

3.2 Publicity function of LR

The argument of some commentators that LR guarantees security is premised on the publicity function of LR described above. For instance, according to Scott (1981), during the earliest days of colonization there was a clear understanding of the essential role of "public record" as a means of establishing ownership security of private interests in landholdings. Larsson (1991) on his part argues that the publicity function of LR prevents the occurrence of land disputes, which significantly reduces the work of State-sponsored courts.

However, the above argument is unconvincing. This is because publicity of land ownership and transactions is about making people aware of somebody's ownership or RE transactions – but that is completely different from society recognizing a person's ownership, allowing the owner to exercise his RE rights peacefully, and there being legal recognition of such rights, which is what ownership security is all about.

Even though, premised on this publication function of LR, Larsson argues that LR prevents the occurrence of disputes on registered land, the evidence adduced in the various studies cited earlier sharply debunks such an argument. Indeed, in a human society land disputes are bound to occur and so it is unimaginable for anybody to argue that LR prevents the occurrence of land disputes.

3.3 Legality of ownership

The third argument that equates LR to security is based on the legal conception of ownership security. However, under the fundamental principles of the DR system, societal and legal recognition of ownership and its protection arise upon conclusion of the agreement between the RE grantor and grantee. Ownership security does not therefore emanate from the fact of registration. Entry of the key contents of the agreement into the public registry is only to provide public notice of the existence of the landed property rights, and challenges to such rights will be handled through civil litigation. This clearly shows that the DR system does not play a role in the resolution of ownership disputes in the law courts, let alone guarantee security.

Obviously, in resolving RE ownership disputes the courts would of necessity rely on the contract or agreement that is entered into between the RE grantor and grantee, and not the fact of registration.

Admittedly, premised on the theoretical principles that underpin the Torrens system, it constitutes evidence of legal ownership. It can therefore be argued that it plays a role in the resolution of RE ownership disputes in State-sponsored courts as registered land title can be tended as evidence of legal ownership. Consequently, the system may be described as a determinant of security. Notwithstanding this, it has to be noted that the Torrens system "being a determinant of ownership security" cannot in any way be equated to "guaranteeing security" – these are completely two different things. If, for example, "X contributes to the occurrence of event Y or X is one of the determinants of event Y" it is not the same as saying that "X guarantees the occurrence of event Y". The former means that X is one of the factors that would make event Y occur – in effect, there are other factors; whilst the latter means that the occurrence of event Y is dependent on only X. The thrust of this chapter is, however, about the latter and not the former.

Also, TR is not the only source of legal ownership. Although proof of traditional landownership is not based on registration or any form of documentation, such ownership is recognised by most jurisdictions in Africa. In Ghana, for instance, the traditional landownership systems are recognised by the 1992 Constitution, the supreme law of the country, and by the Conveyancing Decree of 1973 (NRCD 175), and are therefore admissible in State-sponsored courts as evidence of legal ownership in times of challenges (Abdulai and Ndekugri, 2007). This recognition in Ghana dates back to the colonial era when Native Courts were established in 1925. Similarly, in Nigeria, despite the nationalisation of land under the Land Use Decree of 1978, traditional landownership is recognised by the State-sponsored courts (Ikejiofor et al., 2004), whilst in Mozambique, customary land rights are recognised and protected regardless of whether they have been registered or not (Cotula et al., 2006).

Consequently, if the Torrens system is a determinant of security and it is equated to guaranteeing security, then the customary landownership systems equally guarantee security as they are socially and legally recognised. Thus, from the legal concept of security, LR is not needed in the customary land ownership sector for security to be established. It is therefore astounding that the customary land ownership systems are considered to be purely insecure albeit they are legally recognised; however, regarding the Torrens system the argument is that it guarantees security because it is a source of legal ownership.

Allied to the legal conception of security is the issue of documentary evidence of ownership. Customary landed property rights are described as purely insecure because ownership is not documented. Admittedly, when ownership is documented, it is normally easy to prove in the event of dispute. However, this does not in any way imply that ownership security is guaranteed. In resolving disputes in State-sponsored courts, what has to be established to the requisite standard is the truth about who actually owns the RE, to ensure that there is no miscarriage of justice – delivery of judgments is not merely contingent upon proof of documentary evidence of ownership or registered ownership. Indeed, documents are not sacrosanct as they can be phony documents and so the mere fact that ownership is evidenced by a document does not mean that the evidence is conclusive. It is also argued as if documentation is achieved via only land registration. If, for the sake of argument, it is assumed that documentation guarantees security, there can be documentation without LR and so it is surprising that the focus is on land registration. It is common knowledge that there is documentary evidence of various things but such evidence is not registered or recorded in a central system controlled by the State.

Furthermore, physical possession and occupation as evidence of ownership in the customary RE sector is legally recognised, albeit it is not based on documentation. The potency of this form of evidence is amply demonstrated by the operation of limitation or prescription laws. Under such laws a true owner of RE can be dispossessed of his RE via a reasonable period of occupation by a squatter. In Ghana, under the Limitation Decree of 1972 (NRCD 54), a trespasser dispossesses the true owner of RE if the trespasser occupies the RE for 12 years and within such period the true owner fails to assert his ownership. At the end of the prescription period, the true owner loses his ownership and the right to sue is extinguished. Even where the RE has been registered by the true owner, he still forfeits the RE.

Similarly, under the French Civil Code, the prescription period is 20 years where the true owner is resident outside the territory in which the RE is located; if the true owner lives within the territory where the RE is located, the limitation period is 10 years. In England and Wales the relevant law is the Limitation Act (1980) and the limitation periods are 12 and 10 years for unregistered and registered land respectively. It is not only the customary evidence of RE ownership that is legally recognised even though it is not based on documentation. It is common knowledge that in other cases oral evidence from witnesses is admissible in the State-sponsored courts as sufficient proof of one's case and it is possible for people without documentary proof of evidence to obtain judgment in their favour.

A system that can potently guarantee RE ownership security is ownership or title insurance, which has been extensively treated in the work of Abdulai and Owusu-Ansah (2014) but is beyond the scope of this chapter. Based on the preceding discourse, what then is the right role of LR in the economies of nations? The answer to the question is the preoccupation of the next section.

4. LR and economic development

There is abundant empirical evidence, supported by theoretical arguments and set out in the sections above, that LR per se cannot guarantee RE ownership security. Thus, to argue that LR is the panacea to the problem of poverty and economic underdevelopment, on the basis that it establishes ownership security and consequently guarantees accessibility to investment capital, amounts to a wrong prescription for the wrong malady, with the role of LR being misunderstood. LR is simply a record-keeping system which creates an ownership database that serves various purposes and an economy. RE ownership registration can play a critical role in the economic development of nations.

Notwithstanding that funds, equity, life insurance and other types of investments can be used as additional collateral for investment loans (as well as the income-earning capacity or income level of the investor, which is used to amortize the loan and also serves as a form of security), RE remains the dominant form of collateral and is therefore normally required for mortgage purposes. The importance of mortgaging one's RE to banking institutions cannot be over-emphasised from the perspective of both the mortgagor and the mortgagee. This is because collateral arrangements shift the risks of loan loss from the mortgagee to the mortgagor. Default on the part of the mortgagor will trigger the loss of his mortgaged RE. Thus, for the mortgagee, the collateral provides an additional form of protection or insurance against the loss of the loan, since the law normally makes provision for the mortgage to be foreclosed where the mortgagor defaults. At the same time, the prospect of losing one's RE is an incentive for the mortgagor to be committed regarding the repayment of the loan granted.

The pledging of RE, accompanied by registration of ownership and mortgage transactions, tremendously overcomes the problems of asymmetrical information and moral hazard. When information is recorded in a central system which is accessible to the public, it makes it easy for such data to be accessed for various purposes; thus the importance of any record-keeping system cannot be over-emphasised. Any form of record-keeping system facilitates activities or transactions thereby reducing transaction costs. Thus, LR as a record-keeping system creates a RE ownership database which facilitates RE transactions, thereby reducing transactions costs. The ownership database, for example, facilitates title searches for RE trading and mortgage purposes as ownership details can easily be accessed; this can, in turn, save participants in RE markets considerable search time that can be put to other economic uses.

Larsson (1991) alludes to this role of LR when he explains two basic historical reasons for landownership record-keeping – the need for: (a) the State to know all parcels of land for taxation or other fees; and (b) prospective land purchasers to get publicity for their acquisition of land. It is the same purpose that de Soto (2000) refers to when he emphasises the role of LR in facilitating communication, information sharing, networking and transactions. LR and RE markets are inextricably linked: LR facilitates the RE transfer processes and subsequently ensures the transparency of transactions and further provides records for RE market operations (Farvacque and McAuslan, 1992).

According to Larsson (1991) and Dowson and Sheppard (1956), the earliest evidence of official landed property ownership record-keeping for purposes of taxation and other services dates as far back as 3,000 BC in ancient Egypt, where such records were kept in the royal registry. Thus RE ownership record-keeping, particularly for taxation purposes, has been in existence since society adopted sedimentary agriculture – as evidenced by the Babylonians when they occupied the lands between the Tigris and the Euphrates and the Egyptians cultivated the fertile Nile regions (Dale and McLaughlin, 1988). Similarly, for taxation purpose, LR was introduced in ancient Rome during the third century AD, in China around 700 AD based on crop yields, in South India around 1,000 AD (Larsson, 1991) and in England and Wales in 1086 via the famous Domesday survey ordered by William the Conqueror (Plucknett, 2007). King Gustav of Sweden in 1540 also ordered the registration of all farms for taxation purposes (Larsson, 1991). Toulmin (2008) observes that LR provides governments with information on land ownership and plot sizes as well as the foundation for a RE tax system. Taxation of RE is a major contributor to revenue of governments and is used for various developmental projects and other interventions.

Generally, in the absence of a LR system, legal experts are often commissioned to conduct searches to trace the root of title in RE transactions in order to verify that the RE is not subject to any undisclosed obligations. This is normally a long process, time consuming and expensive. In England and Wales, for example, in those cases where RE is not registered and is to be traded, the law requires that the root of title has to be investigated 15 years back; in the case of France it is 30 years. In the African context, in the absence of LR or a RE ownership database, the only way of verifying the RE ownership information provided by owners is to go around asking members of the communities in which the RE is located; this is obviously a Herculean task, albeit not impossible. Thus any LR system significantly enhances the smooth operation of RE markets as well as reduces transaction costs.

In terms of mortgage registration, the common law position is that if a mortgage transaction is not registered and the mortgagor clandestinely transfers the mortgaged RE via sale to a bona fide purchaser for value who is unaware of the existence of the mortgage transaction, the mortgage as an encumbrance will not be binding on such a purchaser. However, if the mortgage transaction is registered, the purchaser cannot defend himself on the basis that he is a bona fide purchaser for value without notice since the existence of the mortgage will be a public record and, based on the principle of caveat emptor, it will be his responsibility to check the public records. Thus, the registration of mortgage transactions protects lenders from the activities of unscrupulous mortgagors as it makes it impossible for any purchaser of mortgaged RE to plead bona fide purchaser for value without notice. Therefore, apart from the interest charged on loans, registration of mortgages provides an additional incentive for banks to grant loans for investment purposes that can lead to wealth creation and development.

Thus, RE and its ownership registration can play a role in the economic development of nations, albeit that it plays such a role via a completely different route and not through the argument that LR guarantees RE ownership security. Other determinants of economic development include prudent management of a nation's resources and good governance; deficiencies in these have been identified by Abdulai (2011) to be among the major causes of poverty and economic underdevelopment in Africa.

Electoral disputes leading to civil strife are a manifestation of mismanagement of a nation's resources and poor governance, Abdulai (2011) argues, with devastating consequences such as the displacement and impoverishment of many people, loss of human resources via deaths, and destruction of a country's infrastructural base and assets worth billions of US$. These normally reverse the clock of development, as earlier noted. Greed and corruption constitute a second manifestation, it being estimated that 40–60% of national resources are lost through corrupt practices in Africa (Food and Agriculture Organisation, 2005; Abdulai and Ndekugri, 2007; Abdulai, 2010). In support of this observation, Lumumba (2015) attributes Africa's problem to "mis-governance". According to Lumumba, the tragedy of Africa is that, Africa's creed is greed. He concludes that many African countries are kleptocracies; that is, governments where thieves have come together to conspire to rob their countries. In Ghana, for instance, the issue of corruption has been vividly captured by Ofori-Atta (2009), who observes that corruption is inimical to the economic development of any nation and calls on the government to commence the crusade against corruption from its ranks in order to have the moral authority to stamp out the national canker without fear or favour. He also states that when corruption is allowed, either conscientiously or inadvertently, to operate in a free environment at whatever level, it becomes very destructive. Furthermore, he opines that democracy goes with leadership and accountability, adding that good leadership reflects the quality of life of the people and not necessarily policies.

5. Conclusion

There is a longstanding argument that RE market management information via LR guarantees RE ownership security in developing countries. The argument that equates LR to ownership security has been extended by asserting that LR makes RE suitable collateral and thus guarantees access to formal capital for investment, which leads to poverty reduction and economic development. This has triggered the conduct of empirical research that has investigated the extent to which this assertion is true. Such studies have not been able to establish any discernible link between LR and ownership security.

This chapter has critically examined the underpinning theoretical principles of LR systems in order to explain why LR per se cannot guarantee RE ownership security. The chapter establishes that although LR can contribute to economic development, it contributes via a completely different route and not through the commonly-held view about the purpose land LR serves – guaranteeing RE ownership security. LR is a record-keeping system that creates a database of RE owners which can be used for various purposes such as the facilitation of RE taxation and other RE related transactions, thereby reducing transaction costs. The chapter has also defined other determinants of economic development as good governance and prudent management of a nation's resources.

Note

1 RE is an American term but globally used; it refers to land and/or the developments or buildings attached to the land. The other terminologies for RE are "real property" and "landed property".

References

Abbott, D. (2005) *Encyclopaedia of real estate terms*. London: Delta Alpha Publishing.
Abdulai, R. T. (2006) Is land title registration the answer to insecure and uncertain property rights in sub-Saharan Africa? RICS Research Paper Series, 6, 6.
Abdulai, R. T. (2010) *Traditional landholding institutions in sub-Saharan Africa – the operation of traditional landholding institutions in sub-Saharan Africa: A case study of Ghana*. Saarbrücken: Lambert Academic Publishing AG & CO KG.

Abdulai, R. T. (2011) Land registration and poverty reduction in Ghana, in: Home, R. (Ed.) *Local Case Studies in African Land Law*, Pretoria: Pretoria University Law Press, pp. 157–170.

Abdulai, R. T. and Domeher, D. (2012) *Why real estate ownership security cannot be assured via land registration in sub-Saharan Africa*. New York: Nova Science Publishers.

Abdulai, R. T. and Ndekugri, I. E. (2007) Customary landholding institutions and housing development in Ghana: Case studies of Kumasi and Wa. *Habitat International,* 31(2), pp. 257–267.

Abdulai, R. T. and Owusu-Ansah, A. (2014) Land information management and landed property ownership security: Evidence from state-sponsored court system. *Habitat International*, 42, pp. 131–137.

Allanic, B. (2003) *La nouvelle coutume urbaine: Evolution compare des filieres coutumieres de la gestion fonciere urbaine dans les pays d'Afrique subSaharienne, Le cas de Mandela Village*. South Africa: ISTED, DFID.

Andre, C. and Platteau, P. (1998) Land relations under unbearable stress: Rwanda caught in Malthusian trap. *Journal of Economic Behaviour and Organization*, 34, pp. 1–47.

Angel, S., Brown, E., Dimitrova, D., Ehrenberg, D., Heyes, J., Kusek, P., Marchesi, G., Orozco, V., Smith, L. and Ernesto, V. (2006) Secure tenure in Latin America and the Caribbean: Regularization of informal urban settlements in Peru, Mexico and Brazil. Woodrow Wilson School of Public and International Affairs, Princeton University.

Antwi, A. Y. (2000) Urban land markets in sub-Saharan Africa: A quantitative study of Accra, Ghana. Unpublished PhD thesis, Napier University.

Atwood, D. A. (1990) Land registration in Africa: The impact on agricultural production. *World Development*, 18(5), pp. 659–671.

Awuah, K. G. B. and Hammond, F. N. (2013) Prognosis of land title formalization in urban Ghana: The myth and reality of awareness and relevance. *African Studies Quarterly*, 1(2), pp. 57–75.

Banerjee, B. (2002) Background note prepared for the design of Kolkata Urban Services for the Poor Project, DFID Mimeo.

Barrows, R. and Roth, M. (1990) Land tenure and investment in African agriculture: Theory and evidence. *The Journal of Modern African Studies*, 28, pp. 265–297.

Bell, K. C. (2006) World Bank support for land administration and management: Responding to the challenges of the Millennium Development Goals. Paper presented at the 23rd FIG Congress, 8th–13th October, Munich, Germany.

Besley, T. (1995) Property rights and investment incentives: Theory and evidence from Ghana. *Journal of Political Economy*, 103, pp. 903–937.

Bloch, P. (2003) Economic impact of land policy in the English speaking Caribbean, in: Williams, A. (Ed.), *Land in the Caribbean: Issues of policy, administration and management in the English speaking Caribbean*, pp. 13–30. Madison, WI: Land Tenure Centre, University of Wisconsin-Madison.

Brasselle, A., Gaspart, F., Platteau, J. (2001) Land tenure security and investment incentives: Puzzling evidence from Burkina Faso. Centre De Recherche En Economie Du Développement (Cred), Faculty of Economics, University of Namur, Belgium.

Bromley, D. W. (2008) Formalising property relations in the developing world: The wrong prescription for the wrong malady. *Land Use Policy*, 26(1), pp. 20–27.

Bruce, J. W. (1998) Review of tenure terminology. Tenure Brief No. 1. Madison, WI: Land Tenure Centre, University of Wisconsin-Madison.

Bruce, J. W. and Migot Adholla, S. (Eds) (1994) *Searching for land tenure in Africa*. Dubuque: Kendall/Hunt.

Bullard, R. and Waters, H. (1996) Land tenure, the root of land ownership conflicts in Southern Africa, past, present and future, in: *Roots 96: The proceedings of the 1996 rural practice research conference of the Royal Institution of Chartered Surveyors*, London, UK: RICS.

Clay County (2004) Torrens [online]. Available at: http://claycountymn.gov/619/Torrens-Title [Accessed 21 July 2015].

Collier, P. and Hoeffler, A. (1998) On the economic causes of civil war. Oxford Economic Papers, No. 50.

Conroy, C. Rai, A., Singh, N. and Chan, M. (1998) Conflicts affecting participatory forest management: Some experiences from Orissa. Chatham: Natural Resources Institute.

Cotula, L., Toulmin, C. and Quan, J. (2006) Policies and practices for securing and improving access to land. Paper presented at the 18th International Federation of Surveyors (FIG) Congress, 8–13 October, Munich, Germany.

Dale, P. F. and McLaughlin, J. D. (1988) *Land Information Management*, Oxford: Clarendon Press.

Dark, K. (2000) *Britain and the end of the Roman Empire*. London: Tempus.

De Janvry, A., Platteau, J.-P., Gordillo, G. and Sadoulet, E. (2001) Access to land policy reforms, in: de Janvry, A., Gordillo, G., Platteau, J.-P. and Sadoulet, E. (Eds), *Access to land rural poverty and public action*, pp. 1–26. Oxford: Oxford University Press.

De Soto, H. (2000) *The mystery of capital: Why capitalism triumphs in the West and fails everywhere else*. London: Bantam Press.

De Souza, F. A. M. (2001) Perceived security of land tenure in Recife, Brazil. *Habitat International*, 25, pp. 175–190.

Deininger, K. (2003) Land policies for growth and poverty reduction. Policy Research Report. Washington DC: World Bank.

Deininger, K. and Castagnini, R. (2002) Incidents and impact of land conflict in Uganda. Discussion Paper, Washington DC: World Bank.

Deininger, K. and Feder, G. (2001) *Land Policy in Developing Countries*. Development Note 3, Washington DC: World Bank.

Derban, W. K., Derban, D. K., Ibrahim, G. and Rufasha, K. (2002) Microfinance for housing for low/moderate-income households in Ghana. Paper presented at a Conference on Housing and Urban Development for Low Income Groups in sub-Saharan Africa.

Dowson, E. and Sheppard, V. L. O. (1956) *Land Registration*, London: [s.n.].

Durand-Lasserve, A. and Payne, G. (2006) Evaluating the impacts of urban land titling: Results and implications: Preliminary findings. Draft report prepared for Geoffrey Payne and Associates, London.

Enemark, S. (2005) A cadastral tale: Week on geomatics, 8–13th August, Bogota, Colombia.

Farah, I. (2004) Land tenure and conflict in Somalia. *Summary report of the conference on land tenure and conflict in Africa: Prevention, mitigation and reconstruction, 9th–10th December*. Nairobi: African Centre for Technology Studies Press.

Farvacque, C. and McAuslan, P. (1992) Reforming urban land policies and institution in developing countries. Urban Management Program Policy Paper No. 5.

Fitzpatrick, D. (2005) Best practice options for the legal recognition of customary tenure. *Development and Change*, 36(3), pp. 449–475.

Food and Agriculture Organisation (FAO) (2005) Access to rural land and land administration after violent conflicts. FAO Land Tenure Studies, No. 8, Rome.

Goldsmith, A. A. (1995) Democracy, property rights and economic growth. *Journal of Development Studies*, 32(2), pp. 157–174.

Hogg J. E. (1920) *Registration of title to land throughout the empire: A treatise on the Law relating to warranty of title to land by registration and transactions with registered land in Australia, New Zealand, Canada, England, Ireland, West Indies, Malaya etc. A sequel to "The Australian Torrens system"*. Toronto: The Carswell Company Ltd.

Honoré, A. M. (1961) Ownership, in: Guest, A. G. (Ed.), *Oxford essays in jurisprudence first series*, pp. 107–147, Oxford: Oxford Clarendon Press.

Huggins, C., Musahara, H., Mbura, K. P., Oketch, J. S. and Vlassenroot, K. (2005) Conflict in the Great Lakes Region – How is it linked with land registration? Paper No. 96, Natural Resource Perspectives. London: ODI.

Ikejiofor, C. U. Nwogu, K. C. and Nwanunobi, C. O. (2004) Informal land delivery process and access to land for the poor in Enugu, Nigeria. Working Paper No. 2. Birmingham: School of Public Policy, University of Birmingham.

Jacoby, H. and Minten, B. (2005) Is land titling in sub-Saharan Africa cost effective? Evidence from Madagascar. Washington DC: World Bank and Cornell University.

Kagawa, A. and Turkstra, J. (2002) The process of land tenure formalisation in Peru, in: Payne, G. (Ed.), *Improving tenure security for the urban poor*. London: ITDG Publications.

Kironde, L. (2006). The regulatory framework, unplanned development and urban poverty: Findings from Dar es Salaam, Tanzania. *Land Use Policy*, pp. 460–472.

Kvitashvili, E. (2004) Land and conflict: A toolkit for intervention. Washington DC: USAID Office of Conflict Management and Mitigation.

Larsson, G. (1991) *Land registration and cadastral systems: Tools for land information and management*. London: Longman Scientific and Technical.

Li, G., Rozelle, S. and Brandt, L. (1998) Tenure, land rights and farmer investment incentives in China. *Agricultural Economics*, 19, pp. 63–71.

Lumumba, O. (2004) Land related conflicts in Kenya: Policy and legal implications. *Summary report of conference on land tenure and conflict in Africa: Prevention, mitigation and reconstruction, 9th–10th December*. Nairobi: African Centre for Technology Studies Press.

Lumumba, P. L. O. (2015) Good governance: Whither Africa. Lecture delivered at the 2015 PAVA Forum, 28 August, Accra, Ghana.

Lund, C. (2008) *Local politics and dynamics of property in Africa*. Cambridge: Cambridge University Press.

MacGee, J. C. (2006) Land titles, credit markets and wealth distributions. United Nations University/World Institute for Development Economics Research, Research Paper No. 2006/150.

McAuslan, P. (2000) Only the name of the country changes: The diaspora of "European" land law in Commonwealth Africa, in: Toulmin, C. and Quan, J. (Eds), *Evolving land rights, policy and tenure in Africa*. London: DFID/IIED/NRI, pp. 75–95.

Migot-Adholla, S. E., Place, F. and Oluoch-Kosura, W. (1994) Security of tenure and land productivity in Kenya, in: Bruce, J. W. and Migot-Adholla, S. E. (Eds), *Searching for land tenure security in Africa*. Debuque, Iowa: Kendall/Hunt Publishing Company.

Ofori-Atta, G. S. (2009) Corruption is inimical to national development – Archbishop [online]. Available at: www.news.myjoyonline.com/news/200908.asp [Accessed 31 August 2009].

Payne, G. Durand-Lasserve, A. and Rakodi, C. (2007) Urban land titling programmes, in: Brother, M. E. and Solberg, J.-A. (Eds), *Legal empowerment – a way out of poverty*. Norway: Norwegian Ministry of Foreign Affairs.

Payne, G. Durand-Lasserve, A. and Rakodi, C. (2009) The limits of land titling and home ownership. *Environment and Urbanisation*, 21(2), pp. 443–462.

Pemberton, L. (1992) *HM Land Registry: An historical perspective*. London: HM Land Registry.

Place, F. (2009) Land tenure and agricultural productivity in Africa: A comparative analysis of the economics literature and recent policy strategies and reforms. *World Development*, 37(8), pp. 1326–1336.

Place, F., Roth, F., and Hazell, P. (1994) Land tenure security and agricultural performance in Africa: Overview of methodology, in: Bruce, J. and Migot-Adholla, S. E. (Eds), *Searching for land tenure security in Africa*. Dubuque, Iowa: Kendall/Hunt Publishing Company, pp. 15–39.

Platteau, J. P. (1996) Physical infrastructure as a constraint on agricultural growth: The case of sub-Saharan Africa. *Oxford Development Studies*, 24(3), pp. 189–219.

Plucknett, T. F. T. (2007) *A concise history of the common law* (5th ed.). New Jersey: Lawbook Exchange Ltd. Union.

Powelson, J. P., and Stock, R. (1990) *The peasant betrayed*. Washington DC: The CATO Institute.

Ramirez, C. Daniel, G. and Riofrio, G. (2005) Land titling a path to urban inclusion? Policy and practice of the Peruvian model. Paper presented at the N-AERUS Conference, Barcelona, Spain.

Richard, P. (2003) Controversy over recent West African wars: An agrarian question. Unpublished notes for an ISS Seminar, 6th October, ISS, The Hague.

Rosenberg, N., and Birdzell, L. E. (1986) *How best the West grew rich: The economic transformation of the industrial world*. London: I. B. Tauris and Co. Ltd.

Roth, M. and Haase, D. (1998) Land tenure security and agricultural performance in southern Africa. Land Tenure Centre Paper. Madison, WI: University of Wisconsin-Madison.

Schlager, E. and Ostrom, E. (1992) Property-rights regimes and natural resources: A conceptual analysis. *Land Economics*, 68, pp. 249–262.

Scott, R. (1981) The multipurpose land registration system in the public sector community of the North American Continent. Paper presented at the International Federation of Surveyors (FIG) Congress, 8–11 August, Montreux.

Simpson, S. R. (1984) *Land law and registration*. London: Surveyor's Publications.

Sims, D. (2002) What is secure tenure in Egypt? In: Payne, G. (Ed.), *Land rights and innovation: Improving tenure security for the urban poor*. London: ITDG, pp. 79–99.

Sjaastad, E. and Bromley, D. W. (1997) Indigenous land rights in sub-Saharan Africa, appropriation, security and investment dynamics. *World Development*, 25(4), pp. 549–562.

Stamm, V. (2000) Land tenure policy: an innovative approach from Cote d' Ivoire. Drylands Issue Paper 91. London: IIED.

Torstensson, J. (1994) Property rights and economic growth: An empirical study. *Kylos*, 47, pp. 231–247.

Toulmin, C. (2008) Securing land and property rights in sub-Saharan Africa: The role of local institutions. *Land Use Policy*, 26, pp. 10–19.

Toulmin, C. and Longbottom, J. (2001) *West African land: Rights, poverty and growth*. Edinburgh: IIED Drylands Programme.

United States Agency for International Development (USAID) (2005) *Land and conflict: A toolkit for intervention*. Washington DC, USA: USAID.

Wannasai, N. and Shrestha R. P. (2007) Role of land tenure security and farm household characteristics on land use change in the Prasae Watershed, Thailand. *Land Use Policy*, 25, pp. 214–224.

West, H. W. (2000) *On African landholding: A review of tenurial change and land policies in Anglophone Africa*. Lewiston: Edwin Millen.

World Bank (2005) *Doing business in 2005: Removing obstacles to Growth*. Oxford: Oxford University Press.

World Bank (2006) Afghanistan urban land policy notes. Series 5.2. Washington DC: World Bank.

World Bank (2007) World Bank assistance to agriculture in sub-Saharan Africa: An IEG review. Washington DC: World Bank.

Zevenbergen, J. (2002) *Systems of Land Registration: Aspects and Effects*. Delft, The Netherlands: Netherlands Geodetic Commission.

Zevenbergen, J. (2004) A systems approach to land registration and cadastre. *Nordic Journal of Surveying and Real Estate Research*, 1(1), pp. 11–24.

11 Informal real estate brokerage as a socially-embedded market for economic development in Africa

Franklin Obeng-Odoom

1. Introduction

New institutional economists have consistently argued that informal real estate (RE) must be formalised, organised and traded. Formalisation, the argument goes, reduces transaction costs and hence frees up the market for more vibrant trade in RE and property rights. Additional incentives for formalisation, advocates argue, are the easy access to credit, provision of incentives to better maintain RE and the provision of security of RE ownership. It is important to consider these claims in the light of the aims of this book. A massive literature has developed to defend (Anderson and Libecap, 2014), extend (de Soto, 2000) or contest (Domeher and Abdulai, 2012a, 2012b; Gilbert, 2012) these views since they were first made by Coase (1937, 1959, 1974, 1984, 1998), North (1991) and Alchian and Demsetz (1973), the grandfathers of the new institutionalists.

Largely overlooked in the literature, however, is the role of "informal" RE brokers working in RE markets in Africa where there are high information costs. While there is a large literature on formal brokers and brokerage services in the advanced and industrialised countries such as the UK and USA where agents or brokers are typically registered with the government, even if they are not members of professional bodies that issue licenses (Antoniades, 2013), in Africa – where registered RE brokers exist, but alongside informal ones – there is scant research to ascertain the claim of the new institutional economists. It is not that there is no research on informality. On the contrary, Africa is one of the birthplaces, if not *the* birthplace, for conceptual and empirical advances about informality intensifying in the 1960s and 1970s (Meagher, 2007) during which Hart's (1973) pioneering paper on the concept was born. Rather, even in the famed Hart paper, it is non-RE brokerage activities that receive attention. Since then there has been a vast and growing literature on informal economy in this region, culminating among others in a special issue on "informal institutions and development in Africa" published in "Africa Spectrum" and edited by Meagher (2007), who has written widely on informality in Nigeria since editing the special issue (see, for example, Meagher, 2009, 2011).

However, while "informal" RE agents or brokers and their activities have received some attention in Africa (for example, Obeng-Odoom, 2011a), such studies are hardly framed in the wider debates about informal institutions on the continent and their role in the process of economic development (UN-Habitat, 2011). Yet as the number of such agents is substantial, growing and crucial to the operation of the RE sector (Obeng-Odoom, 2011a; UN-Habitat, 2011; Obeng-Odoom and Ameyaw, 2014), it is important to study informal RE brokerage in Africa.

The formal/informal classification can be blurry but a mechanical distinction is typically attempted in terms of: (i) the process of becoming an agent; (ii) outward appearance of agents as expressed in how they are attired; and (iii) how they perform their roles. In this nomenclature, formal brokers usually have formal training, are typically registered or licensed and are usually members of professional bodies or associations, whereas, generally, informal agents are held to be less formally and rigorously trained and regulated. Even in terms of appearance, while formal agents tend to be suited or dressed formally in business attire, the informal agents appear quite casual. Work-wise, too, informal agents adopt informal processes. The small literature on the topic (Mahama and Antwi, 2006; Oladokun and Ojo, 2011; Agboola et al., 2010), typically claims that informal brokers are problematic and emblematic of the stresses in RE markets in Africa. In turn, such brokers must be formalised.

However, these studies are not conclusive. They tend to be connected to a historic push among professional RE bodies to assimilate or exterminate informal brokers from participating in the RE markets to create monopolistic advantages for themselves. Or, as with participants of other informal economies, informal brokers are often the victims of scapegoating when there is any social problem (Asiedu and Agyei-Mensah, 2008) or spatial dispossession when there is a social programme for modernisation, often called "decongestion" in Ghana (Obeng-Odoom, 2011b). Cast more widely from the perspective of new institutional economics, the existing studies can be interpreted as joining a long crusade of eradicating inefficient institutions (Meagher, 2007). Coupled with how little engagement there is in such literature with the economic development discourse (UN-Habitat, 2011), it is important to revisit the topic and within the context of this book to: (i) investigate existing RE brokerage practices; (ii) assess the contribution of informal RE brokerage to economic development; and (iii) examine the argument that regulating informal RE brokerage is the panacea to the problems bedevilling the operation of RE markets.

To do so, the chapter draws selectively on Polanyi's ideas of "embeddedness", "disembeddedness" and "social dislocation" taken from *The Great Transformation* (1944 [2001]) and *Trade and Market in the Early Empires* (edited with Arensberg and Pearson, 1957). Together, these concepts form an analytical framework for studying market activity in economies that are not industrialised and are only partially organised by markets. This Polanyian framework is useful because of the chapter's focus on Africa – a non-industrialised region.

The rest of the chapter is divided into three main sections. It begins with how informal agents are perceived before considering what role informal RE agents play and how policy can be reformed – in that order.

2. Transaction costs, RE agents and the market for lemons

Transactions entail costs, and so to facilitate them, transaction costs must be reduced. Institutions, according to this view, must exist to reduce transaction costs in the form of uncertainty, search times and all other time and resources spent in an exchange. In new institutional economics, North's (1991) definition of institutions has become a classic one. According to North (1991, p. 97), "Institutions are the humanly devised constraints that structure political, economic and social interaction. They consist of both informal constraints (sanctions, taboos, customs, traditions and codes of conduct) and formal rules (constitutions, laws, property rights)." Institutions, North (1991, p. 97) continues: "provide the incentive structure of an economy; as that structure evolves, it shapes the direction of economic change towards growth, stagnation or decline". From this perspective, greater marketisation through private RE is conducive for economic development. It is an argument that has been made extensively in relation to Africa where much RE is regarded as "dead capital" primarily because it is not traded or traded in the same way as in the advanced industrialised countries (de Soto, 2000). For new institutional economists, therefore, transactions are important and it is through them that the process of economic development is completed.

In turn, they look favourably at agents whose duty is to enhance market processes. Agents help with the cost of managing complexity in the RE market, and that can in turn save the buyer considerable search time which can be put to other uses. It is this crucial role played by agents that has made them fundamentally relevant even during the advent of ICT-based advertisement (Sawyer et al., 2003). From this perspective, RE agency arises and exists primarily to reduce the transaction costs in trading RE. So, even if hiring a RE agent entails a transaction cost, the overall effect of hiring them is to reduce cost and facilitate transactions. Wallis and North (1986, p. 101) note:

> We want to treat all of the resources – that is, the total value of the inputs used by intermediaries – as a part of the transaction sector. The problem, of course, is to determine, which firms (industries) are properly classified as intermediaries or what we call "transaction industries". Three cases that seem clear are real estate and finance whose role is primarily to facilitate the transfer of ownership, banking and insurance whose role is to intermediate in the exchange of contingent claims, and the legal profession whose primary role is to facilitate the coordination, enactment and monitoring of contracts.

Yet in practice, RE agents have been branded "fraudsters", "confidence tricksters" and "scammers". Even in countries such as Ghana and Nigeria with particularly high transaction costs, RE agents tend to be despised. According to Mends (2006), RE agents are solely responsible for the stresses of the RE market: they distort market information, they extort money from their clients and conduct valuations that they are not trained to do. So bad is their reputation in Ghana that it is not uncommon to see people advertising for the sale/rent of their houses with the inscription "No agents needed" or "No agents please". Similar concerns apply in Nigeria. RE agents are regarded as performing "illegal services", anti "scientific knowledge", inexperienced in RE transactions and greedy; they are rogues and are out to dupe rather than serve those that consult them. These are simply the "basic issues about quackery" in Nigeria (Oladokun and Ojo, 2011, p. 307).

Such depictions are not limited to informal agents. In jurisdictions where agents are formal, RE agents do not typically have a good reputation either. As a BBC (2004) news item once noted, "Estate agents are the butt of many a dinner party joke, but the estimated 24,000 homebuyers who complain each year about poor service levels are not laughing." RE agents have to live with such a poor public image that they themselves have concluded that they are the most despised people on earth (Hill, 2002). So whether formal or informal, RE agents are deemed fraudulent, dishonest and incompetent (Ratcliffe, 1978; Lim, 2007; Mends, 2006; Mahama and Antwi, 2006; Mahama and Dixon, 2006). For Brinkmann (2000), using RE agents exposes clients to risks and vulnerabilities and so the public should be wary of them.

Such categorical views about RE agents have fuelled much cynicism about the entire institution of brokerage. Many authors have advocated doing away completely with agents. A host of books have been published to demonstrate how this can be done. Among the lot, Glink (2003) has published *Simple Steps You Can Take to Sell Your Home* while Shenkman and Boroson (2001) have authored *How to Buy a House With No (or Little) Money Down*. Joseph Diblasi (2005) has published *Sell Your Home Without a Broker*.

How can this backlash against RE agents (who are widely regarded as crucial to reducing transaction costs) be explained and resolved? The Nobel Prize-winning ideas of the economist Akerlof (1970) are often utilised to explain this paradox and show how the problem may be resolved. According to Akerlof (1970) consumers find it extremely difficult to separate good from bad quality goods. It is this "quality uncertainty" that generates and sustains a vibrant market for "lemons" (poor quality goods), which in turn tend to drive out superior quality goods (as consumers turn to lemons). The cost of this problem includes, but is not limited to, how much is lost in buying a bad product. What is more important to Akerlof (1970) is to drive out good products and hence deprive the entire society of quality goods. Akerlof (1970) offered several approaches to resolve the lemons problems: using brand names and licenses were among the most important.

More recently, a huge literature has been produced to try to explain this negative perception. Apart from a few that blame individual agents for being dishonest (for example, Dabholkar and Overby, 2006), the rest support the analysis of Akerlof (1970). For instance, Rudolf (1998) suggests that information asymmetries, namely adverse selection (the lemons argument) and moral hazard (how to verify the claims by agents), could explain the poor reputation of RE agents. Bishop (2004) famously linked the "despised, slippery and untrustworthy" reputation of RE agents to the very reason why agencies are needed: inherent information asymmetry in the RE market and the cost of correcting it. Others put the blame on dishonest individuals in agency. Bishop (2004) finds that the ease of entry into and exit from the profession creates in consumers' minds the perception that there are RE agents who can just "disappear" as and when they wish.

Like Akerlof (1970), these studies point to the need for licensing. Advocates argue that licensing, especially when administered by professional bodies, will be a form of guarantee to consumers. A license is insurance that guarantees that RE agents who possess it will put in their best effort to accomplish the purpose for which they have been hired. Professional bodies assure this, and there are codes of practice and bye-laws to guide such members. Professional licensing is then supposed to solve both the problem of adverse selection and moral hazard (Mends, 2006; Mahama and Dixon, 2006; Oladokun and Ojo, 2011; Agboola et al., 2010; Antoniades, 2013). This view feeds into a bigger body of work on regulation as being in the public interest (Viscusi et al., 2005, p. 314). From this perspective, consumers will prefer licensed agents who are more professional, more knowledgeable, more secure to use and better positioned to serve their needs. Informal arrangements, on the other hand, should atrophy and where there is a clear choice between a market good and a lemon, consumers must opt for the market good, the professionally licensed agents, at the expense of the lemons, informal agents, such that with time, licensed agents will proliferate while unlicensed agents diminish and eventually leave the market.

It is important to consider these arguments and aspirations carefully.

3. Socially-embedded market transactions

Informal brokers have long played a role in Africa where especially in the Arabic parts such as those in East Africa they are called "dellali". In most cases, they have been informal, sometimes, connecting buyers to cattle traders who themselves might work as "dellali" as seen in the work of Hill (1966). In the housing system, similar issues arise. For instance, in her classic work "Landlords and Lodgers", the economic anthropologist Pellow (2002, pp. 216–218) shows how individuals within communities bring landlords and lodgers together for accommodation purposes. These brokers are informal in the sense that they have no formal training in providing brokerage services: they are brokers because the landlords know them and community members know them and so information about housing for rental purposes is routed through them to lodgers but also to landlords.

More people are using the services of informal RE agents. Oladokun and Ojo's (2011) study provides the transcript of interviews with professionally licensed RE agents working in 172 registered firms in Lagos, Nigeria. Of those interviewed, 78% of reported that there is "frequent" or "very frequent" use of unlicensed RE agents in Nigeria. The pioneering *Ghana Housing Profile* prepared by the UN-Habitat (2011) also documents the growth in the numbers of informal agents.

The question of why such RE brokerage practice is on the rise is even more interesting. Generally, the expansion of the informal system in Nigeria has been ascribed to job cuts or freeze in the public service as part of the wave of structural adjustment policies implemented in the 1980s and the continuing embrace and expansion of neoliberal policy making, which make it extremely difficult to access formal jobs (Meagher, 2006, 2010, 2011). Indeed, it seems even those at the brink of falling into informal jobs were pushed into informality as structural adjustment programmes pulled away formal and public support services from under their feet; in turn, deregulation or reregulation to expand markets are at the heart of the widening of informality in Nigeria (Meagher, 2006, 2010, 2011). In the RE sector, however, there seems to be other drivers. Oladokun and Ojo's (2011) survey found that the incentive for higher rewards is the most highly ranked driver, followed closely by the supply of unsatisfactory services by professionally licensed agents. The first reason is interesting but incomplete without noting the public actually sets out to seek the services of informal agents. The second reason emerging from the interview is consistent with the findings of other studies. Findings from a survey about the ethics of professionally licensed agents in Nigeria (Agboola et al., 2010) showed that most of the respondents in the 125 firms randomly selected in Lagos would prioritise commercial consideration over ethical issues any day. Not surprisingly, 61.3% of consumers interviewed consider the level of ethical behaviour among registered agents to be average, a trend 56.7% of the clients interviewed consider to have remained the same over the years.

What about the effect of licensing on "agency problems"? Again Agboola et al.'s survey is revealing. When professionally licensed agents were asked about the effect of licensing on their ethical behaviour, nearly half (47.1%) said it does not matter. Outside Nigeria, it has been established in Ghana that most licensed agents are the nests and incubators of cankerous corruption in the RE market in Ghana (Obeng-Odoom, 2011a). At the global level, too, registered agents have typically not shown any superior ethical behaviour. One of the oldest formal agency regulatory regimes is in Australia where RE agents have been subjected to formal regulation since the 1800s. Yet, the research of Antoniades (2013) shows widespread dissatisfaction with such licensed agents. Indeed, since 2005, there have been varying levels of increases in consumer complaints, including an increase of 33.3% in 2012 (Antoniades, 2013). Together with the evidence from Africa, the argument for formal regulation begins to look shaky.

On the other hand, if the widespread disdain for informal agents (which almost always comes from affiliates and self-appointed spokespersons of professional bodies) is set aside, members of the general public who have used the services of informal brokers offer more positive ratings. In one survey in Ghana by Obeng-Odoom (2011a), customers expressed more interest in informal agents because they are more accessible, relate better to customers and are more likely to give discounts. A more recent survey by Obeng-Odoom (2014) suggests that formal RE agents in the oil cities in Ghana are starting to take informal brokers more seriously; new alliances across tiers of informal agents on the one hand, and between formally licensed agents on the other hand, have been formed. Typically, the formal agents have started seeking out informal agents for guidance and direction about RE activities. It is still early days yet but it is rather the formal agents that are trying to follow in the steps of the informal agents, not the informal agents trying to become formal agents. This expression of confidence in informal brokerage, especially in informal practices of information gathering, sharing and use, goes to show the resilience of informal brokers. Thus, it is important to consider the activities of informal agents more carefully.

Informal agents in Africa concentrate on RE agency. They typically stay away from technical services such as valuation (UN-Habitat, 2011) and until recently, RE management was outside their purview too (Oladokun and Ojo, 2010). Most of them are not as highly lettered as formal agents; however, apart from a few, most are functionally literate (Obeng-Odoom, 2011a). They can read and write and have basic understanding of measurements they need to describe RE – most of them are fluent in local languages and conversant with the English language (Obeng-Odoom, 2011a). Thus, while most of them are not degree or diploma level qualified in RE courses (Obeng-Odoom and Ameyaw, 2011), informal agents are not illiterate. Indeed, both formal and informal RE agents perform at least four functions: listing, matching, negotiation and closing phases (Turnbull and Dombrow, 2007). However, the mechanisms, instruments or processes by which these functions are performed differ.

The findings of research conducted by Obeng-Odoom (2011) and UN-Habitat (2011) in Ghana shows that in terms of listing, informal agents use mini sign-boards, which may read: "Two-Bedroom House for Sale", "Three Bedroom Houses for Rent" and "Land for Sale". Usually at the topmost part of the sign-board is the name of the practice. According to these authors, this listing practice is unlike the websites and viewing windows in Western cities and unlike the paper-based systems of the formal agents in the global South, but they mimic the informal advertisements posted on noticeboards and traffic lights in the advanced regions. Yet, they differ from these too because informal brokers typically invoke or adopt names that inspire confidence that their activities are genuine. Monikers such as "peace", "hope", and "genuine" are common and are often hand-written in chalk or paint. At the bottom of the signboards or trees are usually the words "Contact" or "Please Call" followed by a personal mobile phone number of the agent. The simplicity of the materials used for listing, then, gives a distinct flavour of informality (Obeng-Odoom, 2011; UN-Habitat, 2011).

Nevertheless, these agents are still able to match those intending to buy, lease or rent and those in the market to sell, lease or let out. As operating informally can connote working outside of formalised offices, informal brokers called via listing advertisements may invite a potential client for a meeting in very informal settings such as a bar or under a shade or to their house. It is not common practice to be directed to an office even though some RE agents have started opening one room/ container offices. The client is at liberty to sign a contract of agreement with the agent, or agree orally to the terms of engagement. Agents are able to advise clients on how difficult or easy it is to get what they require given how much they want to pay. If how much the clients want to pay is below the market rates, there would usually be some haggling. When they agree on the reasonableness of the amount to be paid by the client, the client is taken through the stock of properties available (Obeng-Odoom, 2011; UN-Habitat, 2011).

It is rare and almost impossible to come across an agent who has photographs of available houses or plots of land; the oil cities have welcomed some agencies with such information, but they are the exception (Obeng-Odoom, 2014). Given the absence of pictures, the client has to follow the agent around town to inspect the promised unit of RE. Should it happen that the client is dissatisfied with the house, there is another tour of houses to let. This process goes on until such a time that a house that appeals to the client is found. It must be stressed that in this entire journey around town, the client has to foot the cost of transportation. Should it be the case that the client is a wealthy one with a car, he has to drive the RE agent around. If the house is just in the neighbourhood, then both client and agent do the walking around. Once a property that meets the client's requirement is found, a day is scheduled during which the client and the RE agent have to meet the landlord of the property (Obeng-Odoom, 2011; UN-Habitat, 2011).

Negotiation follows when RE agents successfully match complementary interests in the market. During this time, the agent formally introduces the client to the landlord. The landlord tells the client how much rent is to be paid (Obeng-Odoom, 2011a; UN-Habitat, 2011). The rent advance should be a maximum of six months according to the Rent Act in Ghana (UN-Habitat, 2011). However, in practice, three years to four years' rent can be taken in advance. Often times, these terms will have already been conveyed to the client by the agent but it appears to be ceremonial or perhaps traditional for the landlord to reiterate these terms. There is "closing" when the rent is agreed. It is at this point that the landlord recites the rules of tenancy verbally (or in a usually one-page statement) and then the client has the option of agreeing to or disagreeing to these terms (Obeng-Odoom, 2011; UN-Habitat, 2011).

Most of these processes are socially embedded. Most agents will have lived for at least 10 years in the community in which they practise and for this same reason tend to be more accessible to the community people, most of whom will use an agent they know personally through friendship, familial connections or recommendations by friends (Obeng-Odoom, 2011a). While formal agents commonly ascribe dishonesty, failure to deliver service and fraud to informal brokerage, in practice these problems are the least of consumer concerns. In a survey of 50 clients of agents in Ghana, only five mentioned dishonesty as a problem while three had issues with service delivery and two had cases of fraud against informal agents (Obeng-Odoom, 2011a). The low levels of "agency problems" can be explained by strong community regulation. Thus, not only do social relationships mould transactions, they also act as social regulation to tightly regulate market transactions, which stresses that informal brokerage activities are not "free", but rather "social markets". More than this empirical finding, this argument is logically plausible (Paton, 2010, 2012; Somers and Block, 2014) and consistent with other studies elsewhere (DiMaggio and Louch, 1998). One tends to trust that people will not be duped by friends. People are more careful in how other people behave when those around them might be hurt and people tend to give of their best to friends, school mates, kith and kin.

Singling out informal RE agents for reprimand is therefore problematic, not only because it amounts to scapegoating, but also because such agents are vulnerable. In Ghana, informal RE brokers may work without pay, as some clients and landlords collude to finalise transactions without the knowledge of the agents who brought them together (so that both can avoid the payment of RE agency fees), and are exposed to litigation arising from landlord misbehaviour (such as promising to give one house to two people, going back on their word and illegal ejection of tenants) (Obeng-Odoom, 2011a). So in this sense, RE agents suffer the problem of vicarious penalty in which they shoulder the mistakes of landlords. With a huge and growing housing deficit in Ghana, landlords have become a law unto themselves. As noted in *Oiling the Urban Economy: Land, Labour, Capital and the State in Sekondi-Takoradi, Ghana* (Obeng-Odoom, 2014), it is landlords who "run the show" in the RE market in Ghana. They tend to vary rent at will and evict tenants at will in cahoots with rich oil workers, often migrants from bigger cities or temporary migrants from some advanced industrialised countries whose substantially higher incomes secure them accommodation spaces in prime areas. A similar problem has been reported in Port Harcourt in Nigeria (Ekpenyong, 1989) and in urban Nigeria generally (Lawanson and Oduwaye, 2014). As housing has long been commodified and housing is offered in what is clearly a suppliers' market, even if landlords are known in their communities, landlordism has become an albatross around the neck of communities in Africa.

3.1 Making sense of it all: commodification, competition and confusion

Polanyi's ([1944] 2001) work can help to understand the poor reputation and problems that can be found in the formal brokerage services. In *The Great Transformation*, Polanyi ([1944] 2001) established four pillars of an enduring framework based on the concepts of embeddedness, disembeddedness (commodification), social dislocation, and countermovement. Collectively, these notions may be understood as the "double movement" in capitalist societies. Applying this framework to RE brokerage activities for diagnostic purposes, it can be argued that the attempt to commodify housing and land constitutes the making of a "fictitious commodity". The conversion of land into a commodity is the original sin. As with all fictitious commodities, this process is also the first disintegrating force to "attempt" to disembed market from social relations. But it is not the only one. By privileging the landlord class with a power to say "this is mine" and that is "yours", the very basis of society, a natural connection to the land is lost, leaving humans orphaned.

As noted by Polanyi, all such attempts lead to conflict and social dislocation. In the case of RE brokerage, impersonal, asocial and antisocial agents arise as a professional and professionalised cadre whose activities, while seen as important, are not rooted in the mores and norms of society. This is the social basis of the problematic reputation problems of depersonalised agents and the fountain from which informal RE agents derive their social legitimacy. Thus, with more co-operation, less commodification and competition will also come reduced confusion.

While Polanyi's work in *The Great Transformation* focused mainly on England, where Polanyi clearly suggested that markets can be disembedded, in his essays in *Trade and Market in the Early Empires* (Polanyi et al., 1957), which are focused on Africa, Polanyi argues a rather different point: markets can never be fully disembedded from the State or the polity "in practice" or "substantively".

Polanyi's (1957) idea of "the economy as instituted process" exemplifies this "substantivist view" of markets and economies. From a Polanyian perspective, then "free market" is not the one without regulation (Paton, 2010, 2012). It simply means "freeing" or extending market processes to let them work in the interest of capital. Regulation then is not an attempt to embed, for the market is "always embedded" (Cahill, 2012, pp. 114–122); that is, it is always built on other relationships that make it work.

The point about formal brokerage then, is not that it has been robbed of its social basis completely, but that statism, professionalism and careerism have robbed agency of communal linkages within which informal brokerage is embedded. Such linkages are the mechanisms of stability and they revolve around reciprocity, redistribution and exchange, not only at the level of the individual but also at the level of social groups, kith and kin, and webs of social networks. In this sense, externally imposed top-down asocial and antisocial transaction processes roll back the three processes of "reciprocity, redistribution and exchange" (Polanyi, 1957) that typify informal brokerage. Formal brokerage processes swing brokerage from a "gemeinschaft" to "Gesellschaft" or from a personalised community to an impersonalised market society (Tönnies, 1887 [1957]).

Within this context, the use of regulation, professionalism and professionalisation – the expressed interest of which is to control the market and to protect the public interest – is limited. Indeed, more and more statist and professional regulation ends up freeing the market as impersonal, antisocial, and asocial agents arise or better still are imposed to re-embed the market in society. In principle, as people have seen, the market is "always embedded" (Cahill, 2012) but the "faux" thinking leads to a widening of the market vis-à-vis society and hence sets society on the path of becoming an accessory of the market.

Herein lies the root cause of the problem: the fictitious commodification of the commons and the expansion of the depersonalisation of the market in the form of professionalism of brokerage services. Were land and housing socially managed and reproduced and professionalism not so much elevated to the rank of an emperor, the agent would still be needed but, like the informal broker, the transactions would be more tightly covered by social relationships. Besides, it would not elevate a landlord class over and above the community. The solution, then, is not to regulate to further enhance markets in land but rather to more tightly embed the market within society and community.

Thus, in terms of policy, there is clearly a role for the State to support and to nourish rather than to attack and to undermine informal systems. More fundamentally, there is a role for the State to deploy its development planning towards RE for human need, not profit. Therefore, this analysis is distinct from prevailing mainstream claims about informality in several important ways. First, the analysis here differs from praise-singing studies about how "self-help" activities in Africa are resilient, and the need to marginalise the "failed", corrupt or weak States in Africa and hence expand the market for over autonomous and over "agencied" informal actors (Obeng-Odoom, 2011b). Second, the analysis parts company with "social capital" studies, which essentially make the business case for social networks in

Africa about which much has been published in the *Journal of Developmental Entrepreneurship*. It is also distinct from arguments about cultural primitivity in Africa where ethics and ethos are frozen in time and fictionalised (for an extended discussion of these insular perspectives on social networks, see Meagher, 2006, 2007, 2009, 2010, 2011). Aside the different resulting policies on how to organise the State, market and community, the analysis in this chapter is methodologically distinct from the mainstream, which tends to be ahistorical, deductive and rationalist. The case for informal RE brokerage as a socially-embedded market for economic development in Africa bridges existing formal–informal, micro–macro and agency structure dichotomies and transcends them into a whole and holistic description and a prescription for action.

In this sense, informal RE agency constitutes a socially embedded market with much promise for economic development. While there are "agency problems", they hardly arise because of informality. Instead, agency problems arise from the attempt to dis-embed RE brokerage from its socially moulded relationships. From this perspective, "statist" regulation can only be a "fictitious decommodification" (Standing, 2009, p. 32) instrument that acts as a "faux" and corrupted version of an original social moulding that serves to divide and weaken the class of brokers into a "labour aristocracy" with few workers (those with licenses) attempting to monopolise the RE brokerage market. This perspective has important ramifications for how to organise market, state and community and in what ways informal RE agency can contribute to economic development in Africa.

4. Conclusion

Drawing selectively on a Polanyian framework supported by examples from Africa, this chapter has shown that informal RE agency constitutes a socially embedded market with much better promise for economic development than current forms of formal agency. It accepts that there are some challenges with the informal RE brokerage system. However, it shows that such difficulties pale in the face of mounting dissatisfaction with current forms of RE agency, which arise because of the attempts to dis-embed brokerage from its socially moulded relationships. More fundamentally, the chapter shows that it is the result of the fictitious commodification of land and housing that leads to the rise of a class of unaccountable landlords and impersonal cadres of professionals, both of whom serve the market and are served by overbearing markets.

The implication of the analysis is that "statist and professional" regulation is inferior to "community regulation" steeped in norms and mores. The antisocial and asocial instrument of professional and State regulation acts as a faux and corrupted version of original social moulding and serves to alienate brokers from the public. In this process of over extending the market, society stands dislocated, leading to a backlash of complaints from consumers in what Polanyi called a "double movement".

This perspective is a major departure from existing studies on RE agency, be they from the new institutional economics fold or elsewhere, and has important ramifications for how to organise the market, State and community. There is an important role for the State, not of inaction but of action; to deploy its development planning towards RE for human need, not profit. The case for informal RE brokerage as a socially-embedded market for economic development in Africa then bridges existing formal–informal, micro–macro and agency structure dichotomies and transcends them into a grounds up, comprehensive description and prescription for action. In the light of the debates about brokerage and economic development, the thesis canvassed here is that only a socially-embedded informal brokerage market can foster economic development in Africa.

Acknowledgements

Many thanks to Dr Joy Paton of the Department of Political Economy at the University of Sydney in Australia for helpful comments on an earlier draft of the paper and to Dr Raymond T. Abdulai for editorial feedback.

References

Agboola, A. O., Ojo, O. and Amidu, A. R. (2010) The ethics of real estate agents in emerging economies. *Property Management*, 28(5), pp. 339–357.
Akerlof, G. A. (1970) The market for "lemons": Quality uncertainty and the market mechanism. *The Quarterly Journal of Economics*, 84 (3), pp. 488–500.
Alchian, A. A. and Demsetz, H. (1973) The property right paradigm. *The Journal of Economic History*, 33(1), pp. 16–27.
Anderson, T. L and Libecap, G. D. (2014) *Environmental Markets: A Property Rights Approach,* Cambridge: Cambridge University Press.
Antoniades, H. (2013) The national occupational licensing system: Codifying real estate agents' trust accounts. *Australia and New Zealand Property Journal*, 4(4), pp. 382–391.
Asiedu, A. and Agyei-Mensah, S. (2008) Traders on the run: Activities of street vendors in the Accra Metropolitan Area, Ghana. *Norwegian Journal of Geography*, 62(3), pp. 191–202.
BBC (2004) Estate agent service no joke [online]. Available at: http://news.bbc.co.uk/2/hi/business/3664703.stm [Accessed 10 October 2014].
Bishop, P. (2004) Despised, slippery and untrustworthy? The brokerage industry: An economic analysis. *Journal of Real Estate Research*, 1/2, pp. 31–47.
Brinkmann, J. (2000) Real-estate agent ethics: Selected findings from two Norwegian studies. *Business Ethics: A European Review*, 9(3), pp. 163–173.
Cahill, D. (2012). The embedded neoliberal economy, in: Cahill, D., Edwards, L. and Stilwell, F. (Eds), *Neoliberalism: Beyond the Free Market*. Cheltenham: Edward Elgar Publishing, pp. 110–127.
Coase, R. H. (1937) The nature of the firm. *Economica*, 4(16), pp. 386–405.
Coase, R. H. (1959) The Federal Communications Commission. *Journal of Law and Economics*, 2(1), pp. 1–40.
Coase, R. H. (1974) The market for goods and the market for ideas. *American Economic Review (Papers and Proceedings)*, 64 (2), pp. 384–391.
Coase, R. H. (1984) The new institutional economics. *Journal of Institutional and Theoretical Economics*, pp. 229–231; also (1998) *The American Economic Review*, 88(2), pp. 72–74.

Dabholkar, P. A. and Overby, J. W. (2006) An investigation of real estate agent service to home sellers: Relevant factors and attributions. *The Services Industries Journal*, 26(5), pp. 557–579.

De Soto, H. (2000) *The Mystery of Capital: Why Capitalism Triumphs in the West and Fails Everywhere Else*. New York: Bantam Press.

Diblasi J. P. (2005) *Sell Your Home Without a Broker: Insider's Advice to Selling Smart, Fast and for Top Dollar*. Naperville, Ill, Sphinx Pub.

DiMaggio, P. and Louch, H. (1998) Socially embedded consumer transactions: For what kinds of purchases do people most often use networks? *American Sociological Review*, 63(5), pp. 619–637.

Domeher, D. and Abdulai, R. (2012a) Access to credit in the developing world: Does land registration matter? *Third World Quarterly*, 33(1), pp. 161–175.

Domeher, D. and Abdulai, R. (2012b) Land registration, credit, and agricultural investment in Africa. *Agricultural Finance Review*, 72 (1), pp. 87–103.

Ekpenyong, S. (1989) Housing, the state and the poor in Port Harcourt. *Cities*, February, pp. 39–49.

Geoghegan, T. (2008) Should we feel sorry for estate agents? *BBC News Magazine* [online]. Available at: http://news.bbc.co.uk/2/hi/uk_news/magazine/7338512.stm [Accessed 10 July 2008].

Gilbert, A. (2012) De Soto's *The Mystery of Capital*: Reflections on the book's public impact. *International Development Planning Review*, 34(3), pp. v–xvii.

Glink, I. (2003) *Simple Steps You Can Take to Sell Your Home*. New York: Three Rivers Press.

Hart, K. (1973) Informal income opportunities and urban employment in Ghana. *Journal of Modern African Studies*, 11(1), pp. 61–89.

Hill, A. (2002) You'll be one of the most despised people on earth. *The Observer*, 7 July and *Journal of Economics*, 84, pp. 488–500.

Hill, P. (1966) Landlords and brokers: A West African trading system (with a note on Kumasi butchers). *Cahiers d'études africaines*, 6(23), pp. 349–366.

Lawanson, T. and Oduwaye, L. (2014) Socio-economic adaptation strategies of the urban poor in the Lagos metropolis, Nigeria. *African Review of Economics and Finance*, 6(1), pp. 139–160.

Lim, V. (2007) Licensed real estate agent vs. bogus real estate agent [online]. Available at: http://www.iproperty.com.my/articles/licensedvsbogusagents.asp [Accessed 30 November 2007].

Mahama, C. and Antwi, A. (2006) Land and property markets in Ghana. A discussion paper prepared for the Royal Institution of Chartered Surveyors for presentation at the 3rd World Urban Forum III, 19–23 June, Vancouver, Canada.

Mahama, C. and Dixon, M. (2006) Acquisition and affordability of land for housing in urban Ghana: A study in the formal land market dynamics. *RICS Research paper series* 6(10).

Meagher, K. (2006) Social capital, social liabilities and political capital: Social networks and informal manufacturing in Nigeria. *African Affairs*, 105(421), pp. 553–582.

Meagher, K. (2007) Introduction: Special issue on "informal institutions and development in Africa". *Afrika Spectrum*, 42(3), pp. 405–418.

Meagher, K. (2009) The Informalization of belonging: Igbo informal enterprise and national cohesion from below. *Africa Development*, 34(1), pp. 31–46.

Meagher, K. (2010) Identity Economics: Social Networks and the Informal Economy in Nigeria. African Issues series. Suffolk: James Currey.

Meagher, K. (2011). Informal economies and urban governance in Nigeria: Popular empowerment or political exclusion? *African Studies Review*, 54(2), pp. 47–72.

Mends, T. (2006) Property valuation in Ghana: Constraints and contradictions. Paper presented at the 5th FIG Regional Conference, 8–11 March, Accra, Ghana.

North, D. C. (1991) Institutions. *The Journal of Economic Perspectives*, 5(1), pp. 97–112.

Obeng-Odoom, F. (2011a) Real estate agents in Ghana: A suitable case for regulation? *Regional Studies*, 45(3), pp. 403–416.

Obeng-Odoom, F. (2011b) The informal sector in Ghana under siege. *Journal of Developing Societies*, 27(3&4), pp. 355–392.

Obeng-Odoom, F. (2014) *Oiling the Urban Economy: Land, Labour, Capital and the State in Sekondi-Takoradi, Ghana*. London: Routledge.

Obeng-Odoom, F. and Ameyaw, S. (2011) The state of surveying in Africa: A Ghanaian perspective. *Property Management*, 29(3), pp. 262–284.

Obeng-Odoom, F. and Ameyaw, S. (2014) A new informal economy in Africa: Case study of Ghana. *African Journal of Science, Technology, Innovation and Development*, 6(3), pp. 10–20.

Oladokun, T. T. and Ojo, O. (2011) Incursion of non-professionals into property management practice in Nigeria. *Property Management*, 29(3), pp. 305–320.

Paton, J. (2010) Labour as a (fictitious) commodity: Polanyi and the capitalist market economy. *The Economic and Labour Relations Review*, 21(1), pp. 77–88.

Paton, J. (2012) Neoliberalism through the lens of embeddedness, in: Cahill, D., Edwards, L. and Stilwell, F. (Eds), *Neoliberalism: Beyond the Free Market*. Cheltenham, UK: Edward Elgar Publishing, pp. 90–109.

Pellow, D. (2002) *Landlords and Lodgers: Socio-Spatial Organisation in an Accra Community*. London: Praeger.

Polanyi, K. (1957) The economy as instituted process, in: Polanyi, K., Arensberg, C. M. and Pearson H. W. (Eds), *Trade and Market in the Early Empires*. New York: The Free Press and London: Collier-Macmillan Ltd., pp. 243–270.

Polanyi, K. ([1944] 2001) *The Great Transformation: The Political and Economic Origins of Our Time*. Massachusetts: Beacon Press.

Polanyi, K., Arensberg, C. M. and Pearson, H. W. (1957) *Trade and Market in the Early Empires*. New York: The Free Press and London: Collier-Macmillan Ltd.

Ratcliffe, J. (1978) *An Introduction to Urban Land Administration*. London: The Estates Gazette Limited.

Rudolph, M. P. (1998) Will mandatory licensing and standards raise the quality of real estate appraisals? Some insights from agency theory. *Journal of Housing Economics*, 7(2), pp. 165–179.

Sawyer, S., Crowston, K., Wigand, R. T. and Allbritton, M. (2003) The social embeddedness of transactions: Evidence from the residential real-estate industry. *The Information Society*, 19(2), pp. 135–154.

Shenkman, M. M. and Boroson, W. (2001) *How to Buy a House With No (or Little) Money Down*. New York: John Wiley and Sons Inc.

Somers, M. and Block, F. (2014) The return of Karl Polanyi. *Dissent*, 61(2), Spring, pp. 30–33.

Standing, G. (2009) *Work after Globalization: Building Occupational Citizenship*. Cheltenham: Edward Elgar.

Tönnies, F. (1887) *Gemeinschaft und Gesellschaft*. Leipzig: Fues's Verlag (Translated, 1957 by Loomis, C. P. as *Community and Society*). East Lansing: Michigan State University Press.

Turnbull, K. T. and Dombrow, J. (2007) Individual agents, firms, and the real estate broker-age process. *Journal of Real Estate Finance and Economics*, 35(1), pp. 57–76.

UN-Habitat (2011) *Ghana Housing Profile*. Nairobi: UN-Habitat.

Viscusi, W. K., Harrington, J. and Vernom, J. M. (2005) *Economics of Regulation and Antitrust* (4th ed.). Cambridge: MIT Press.

Wallis, J. J. and North, D. C. (1986) Measuring the transaction sector in the American econ-omy, 1870–1970, in: Engerman, S. L. and Gallman, R. E. (Eds), *Long-Term Factors in American Economic Growth*. Chicago: University of Chicago Press, pp. 95–162.

12 Student housing investment and economic development

Eric Yeboah

1. Introduction

The real estate (RE) sector is increasingly being regarded as an integral component of the economies of nations. It has been established that investment in RE has direct impact on economic development. The recent global financial crisis helps to illustrate this linkage. The period preceding 2008 was characterised largely by lax underwriting standards and aggressive selling of mortgages and other RE products to sub-prime market in the United States of America (Doling et al., 2013). The overall impact of this was far reaching. It sparked off the recent financial crisis in America, which spiralled across the world (Doling et al., 2013). Conversely, the economic renaissances of China and Hong Kong, for example, are underpinned by significant investment in the RE sector (Hongyu et al., 2002; Chui and Chau, 2005). Indeed, it has been argued that one indicator of the extent of economic advancement is the extent of investment in the RE sector (Hongyu et al., 2002).

The RE industry is diverse and complex. It includes the more tangible aspects such as housing as well as the less visible components such as RE Investment Trusts (REITs) and other types of securitisation (Capozza and Seguin, 2003). As a result of the diversity of the RE sector, it is often important to disaggregate the various sectors in order to effectively map out their respective contribution to economic development. Even when RE is conceptualized broadly as land and/or the buildings attached to the land, there are several cleavages such as commercial, industrial, offices, leisure and residential RE including student housing. Conventional studies have often treated RE as a homogenous subject (Glossop, 2008). In as much as this may be convenient, it comes with its inherent risk of failing to explore the unique contribution of the various sub categories of RE.

A labor force with capabilities in abstract and problem-solving skills is critical to the development of nations. Therefore, Ghana and many other African countries are increasingly making efforts to expand access to education across all levels (Darvas and Balwanz, 2013). Invariably, ensuring such expanded access to education requires a commensurate expansion of educational infrastructure such as student residential facilities. The expansion of such infrastructure is essentially investment in the RE sector, which in itself can be a catalyst for economic development. But the linkages between investment in student housing and economic development remain under researched. Earlier studies have often concentrated on examining housing quality and its contribution to student well-being and performance (Addai, 2013). Such efforts are welcome but further investigations are required to map out the nature of the relationship between student housing and economic development.

This chapter thus attempts to contribute towards addressing this gap by drawing lessons and insights from the Ghanaian situation. Within Ghana, the study focuses on Kwame Nkrumah University of Science and Technology (KNUST) and its immediate neighbouring communities such as Ayeduase Ayigya, Boadi, Bomso and Kotei. KNUST was established in 1952, and it is a leading university in Ghana that continues to attract students across the continent. From a total of about 6,000 students in 1995, the student population has been rising steadily over the years. The student population in 2014 was about 40,000 (Vice Chancellor of KNUST, 2014). This steady rise of student population has not witnessed a commensurate increase in the needed infrastructure by the University authorities and the State. The University has six traditional halls of residence, which together provide accommodation for about 4,800 students. Construction of the last hall was completed in 1968, and this highlights how authorities have struggled over the years to provide the needed infrastructure to keep pace with the rapidly rising student numbers. The acute deficit in providing accommodation to students has effectively created a thriving investment opportunity, and private investors are continuously developing student housing to cater for such increasing demand. Presently, the University authorities have certified 61 such private student housing developments (KNUST, 2014). Beyond these 61 purpose built student hostels, there are units of RE that were originally designed as family residential units but have been converted into student housing units to meet growing demands. This, therefore, provides an ideal context to explore the relationship between student housing investment and economic development.

KNUST is located in Kumasi, the second city of Ghana, which is about 250 km north of the national capital, Accra. Figure 12.1 shows the location of Ghana in Africa and the specific case study area.

The remainder of this chapter is organised in five sections. The next section discusses the choice and justification of the methodology, which was employed for the research. This is followed with a theoretical examination of the meaning of "economic development". The third section then interrogates Ghana's current educational system and the increasing political contestation with issues such as the provision of educational infrastructure. These set the scene for the fourth section that analyses and discusses field data to help develop an improved understanding of the linkages between student housing investment and economic development. The chapter concludes by looking at the policy implications of the study's findings and recommendations, which could help to make investment in student housing more central to the economic development dialogue.

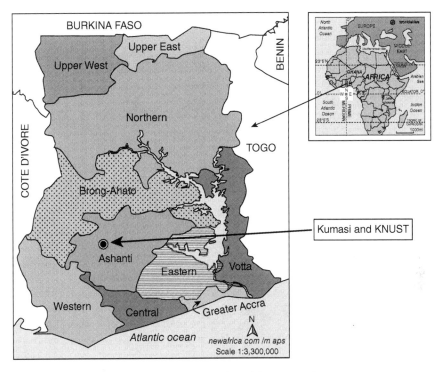

Figure 12.1 Map of Ghana showing the location of the case study area

Source: Adapted from www.newafrica.com (2000).

2. Research methodology

There are two main research methodologies, qualitative and quantitative, although there is mixed methods of research, which are increasingly gaining traction (Johnson et al., 2007). The other names for the mixed methods are mixed methodologies or multi-methodology, which is the use of both the qualitative and quantitative research methodologies in a single research project (Abdulai and Owusu-Ansah, 2014). The choice of a particular research methodology should, however, reflect the nature of the research problem under investigation and the overall research aim and objectives one seeks to address. This chapter seeks to examine the linkages between student housing investment and economic development. Such a task requires a greater exploration in order to develop a deeper understanding and effectively interpret the issues under investigation. In this type of situation, Creswell (2003) argues that the qualitative research methodology is more appropriate.

Within the qualitative research sphere, the case study strategy was employed. Case study is an empirical inquiry that investigates a contemporary phenomenon within its real life context using multiple sources of evidence (Yin, 2003).

The case study area (KNUST and its immediate neighbouring communities speci-fied above) was purposively selected.

The first part of this study involved a review of related literature in order to situate the study in the appropriate existing body of knowledge. The process of engaging the literature also helped to develop the meaning of a conceptual phrase such as "economic development" which eventually became the analytical frame-work for further discussions. The second phase of the research focused on col-lecting and analysing field data. Through the field investigation, a cross section of stakeholders (respondents) was surveyed. The survey participants included stu-dents, landlords/landladies, facilities managers of student housing and officials from the Estate Unit of KNUST, which regulates private student housing provid-ers on behalf of the University. Other respondents included managers of local businesses such as retail outlets, internet cafes and laundries. Respondents were also drawn from the local government authorities who are tasked with local rev-enue generation as well as officials from RE insurance firms.

The study randomly selected 10 student housing provides from a sampling frame of the 61 approved hostels (see KNUST, 2014, for the list of the approved hostels). In each of the 10 student hostels, the facilities managers were purpo-sively selected while two students were sampled using the convenient sampling technique. Other non-probability sampling techniques such as purposive and snowballing approaches were employed in the selection of respondents from the University's Estate Unit, RE insurance brokers, landlords/landladies and revenue officials from the local government authority. With the support of a research assis-tant, all the respondents were surveyed through semi-structured interviews. An interview guide with questions on the benefits of student housing developments to communities and the reasons for encouraging the development of additional stu-dent housing units in communities among others were posed to the respondents. Table 12.1 summarises the stakeholder categories and the number of respondents who were surveyed in the study.

As part of the analysis, the content analysis technique (CAT) as discussed by Scherier (2012) was employed. Even though the study is specific to KNUST and its surrounding communities in Ghana and, therefore, the findings may not be applicable in their entirety to other places, the research design ensures greater exploration of local dynamics and linkages. Thus, the study still provides useful insights towards an understanding of the relationship between student housing investment and economic development. It also provides the basis for similar stud-ies to be conducted in other places.

3. Conceptualising economic development

Like many conceptual terms and phrases in the development literature, "economic development" has been subjected to varied interpretations. The phrase has been conventionally associated with gauging the extent to which nations are prospering (Kanth, 1997). In this regard, issues such as technological advancement and macro-economic indicators such as Gross National Product (GDP) and per capita income

Table 12.1 Categories and number of respondents interviewed

Respondent	Number	Brief comment
Facility managers of student housing	10	These were made up of eight males and two females who are responsible for the day-to-day management of the student housing units. Six of them noted that they were part of the project to develop their respective units of RE from the outset.
Landlords/ladies	4	These consisted of three males and one female who have converted part of their houses into units for students.
Students	20	These consisted of 13 males and seven females.
University's Estate Unit	1	This is the Estate Officer of KNUST, who is responsible for liaising with private student housing providers.
Revenue officials of local government officials	2	These officials are part of those with the mandate to oversee the internal revenue generation within the Kumasi Metropolitan Authority.
Insurance brokers	1	One respondent from an insurance firm was surveyed to develop a better understanding of how investment in student housing impacts on RE insurance.
Managers of local businesses	5	These included owners of three grocery shops and a manager each of an internet café and laundry service provider.
Total	43	

Source: Field survey (2014).

have often been used. But the use of such "hard" yardsticks as the measuring principles of economic development has been criticised over the years by critics of mainstream theories, such as Sen (1999). Such commentators argue that a country may record impressive score on the various indicators but could still be saddled with several real life challenges. According to the latest Human Development Index (HDI, 2014), for example, Trinidad and Tobago has an impressive per capita income of US$25,325, which is almost four times that of Cape Verde's US$6,365. However, the former has a life expectancy of 69.6 years and the latter 75.1 years. High per Gross National Product (GNP) may not, therefore, necessarily guarantee a high expectancy rate. As a result of such variations and inconsistencies, relying solely on macro-economic indicators as the basis of gauging economic development may fail to paint an accurate picture of the state of affairs.

These concerns have resulted in the search for alternative parameters to measure economic development. In this regard, the seminal work of Sen (1999) provides a useful conceptual foundation. This work builds on the conceptual definition of poverty as encompassing material need, income poverty and capacity deprivation (UNDP, 2006). These variants of poverty are mutually reinforcing and often result in poor quality of life, which is characterised by premature mortality, avoidable morbidity, overwhelming illiteracy, unemployment and slums, among others. Sen (1999) conceives these as symptoms of under development. Economic development should liberate people from such ominous conditions. Sen (1999, p. 3) argues that:

> [Economic] development can be seen . . . as a process of expanding real freedoms that people enjoy. Focusing on human freedoms contrast with the narrower views of development, such as identifying development with the growth of gross national product or with the rise in personal incomes or with industrialisation or with technological advancement or with social modernisation.

This world view of poverty traces its roots to the Aristotelian theory of human good. Aristotle posits that there are differences in opinion about what is best for human beings and that to profit from ethical inquiry, one must resolve this disagreement (Broadie, 1991). The inherent difficulty in resolving what is good for humans through very objective parameters prompted the use of proxies that could easily be accepted as the basis for gauging human progress. Two indicators, "*eudaimonia*" ("happiness") and "*euzên*" ("living well") were employed (Woods, 1992). These were grounded in the logic that, to ensure a sense of happiness and well-being, people should be able to access "human goods" such as health care, employment and wealth recreation, among others (Broadie, 1991). The happiness and well-being of the people is, therefore, the desired end of all development interventions.

This Aristotelian approach of conceptualising development is central in contemporary discourse, primarily as a result of the attention it pays to the social ends of the development process. The popular HDI, for example, measures the progress of nations using macro-economic indicators such as GNP and also social ends such as life expectancy and issues relating to education.

Defining economic development using social parameters rather than relying solely on macro-economic outlook has heavily influenced the post-2000 development agenda. The United Nation's coordinated effort to ensure holistic economic growth globally focuses primarily on social considerations such as attainment of universal primary education, reduction of poverty and hunger and reduction of child mortality, among others (United Nations, 2010). Economic development should effectively result in the attainment of social goals, which will foster happiness and greater well-being of the people. This means that an analysis of the linkages between student housing investment and economic development should focus primarily on examining how the former contributes to ensuring social and human good such as employment creation, revenue generation and improvement in the standards of living, which are all prerequisites for the happiness and general well-being of the people.

4. Overview of the educational system in Ghana

Like many other countries, education has long been considered as a critical driver of economic development in Ghana. This is because it provides the pathway for acquiring skills and expertise, which are needed for nation building. As a result, there were mechanisms to facilitate the transfer of various skills like farming and artisanal vocations such as sewing long before the era of colonial rule; however, the current formal educational system in the country is a vestige of colonial rule in Ghana (Foster, 1965). Faced with the need to educate their own children and also to actively involve the local people in trade and missionary work, all colonial authorities who operated in Ghana (that is, Portuguese, Dutch, Danes and British) made efforts to establish formal educational systems, which were modelled along the ideals of their respective countries (McWilliam and Kwamena-Poh, 1975).

The establishment of formal educational institutions were initially concentrated along the coastal belt where the colonisers settled (McWilliam and Kwamena-Poh, 1975). However, the need to extend the activities of both missionaries and merchants across the colony saw rapid expansion of schools into the inland areas. By 1881, there were a total of 139 schools established across the country (Foster, 1965). The expansion of education continued to remain on the agenda of the colonial authorities and, by 1950, about 3,000 primary and secondary schools were established nationwide; student enrolment rose to about 280,000, representing 6.7% of the population (Darvas and Balwanz, 2013).

The expansion of education still remained high on the development agenda after independence. The promulgation of the Education Act 1961 (Act 87) by the then government represented a clear intent to make education much more accessible. Indeed, this enactment sought to make primary education compulsory. Section 2 of the Act provided that:

1 Every child who has attained the school-going age as determined by the Minister shall attend a course of instruction as laid down by the Minister in a school recognised for the purpose by the Minister.

2 Any parent who fails to comply with the provisions of the preceding subsection commits an offence and shall be liable on summary conviction to a fine. . . .

The intention of the State was to heavily subsidize the cost of obtaining formal education. As a result, "no fee, other than the payment for the provision of essential books or stationery or of materials required by pupils for use in practical work, shall be charged in respect of tuition at a public primary, middle or special school" (s. 21(1)). At this stage, the education system consisted of a six-year primary and four-year secondary school. After secondary school, suitable students proceeded to undertake a further two years of study in the sixth form. After this level, qualified students could proceed to undertake a three year programme at the university. Those who did not qualify to be admitted into the university were, however, required to enrol for two years of vocational training.

Ghana has experienced a chequered political history with frequent overthrow of democratically elected governments (Gockling, 2005). The educational system, however, remained largely unchanged despite efforts to introduce some reforms in 1974 (Akyeampong et al., 2007). A major overhauling of the educational system was carried out in 1987 by the government with support from the World Bank and other development partners. The reform resulted in the introduction of the junior and senior secondary schools concept. There was six years of primary education, three years of junior secondary and another three years of senior secondary education (Darvas and Balwanz, 2013). After the senior secondary level, students could proceed to university, polytechnic or teacher or nursing training.

Ghana's return to democracy in 1992 has witnessed renewed attention to make education, particularly basic education, more accessible to the citizenry. As a result, advancing education is one of the cardinal issues, which have been enshrined under the Directive Principle of State Policy. Article 38(1)–(3) of Ghana's 1992 Constitution mandates governments to provide educational facilities and ensure free compulsory universal basic education (FCUBE). As a result of the Constitutional mandate on various governments to prioritize the advancement of education in the country, it is not surprising that it has become very topical in the recent political landscape. Thus, when there was a regime change in 2001, the New Patriotic Party (NPP) government promised to revise the educational system. This culminated in the extension of secondary education from three to four years (Little, 2010). But when the National Democratic Party (NDC) returned to government in 2008, the duration of secondary education was reversed to three years. This was carried out despite evidence that students who had four years of secondary education generally performed better in the West African Senior Secondary Certificate Examination (Dabie, 2013; Djangmah, 2014).

During the 2012 general elections in Ghana, issues relating to education were again firmly placed on the front burner. Opposition parties such as the NPP and the Convention Peoples Party (CPP) had education as their flagship campaign messages. CPP promised to ensure twenty-first century education (CPP, 2012), whereas NPP assured the electorate of free senior high school education (NPP, 2012). The ruling NDC spent considerable time and resources to discredit the NPP's promise of free senior high school, arguing that ensuring quality education must be the priority (NDC, 2012). At the same time, NDC touted its achievements, prominent among which were education-related ones. These included claims of unprecedented provision of infrastructure, which resulted in the elimination of several schools that were hitherto held under trees. There were further claims of the distribution of free exercise and text books as well as uniforms (NDC, 2012). There are indications that the NDC may roll out the free senior high school concept (Government of Ghana, 2014) despite fiercely resisting it in 2012. In the view of some analysts like Braimah (2014), this sudden "U" turn is largely to gain political capital rather than the result of genuine efforts to improve education.

About 7,160,000 pupils were enrolled in primary schools across the country in 2013 (Ministry of Education, 2013). In 2004, at the secondary tier of the educational system, about 850,000 students were enrolled in 828 schools (World Bank, 2014). According to National Council for Tertiary Education (NCTE, 2014), there were 64 public tertiary institutions in 2014. These included nine public universities, 10 polytechnics and 38 colleges of education. There were 94 accredited tertiary institutions as well as specialized institutions, regulatory bodies and other sub-vented organisations in 2014. In 2013, there were about 260,000 students across all tertiary institutions (Ministry of Education, 2013).

4.1 Educational infrastructure provision in Ghana

The government of Ghana has over the years attempted to invest in educational infrastructure to meet the demands of a growing population. In 2003, for example, an amount of GHS534 million (representing 5.6% of GDP) was invested in education; this amount tripled to GHS1.7 billion in 2011 (Darvas and Balwanz, 2013). In 2014, the educational sector represented 30% of the gross expenditure by government (Government of Ghana, 2014). Governments continue to explore additional sources of funding for the expansion of education. The Ghana Education Trust Fund Act 2000 (Act 581), for example, was promulgated to achieve this purpose. Under this Act, the rate of value added tax (VAT) was increased by 2.5%. The additional amount of VAT levied is paid into a special account – the Ghana Education Trust Fund (GET Fund) – for the purpose of financing educational programmes and infrastructure.

Despite such initiatives, financing educational infrastructure remains a challenge in Ghana. For example, several primary schools take place under trees (Darvas and Balwanz, 2013) whilst only about 15% of the student population in the tertiary level are offered accommodation in hostels provided by the State (Addai, 2013; Braimah, 2014). As a result of this deficit, investment in student housing by the private sector, like in many areas of the economy, is highly encouraged by successive governments. The private sector has accordingly become a crucial provider of student residential facilities, especially at the tertiary level. But how does the provision of student housing impact on economic development? The ensuing section attempts to answer this question by examining the state of affairs in KNUST and its neighbouring communities.

5. Data presentation, analysis and discussion

By employing the content analysis technique (CAT), the following were identified as being the dominant and recurring themes from the various interviews with the survey participants, which are employed as analytical parameters for subsequent discussions:

- investment in student housing enhances local communities through employment generation and provision of local infrastructure;
- investment in student housing creates additional sources of revenue for the State through rates and other forms of taxes; and
- investment in student housing strengthens the RE insurance industry.

5.1 Enhancement of local communities through employment generation and provision of local infrastructure

There are several stages in the process of investing in student housing and these include planning, designing, construction and management. At each of these stages, several employment opportunities are either directly or indirectly created. The planning and preparatory stages of the development of student housing create job opportunities for an array of professionals. Carrying out investment appraisal to ascertain the feasibility or otherwise of the proposed investment is a source of employment to professionals such as land economists and investment appraisers. The process of land acquisition also offers job opportunities to professionals such as RE brokers and lawyers. The services of architects, structural engineers and project management specialists are all needed in the development of student housing at the planning and preparatory stage.

At the construction stage of the investment process, the extent of employment opportunities becomes even more profound. All the student housing providers who were surveyed in the study recounted how the construction of hostels or student residential facilities created jobs for different artisans such as masons, plumbers, electricians, carpenters and laborers, among others. Figures provided by the facilities managers of student residential units of RE who participated in the survey estimated the construction phase of the investment process created between 50 and 100 jobs for the various categories of skilled and unskilled labor. It must be admitted that these jobs created are not permanent in nature. However, the frequency with which student housing is being constructed in the vicinity of the University means people with the requisite expertise could have their skills being demanded on regular basis. Investment in student housing continues to generate employment even after the construction phase. For example, each of the purpose-built student accommodation facilities has a manager, caretakers, janitors and security personnel. All these categories of personnel work closely together to create and maintain conducive environments that support students' living. Rendering these supporting services is a source of employment.

Beyond these examples of direct employment, investment in student housing also creates several indirect jobs, which vitalise the local economy. Student housing requires complementary services in order to effectively support the lives and activities of students. These include conveniently located retail outlets, eateries and fitness centres, among others. The six halls of residence within the KNUST campus have 117 service providers who offer various ancillary services. Each of the registered service providers employs other persons to support the operations.

In the neighbouring communities, the local economies are largely dependent on the vibrant student population, which effectively provides a ready market. The view of one respondent, an owner of a retail outlet in Ayeduase, highlights the linkages between the increasing student population, proliferation of student housing and the dynamics within the local economy:

About 99 percent of local businesses in this community are established to meet students' demands and needs . . . when students are away on holidays, there is a major slump in business. . . . The concentration of student housing facilities in this community means we are benefiting from the presence of students . . . these students have effectively become the bloodline of the local economy.

(Respondent 4, retail shop owner)

Local economies, therefore, thrive on the presence of students, just as students also rely on the local economy for complementary services, which enhance their well-being. This creates a symbiotic relationship between investment in student housing and the local economy.

Furthermore, investment in student housing creates demand for building materials, which in turn creates employment. This demand derived from construction and investment in student housing stimulates the respective industry to produce in order to meet the new demand levels. This in itself generates and sustains employment across the value chain. When demand for building materials results in increased production, aspects of the value chain such as distribution and retail will often experience commensurate expansion. Significantly, growth in each stage of the production process creates employment.

Unemployment is a major challenge bedevilling governments of both developing and advanced economies. In the case of Ghana, for example, an estimated 25% of the labor force is unemployed (African Economic Outlook, 2012). This troubling statistic has been recorded at the time when the economy is making steady gains. After a slowdown of economic activity in 2009, the economy picked up in 2010 and grew in real terms by 7.7% (Government of Ghana, 2014). The economy recorded an impressive rate of 13.7% in 2011 as a result of oil revenue and strong export performance of cocoa and gold in volume and prices (ibid., 2014). In effect, the economic gains in recent times have not addressed the unemployment challenge. This means that any investment that has the inherent opportunity to create both direct and indirect jobs is critical to the development of the people. Investment in student housing is, therefore, a catalyst for employment generation, which is key to economic development.

Investment in student housing also results in the expansion and improvement of the local services and infrastructure that are needed to ensure convenience and well-being of the local people, and this was identified by each of the respondents who were interviewed. In particular, all the respondents indicated that the development of new student housing has resulted in the extension of electricity, improved roads and potable water to various neighbourhoods. The proliferation of student housing near the University community has led to competition. Student housing providers are, therefore, increasingly improving services and infrastructure in order to be competitive and such investment in the local infrastructure invariably ends up entirely improving the communities within which they are found. Investment in student housing could, therefore, help to improve the provision of local infrastructure and services by the private sector whilst creating employment.

5.2 Investment in student housing as a source of revenue mobilisation through taxation

Governments, both local and central, are increasingly faced with the challenge of raising adequate revenue for the purpose of providing needed services and infrastructure, among other requirements. Taxation represents one important source of revenue in this regard, and investment in student housing opens up different opportunities for taxation. As it has been established, investment in student housing provision creates the multiplier effect of creating jobs at different levels. This includes the production and distribution of building materials and the opening up of the retail and service sectors to support student living. As required by the Ghana Revenue Authority, the profits of all such businesses are treated as income and, therefore, are taxable. Furthermore, all those who are employed are expected to pay their personal income tax. Together with taxes on businesses, the Local Government Act 1993 (Act 462) empowers all local government agencies to impose a business operating permit levy on all businesses within their jurisdiction. Business operating fees are annual payments and this means businesses established to support student living become a vital source of revenue to the local government on yearly basis. The various businesses that spring up to support students' living are all mandated to pay the business operating fee. This source of revenue is becoming an important source of internally generated revenue. In 2013, for example, the fee amounted to 3% of internally generated revenue for Kumasi Metropolitan Authority. Together with this, businesses that are established to meet the demands of students are, like any other businesses, subject to a corporate tax of 25%.

Investment in student housing widens the opportunities for revenue generation by the State through indirect taxation. As it has been noted earlier, investment in student housing results in corresponding demand for building materials. Building materials such as cement, iron rods and tiles are value added products. Accordingly, they attract VAT. Since its introduction, VAT has remained an immense source of revenue for government. Currently, the VAT rate in Ghana is 17.5%. Out of this, 2.5% is dedicated to supporting the National Health Insurance to improve health care delivery. Another 2.5% is ring-fenced to support the financing of education through the Ghana Education Trust Fund. As noted by Sen (1999), improvement in the education and health of the people represents an important indicator of economic growth. Therefore, the importance of any investment that creates the opportunity to mobilize revenue through indirect sources such as VAT cannot be underestimated.

RE rates are a major source of revenue to local government authorities. In the Ghanaian case, for example, RE rates can be the means of raising funds to meet the shortfall of projected expenditure of local government authorities. From the previous total of 110 districts, the number has doubled to 216 over the past

26 years, and this has put a strain on central government funding for local government authorities. This situation has heightened the need for local government authorities to raise funds internally in order to provide basic services and other amenities. In this regard, the investment in the development of student housing in Ghana is helping to expand the base for RE rates at the local level.

In Ghana, Act 462 provides that RE rates be imposed on the value of developments and other improvements on land. According to the Estate Unit of KNUST, which is charged with regulating student housing, as many as 47 student housing facilities were constructed between 2007 and 2014. All such facilities are multi-storey buildings, which have been constructed with materials such as sandcrete blocks, tiles, glass, aluminium and steel. The size and the quality of constructional materials used means that modern student housing are often of high rateable value. Table 12.2 summarises the amount of RE rates, which were paid by the student housing providers around KNUST who were surveyed in the study.

The 10 student hostels paid a total amount of GHS9, 680 as RE rates during the 2012/13 fiscal year. For the same period, a district (Sisala East District) in one of the political administrative regions in the country (the Upper West region) generated a total of GHS14,785 from RE rates (Sisala East District, 2014). The contribution of 10 student housing facilities around KNUST effectively represents about 65% of the RE rates that were generated from an entire district. The importance of student housing as a source of revenue to local government authorities can, therefore, not be overstated. These revenues support other sources of income to enable both local and central government to provide services and infrastructure, which together help to improve the quality of life and well-being of the people. Therefore, as a source of both direct and indirect taxes to the State, student housing development contributes the much needed revenue required to expedite economic development.

Table 12.2 RE rates paid by the 10 surveyed student hostels during 2013

Name of student hostel	RE rate paid during 2012/13 fiscal year (GHS)
Splendor Hostel	1,300
Shepherdsville	1,450
Evandy	1,840
West End Hostel	1,500
Nana Adomah Hostel	1,110
Hydes Hostel	900
No Weapon	600
Nyberg Hostel	650
Frontline Hostel	900
Christ the King Hostel	780
Total	9,680

Source: Field survey (2014).

5.3 Investment in student housing and the RE insurance industry

Economies of nations are built on three main sectors: production and manufacturing; services; and the financial sector (Aryeetey and Kanbur, 2007). The financial sector can be further split into two sub-sectors, the banking and insurance industries. A healthy financial sector is pivotal to the economic development agenda (Monin and Jokipi, 2010), and investment in student housing can have an impact on the insurance sub-sector. Insurance is a contract that, subject to the payment of a premium, one will be indemnified in the occurrence of a stated peril (Dionne, 2000). One key contribution of the insurance industry is that it provides the opportunity for capital, in the form of the premium, to be aggregated for the purpose of investing in other ventures. Insurance, therefore, helps in the efficient re-allocation of capital.

In Ghana, the Insurance Act 2006 (Act 724) makes it mandatory for all commercial RE to be insured. Section 184(1) provides that every commercial RE be insured against perils such as fire, flood and building collapse. Section 184(3) defines such categories of RE as those that the public have the right to access for the purpose of recreation, business transactions and residence, under which student housing falls. Student housing faces several risks including fire and building collapse. It is, therefore, important that such RE is insured to spread the associated risk. This provides the assurance that, in the event of a peril, the investment will be reinstated. In Table 12.3, the premiums that were paid in respect of selected student housing facilities in 2013 are provided.

It must be noted that the premiums that are paid to insurers are expected to be invested. The investment portfolio for investors is often broad and may include RE, shares, bonds, equity and stocks. Shares, for example, provide a cheap source of capital because money is raised without incurring interest payments (Parameswaran, 2007). Such capital injection into the activities of businesses and companies could potentially result in improved productivity and growth.

Table 12.3 Insurance premiums paid by the 10 surveyed student hostels in 2013

Name of hostel	Premium paid in 2013 (GHS)
Splendor Hostel	7,000
Shepherdsville	6,520
Evandy	7,600
West End Hostel	
Nana Adomah Hostel	5,000
Hydes Hostel	5,700
No Weapon	4,100
Nyberg Hostel	5,000
Frontline Hostel	6,400
Christ the King Hostel	4,800

Source: Field Survey (2014).

Investment in student housing development can, therefore, result in a more dynamic insurance sector that helps to effectively re-allocate resources through the collection of premiums and their subsequent investment.

6. Conclusion

As a sub-niche of the housing sub-sector of the RE industry, student housing has a broad and complex relationship with economic development, both at the local and national levels. The ability of student housing to reinvigorate local economies through job creation and improved supply of infrastructure and services cannot be overstated. Investment in student housing provision stimulates the production of construction materials, creates jobs along the value chain and ultimately helps to improve student well-being. Investment in student housing also generates revenue through both direct and indirect taxation and such revenue could contribute towards improving local infrastructure and services. But to what extent is State policy facilitating the development of student housing in order to sustain its linkages with both the local and national economies?

Investors in RE currently enjoy a five year tax holiday (Leary and McCarthy, 2013). This is commendable, although much can be done to attract investors into the development of student housing. For example, corporate tax could be lowered from 25% to 15%. Notwithstanding that these could potentially increase investment, it is more likely in the short-term to result in a shortfall in tax revenues from student residential facilities. And as a policy, loans for the development of student housing could be granted on concessionary basis, where the interest rate is fixed below the market rate. All these could help to create an enabling environment for student housing investment. The provision of student housing should be envisaged as a complementary social good that is required for improved teaching and learning while serving as a catalyst to boost both local and national economies.

References

Abdulai, R. T. and Owusu-Ansah, A. (2014) Essential Ingredients of a Good Research Proposal for Undergraduate and Postgraduate Students in the Social Sciences. *SAGE Open*, July–September, pp. 1–15 [online]. Available at: http://sgo.sagepub.com/content/4/3/2158244014548178 [Accessed 16 December 2014].

Addai, I. (2013) Problems of Non-Residential Students in Tertiary Educational Institutions in Ghana: A Micro-Level Statistical Evidence. *Journal of Emerging Trends in Educational Research and Policy Studies*, 4(4), pp. 582–588.

African Economic Outlook (2012) Economic Outlook of Ghana [online]. Available at: http://www.africaneconomicoutlook.org/fileadmin/uploads/aeo/PDF/Ghana%20 Full%20PDF%20Country%20Note.pdf [Accessed 10 June 2014].

Akyeampong, K., Djanmah, J., Oduro A., Seidu, A. and Hunt H. (2007) Access to Basic Education in Ghana: The Evidence and the Issues. Centre for International Education, University of Sussex [online]. Available at: CREATE.http://files.eric.ed.gov/fulltext/ED508809.pdf [Accessed 5 April 2014].

Aryeetey, E. and Kanbur R. (2007) *Economy of Ghana: Analytical Perspectives on Stability, Growth and Poverty*. Suffolk: James Currey.

Braimah, A. I. (2014) The Political Economy of Cost-Free Education in Ghanaian Public Schools: A Critical Analysis of National Resources. *Journal of Education and Practice* 5(25), pp. 19–27.

Broadie, S. (1991) *Ethics with Aristotle*. New York: Oxford University Press.

Capozza, D. R. and Seguin, P. J. (2003) Special Issue: Real Estate Investment Trusts – Foreword from the Guest Editors. *Real Estate Economics*, 31, pp. 305–311.

Chui, L. and Chui, K. (2005) An Empirical Study of the Relationship between Economic Growth, Real Estate Prices and Real Estate Investments in Hong Kong. *Surveying and Built Environment*, 16(2), pp. 19–32.

CPP (2012) Manifesto, Election 2012. Accra, Ghana: CPP.

Creswell, J. W. (2003) *Research Design: Qualitative and Quantitative Approaches*. Thousand Oaks, CA: Sage.

Dabie, Y. (2013) 4 Years SHS Is Better than 3 Years [online]. Available at: http://www. modernghana.com/news/487861/1/4-years-shs-is-better-than-3-years.html [Accessed 12 July 2014].

Darvas P. and Balwanz, D. (2013) *Basic Education beyond the Millennium Development Goals in Ghana*. Washington DC: World Bank.

Dionne, G. (2000) *Handbook of Insurance*. London and Boston: Kluwer Academic Publishers.

Djangmah, J. (2014) Reversal of 4 Year SHS Disappointing [online]. Available at: http:// ghanapolitics.net/Ghana-Education-News/reversal-of-4-year-shs-disappointing-professor-djangmah.html [Accessed 14 October 2014].

Doling, J., Vandenberg, P. and Tolentino, J. (2013) Housing and Housing Finance – A Review of the Links to Economic Development and Poverty Reduction. ADB Economics Working Paper Series, No. 362.

Foster, P. (1965) *Education and Social Change in Ghana*. London: Routledge and Kegan Paul.

Glossop, C. (2008) Housing and Economic Development: Moving Forward Together [online]. Available at: http://www.centreforcities.org/wp-content/uploads/2014/09/08-11-06-Housing-and-economic-development.pdf [Accessed 8 November 2014].

Gockling, R. S. (2005) *The History of Ghana*. Westport, Connecticut and London: Greenwood.

Government of Ghana (2014) Budget Statement and Economic Policy. Accra: Ministry of Finance and Economic Planning.

HDI (2014) Human Development Index and Its Components [online]. Available at: http://hdr.undp.org/en/content/table-1-human-development-index-and-its-components [Accessed 9 November 2014].

Hongyu, L., Yun, W., and Siqi, P. Z. (2002) The Interaction between Housing Investment and Economic Growth in China. *International Real Estate Review*, 5(1), pp. 40–60.

Johnson, R. B., Onwuegbuzie A. J. and Turner L. A. (2007) Toward a Definition of Mixed Methods Research. *Journal of Mixed Methods Research*, 1, pp. 112–133.

Kanth R. K. (1997) *Against Economics: Rethinking Political Economy*. Aldershot: Ashgate.

Kwame Nkrumah University of Science and Technology (KNUST) (2014) Approved Student Hostels [online]. Available at: http://www.knust.edu.gh/students/housing/hostels [Accessed 28 November 2014].

Leary, M. and McCarthy, J. (2013) *The Routledge Companion to Urban Regeneration*. New York: Routledge.

Little, A. W. (2010) Access to Basic Education in Ghana: Politics, Policies and Progress [online]. Available at: http://files.eric.ed.gov/fulltext/ED512117.pdf [Accessed 28 November 2014].

McWilliam H. O. A. and Kwamena-Poh, M. A. (1975) *The Development of Education in Ghana*. London: Longman.

Ministry of Education (2013) Education Sector Performance Report 2013 [online]. Available at: http://www.moe.gov.gh/docs/ESPR%202013%20Final%20August.pdf [Accessed 28 November 2014].

Monnin, P. and Jokipii, T. (2010) The Impact of Banking Sector Stability on the Real Economy [online]. Available at: http://www.ecb.europa.eu/events/conferences/shared/pdf/cfs/monnin.pdf??1be70eb2557b66d02dce61d086c4c581 [Accessed 28 November 2014].

National Council for Tertiary Education (NCTE) (2014) Tertiary Education in Ghana [online]. Available at: http://www.ncte.edu.gh/ [Accessed 19 November 2014].

NDC (2012) Manifesto, Election 2012. Accra, Ghana: NDC.

NPP (2012) Manifesto, Election 2012. Accra, Ghana: NPP.

Parameswaran, S. (2007) *Equity Shares, Preferred Shares and Stock Market Indices*. Noida: Tata McGraw-Hill Education.

Schreier, M. (2012) *Qualitative Content Analysis in Practice*. New York and Newbury: Sage.

Sen, A. K. (1999) *Development as Freedom*. New York: Anchor Books.

Sisala East District (2014) Performance Review Report 2013. Presented to the General Assembly on 19 March, 2014 at Gwollu, Ghana.

United Nations (2010) Millennium Development Goals. New York: United Nations.

United Nations Development Programme (UNDP) (2006) What Is Poverty? Concepts and Measures [online]. Available at: http://www.ipc-undp.org/pub/IPCPovertyInFocus9.pdf [Accessed 9 September 2014].

Vice Chancellor of Kwame Nkrumah University of Science and Technology (KNUST) (2014) Bringing KNUST to Your Door Step. Speech delivered at the commissioning of the Kwabenya Campus, 19 September, Accra, Ghana.

Woods, M. J. (1992) *Aristotle's Eudemian Ethics*. Oxford: Clarendon Press.

World Bank (2014) Secondary School Improvement Project, Project Appraisal Document. Report No: 86520-GH. Washington DC: World Bank.

Yin, R. K. (2003) *Case Study Research: Design and Method* (3rd ed.). London: Sage.

Legislation

Education Act 1961 (Act 87).

Insurance Act 2006 (Act 724).

Local Government Act 1993 (Act 462).

13 Real estate and social inequality in Latin America

Approaches in Argentina, Brazil, Chile and Colombia

Claudia B. Murray

1. Introduction

Cities produce 50% of gross domestic product that manifests in increasing competition to attract real estate (RE) investment and development (World Bank, 2013). However, parallel growth in inequality suggests that this expansion is not benefiting all citizens (Lima, 2006; Bourguignon et al., 2007). Latin America is reportedly the most unequal region in the world and the question that always remains is why a region with such a wide range of natural resources scores so highly in the Gini Index?[1] Policymakers, academics and experts are increasingly concerned over the widening gap between Latin American countries and developed economies (de Soto, 2001; Sacks, 2006; Fukuyama, 2008; Edwards, 2010). Some highlight social inequality as the main reason for this gap (Fukuyama, 2008; González and Martner; 2012) while others claim that globalization is mostly to blame (Stiglitz, 2003; Chomsky, 2010).

It has been stated that the solution lies in security of RE rights (de Soto, 2001; Fukuyama, 2008). De Soto (2001) argues that Latin America's poor sit on dead capital (their homes) due to insecure RE titles, which hinders collateralization of RE to acquire entrepreneurial capital. His views have been challenged (Varley, 1987; Samuelson, 2001; Gilbert, 2002; Frankema, 2006; Van Gelder, 2009) but followed and praised by regional governments (Fernandes, 2002; Clift, 2003). However, the title's legalization theory is based on simplistic assumptions that miss the nuances existing in Latin America's land markets; see, for example, Van Gelder (2009) for an in depth discussion. In addition, it assumes that the underlying planning system in most of these countries is sophisticated enough to secure urban land equity (using here the term in its dual connotation).

Given the importance of RE titles and home ownership that has been stressed so far in academic discussions, this chapter looks into the question of inequality from the perspective of the residential RE sector. It looks at how the development of housing evolved in Argentina, Brazil, Chile and Colombia from the 1970s to current times. It presents an overview of the relation between government housing policies and development/planning regulations, which are shaping the urban landscape. At the lower end of the residential RE sector, it considers informality and social housing, and at the higher end it looks at new luxury gated developments:

Where are they located in comparison to informal developments? Are low and high end markets competing to purchase the same land? How does that competition fare? If so, what redistributive mechanisms do local governments have in order to spread prosperity? The objectives are to: (a) bring to light the counterproductive housing policies that have been pursued over the years to help the poor in order to improve their standards of living and contribute to economic development; (b) reveal the lack of governance and accountability that municipalities have; (c) signal the scarcity of planning regulations to control speculative housing development; and (d) examine the difficulties arising when RE profits are only benefiting a very small percentage of the population.

The information presented in the chapter is collected from official documents of government and international organizations, peer-reviewed academic journals and the latest media articles from the national press of each country. The collected literature is in Spanish, Portuguese and English.

The rest of the chapter is divided into six main sections. Section two provides a brief summary of the region's socio-political background whilst section three explains how informality spread and social housing debt mounted through decades of neglect in the selected countries. The fourth section looks at the process of market liberalization. In the fifth section, building the way out of recession is considered whilst section six discusses the main findings and their implications for economic development. The chapter is concluded in section seven.

2. Socio-political background of the region

Historically, Argentina, Brazil and Chile share the legacy of military governments that scarred the political scene and influenced the welfare systems of their countries. Colombia, on the other hand, never suffered military rule to the scale of the others but suffered, and indeed still suffers, from more than 50 years of internal civil conflicts against illicit drug trafficking; there was a coup d'état in 1953 but democracy was restored in 1958 (Arismenedi, 1983; and for a revision of authoritarian tendencies and their roots in Latin America, see Collier, 1978). Despite the differences in their own struggles and conflicts, all these countries share the transition from tariff-protected manufacturing industries known as import-substitution industrialization (ISI) (Macario, 1964) from the 1940s to the 1980s, followed by the adoption of very important economic recommendations known as the Washington Consensus during the 1990s (World Bank, 1993, 1994; Gilbert, 2002; Stiglitz, 2003).

Broadly speaking, this implied the reduction of State control and a move towards deregulation, privatization and opening of markets. The transition from a heavily protected economy that favoured domestic products towards open competition had a direct impact on RE markets – commercial and residential RE investment was ripe to produce high returns given that the market had been heavily protected and stagnant for decades (Thibert and Osorio, 2014). In addition, privatization of infrastructure services also attracted foreign direct investment, which in the case of motorways shaped the new spatial organization of the cities by encouraging residential ghettoization, a process that is known as fragmentation of the urban space (Bäbr and Borsdorf, 2005).

It must be pointed out that Chile had an earlier start towards liberalization of markets as some of these recommendations were already underway in the country after the military coup of 1973 (Arellano, 1982; Rojas, 2000; Gilbert, 2002; MINVU, 2004; de Soto and Torche, 2004; Garcia de Freitas and Cunha, 2013). Among wider economic aims, the reform intended to attract more private investment and enable the housing market to work more efficiently with the hope that the housing deficit in all sectors of the population could be solved by private investment with the Chilean State acting as a facilitator. But as the Ministerio de Vivienda y Urbanismo (MINVU, 2004) explains the reform took some time to develop and get established. Furthermore, some even claim that its visible results only arrived with the return of democracy in 1990 (Rojas, 2000). This means it took Chile nearly 20 years to adapt to the economic reforms introduced in the 1970s. Argentina, Brazil and Colombia attempted to take that leap in a few years (late 1980s and early 1990s) with mixed reviews on the consequences of these reforms (McKenzie and Mookherjee, 2003; Castañeda, 2006; Stiglitz, 2007). Lack of transparency in the privatization process and lack of a mature institutional framework capable of dealing with the sophistication that international private investment demands have been blamed for the economic crises that affected the region after market liberalization, starting with Mexico's Tequila crisis in 1994, through Brazil's hyperinflation crisis in 1999 and ending with Argentina's default in 2001 (Stiglitz, 2007).

On the other hand, supporters of liberalization claim that in some countries' privatization of utilities such as electricity and telecommunications has led to an increase in the accessibility to these facilities for the poor (McKenzie and Mookherjee, 2003). Notwithstanding, the authors acknowledge the scarcity of data availability as it is very difficult to measure accessibility of public services when most people live in informality and have therefore no legal access to formal connections. Furthermore, they also acknowledge that in most cases they could not measure the quality of the services the population receives.

The last point is particularly important as Latin American governments are demonstrating a concern over the quality of what has been achieved so far in urban development (Murray et al., forthcoming). Qualitative attributes to assess sustainable prosperity are increasingly taking importance over quantitative attributes (Parris and Kates, 2003). As these authors argue, local understanding of sustainability and political, economic and cultural factors affects the multiplicity of indicators that have been developed so far. As a result, and two decades after the experience of market liberalization, Latin America is pursuing its own methods to develop social policies in accordance with regional theories of sustainable growth; these look at reaching economic targets while simultaneously ensuring: (a) equal share of prosperity among all sectors of society; (b) a reduction in carbon emissions; and (c) avoidance of the depletion of natural resources (Economic Commission for Latin America and the Caribbean [ECLAC], 2010, 2012, 2014). Notwithstanding this, some countries' policies towards equitability are seen as too populist (Edwards, 2010) and said to be inflicting further damage to Latin

American economies. This question of populist government and urban equality is central to the selection of countries under study here.

The selection of the countries is based on the political classification presented by political scientist Jorge Castañeda (2006). As he explains, there are two lefts in Latin America: the "reconstructed left", a fusion of communist, socialist and Castroist tendencies, which has evolved by learning from past mistakes and intends to develop new policies that can keep the region as an important player in the global market; and the "populist left", which has a deep attachment to nationalism and tends to be anti-globalization (Castañeda, 2006, pp. 34–35). There are also the centre and centre right groupings that dominated the scene during the 1990s. This political division clashes in the way societies and governments intend to compensate for the years of social neglect that the region's poor have suffered. The clash seems to have little hope for compromise solutions and manifests in recent street protests across all countries: student revolt in Chile, anti-World Cup demonstrations in Brazil, workers' strikes in Argentina and anti-national government protests in Colombia to support a leftist mayor in Bogota. The common denominator in all the countries is the accumulated social debt and the conflicting approaches to trying to solve inequality.

In accordance with Castañeda's political spectrum, the selected countries in this chapter each demonstrate different tendencies, but more relevant are their housing policies and development regulations. These are intended to bring urban equity, which in turn is supposed to contribute to economic development. Argentina, representing the populist left, has seen the coalition of mostly socialist parties (Frente para la Victoria) win successive elections since 2003; the same year that Brazil, representing a "tenuously" reconstructed left (Castañeda, 2006, p. 35) saw the rise of the Partido dos Trabalhadores (workers' party) who are also still in power. Chile, a straight reconstructed left, has followed this tendency for almost two decades, and after a short period with the centre-right (2010–2014), it has also recently returned to where it previously was with a newly formed coalition of leftists parties known as Nueva Mayoría. The exception here is Colombia, which has been mostly moving from the right to the centre-right, with the Unidad Nacional coalition currently in power. Therefore, the selected countries represent a spectrum of policies to tackle inequality from different political angles.

3. The spread of informality and accruing social housing debt

During the 1960s and early 70s, housing policies in Argentina, Brazil, Chile and Colombia were directed to the supply of housing to those in formal employment. The literature on housing policies for these countries is quite vast. The most relevant for this period are: for Argentina, Zanetta (2005), Barreto and Alcala (2008) and Barreto (2012); for Brazil, Valladares (1978), Andrade and Azevedo (1982), Melo (1992) and Bonduki (2008); for Chile, Arellano (1982), Collins and Lear (1995), Kusnetzoff (1987), Rojas (2000) and Silva Lerda (1997); and for Colombia, Ceballos Ramos and Saldarriaga Roa (2008b) and Cuervo and Jaramillo (2009). All these authors agree that the policies failed to target the unemployed and those working in informal or casual jobs, which negatively affected economic development. This, at the time, was a considerable failure given that the informal sector accounted for more than 60% of the population (Silva Lerda, 1997). As a consequence, informality increased, and, with it, governments' policies of slum clearance and relocation. It is worth pointing out that there was also a belief, particularly among military governments, that the situation of the unemployed would reach a breaking point, forcing them to be eventually absorbed into the formal economy. There was also the theory that Marxist activists as well as drug cartels were infiltrating informal settlements, using them as a focus of resistance, indoctrinating residents and recruiting the young into their networks. This is the main reason why governments tended to ignore social policies and continued with slum clearance programmes.

In Chile, the military imposed strict controls over the expansion of informal developments, which generated another problem: the *allegados* – the family's relatives and friends who arrived to share already developed houses, adding over-crowding to already infrastructure-thirsty urban environments (Rojas, 2000). In Argentina, new laws were passed with the clear intention of discouraging informal settlements on outside boundaries. The Ley 8912 of 1977 classifies the periphery of Buenos Aires for recreational use and non-permanent residence and specifically refers to country clubs (second holiday homes) as the preferred typology for the area (Article 7) (Gobierno de Buenos Aires, Legislación, Decreto-Ley 8912/77; Thuillier, 2005). Encouraged by favourable legislation and the gradual open-ing of motorways that linked the periphery to the centre, residential RE invest-ment moved to the suburbs where country clubs for the wealthy were developed. Paradoxically, the very few social housing schemes for those in formal employ-ment and low wages were also being developed in the periphery. Affordable land was attracting all market sectors, making social housing compete for land with speculative projects. As it will be demonstrated later, this process only became more aggravated with the arrival of market liberalization policies in the 1990s. Meanwhile the exodus to the periphery began in all countries under study and the vacated city centres were gradually taken up by squatters or lay empty and neglected (Bähr and Borsdorf, 2005).

The state was at this time the main provider of social housing, responsible for the financing, planning and delivery of the projects. The institutions involved were usually a national mortgage bank and a relevant government institution at the national level that coordinated all actors (local governments, construction companies and beneficiaries). The schemes delivered by these institutions were generally low cost mass-produced housing units with very little infrastructure. The largest example of housing units built during the 1970s in Latin America is located in Sao Paulo and has 84 five-storey housing blocks with no public trans-port, a problem that has not yet been adequately solved: it takes today's residents a whole two hours of commuting to the city centre (Budds et al., 2005).

The landscape of the 1970s and 1980s was therefore of increasing informal-ity and insufficient and low quality social housing stock. In addition, given the high inflation, most State-owned mortgage banks were running large deficits and there were allegations of corruption in the system (Valença, 1992; Zanetta,

2005; Tachópulos Sierra, 2008; for a more positive view on the state's abilities to finance social housing programmes, see, for example, Cuervo and Jaramillo, 2009). Spatially, the tendency of country clubs in the peripheries was not only hollowing out the city centres, but there was also the fragmentation of the society by the polarization of the urban fabric. In these conditions, countries followed the Washington Consensus recommendations with the view that the private sector could help to bridge the social divide.

4. Transition and liberalization of markets

4.1 Social housing policies until the end of the 1990s

As stated in the introduction, Chile's reform started in the mid 1970s. The aim of the reform was to attract more private investment and enable the housing market to work more efficiently with the hope that urban land could yield better investment returns even for social housing schemes (Rojas, 2000). Accordingly, the MINVU was to cease all developing activities and act now as a facilitator on the demand side by subsidizing mortgages to the poor. The World Bank with whom Chile had challenging relations praised the market conditions of Chile in the early 1990s, stating that the country had been able to provide three key elements in its housing programme: (a) specific targeting of the poor; (b) transparency; and (c) private market provision (World Bank, 1993). For an account of the struggles between Chile, the World Bank and other international organizations that were at the time unconvinced of the benefits of subsidies, see Gilbert (2002). The Chilean model was considered as best practice (Gilbert, 1997) and other Latin American countries adopted the subsidized system.

Socio-economic conditions, more than politics, drove the subsidized system in all countries under study, therefore bringing divergences in terms of the percentage of housing units that were subsidized and the availability of credit, which depended on the value of the house. There were also differences on eligibility of applicants according to their minimum salaries (Gilbert, 1997; Chiape de Villa, 1999; Budds et al., 2005; Zanetta, 2005). However, regardless of the particular socio-economic conditions, the system generally consisted of an A-B-C programme: Ahorro, Bono and Crédito (savings, subsidy and credit), whereby families with savings (ahorro) were eligible for a government subsidy (bono) and were better placed to obtain a bank loan (crédito) in order to purchase a home. As the following section shows, this structure was the results of the dissemination of new ideas on the way informal settlers developed their dwellings.

4.2 The influence of John Turner

Around the 1960s and 1970s, the architect John Turner spent time in Peru and other Latin American countries studying informal developments. Turner found that households improved their houses incrementally (Turner, 1977). He argued that their self-help approach was better than the large-scale schemes developed so far by the governments and that a system of progressive building based on families' savings was better suited to solve the housing deficit, particularly for the very poor, thereby improving their standards of living and contributing positively to economic development. His ideas were popularized around the mid 1990s, as well as those of others with similar views on self-build solutions (Pugh, 1994; and for a criticism to Turner see, for instance, Werlin, 1999). Site-and-services schemes that originated after these theories intended to provide serviced land to poor families or group of families that had a plan or community plan and were able to build with technical support from the municipality. Help was provided on condition that benefited families would send their children to school and thus break the poverty cycle.

All the countries under study implemented progressive building or upgrading schemes of some sort as part of their subsidized program. Nearly 20 years later, current reviews of those past programs are mixed. Some claim that municipalities did not welcome upgrading programs as they doubted they could collect service charges from very poor families, and therefore the investment in infrastructure was unlikely to be recovered. This had the effect that municipalities and some-times the government themselves either restricted the land available for these projects or rejected them altogether (Gilbert, 1997; Budds et al., 2005). Others argue that the progressive building strategy lent itself to abuse as it was just an excuse to deliver low quality unfinished units (Chiape de la Villa, 1999). Positive views of upgrading informality and site-and-service plans shows that the scheme has been more successful at targeting the very poor than the massive new built complexes, as in the former the beneficiaries own their houses outright with no mortgage while the latter represents a burden (Gilbert, 1997; Rojas, 2000; Green and Ortúzar, 2002; Boldarini, 2008). Nevertheless, at the end of the 1990s, and for reasons that still need to be researched, the four countries under study have tended to favour the delivery of new homes whilst offering progressively fewer subsidies towards other alternatives.

4.3 Implementation of the subsidized system

Whatever the program, be it progressive building, site-and-services or upgrad-ing, the mechanism whereby the state helps to subsidize the demand for housing units prevailed. This was a major change in countries that had traditionally relied on national governments for the supply of social housing. Understandably, new procedural issues emerged as changes became operational. One of the major chal-lenges was how to create a more attractive market for the private sector. Two main barriers had to be addressed: the reduction in planning bureaucracy and the legal-ization of land titles (Borja and Castells, 1997; Cuenya, 2005; Barreto, 2012). The way to reduce bureaucracy was to decentralize activities, which in Latin America at the time was complicated as provincial governments were dependent on national states with highly centralized administration systems. On the other hand, capital cities had always been under the suspicion that they were too independent from their governments and this is still a current problem; the cases of Bogota and

Buenos Aires are paradigmatic as they have a history of being ruled by the opposition party (Rodriguez-Acosta and Rosenbaum, 2005; de Duren, 2006; Thibert and Osorio, 2014). A fine balance needed to be achieved between relinquishing central powers to the provinces and strengthening their capacity to take on new functions while keeping metropolitan areas under control.

Not surprisingly, the process has received severe criticism in all the four countries. Some argue that the decentralization was heavily politicized and that power was given to provinces in exchange of political support to pursue wider reforms (Coy and Pöhler, 2002; Zanetta, 2005). All countries were simultaneously undergoing an extensive program of privatization of their national infrastructure networks and services and support from local government was needed to implement these changes. Other critics to the decentralization process focus on the lack of coordinated action at the government level and particularly the unregulated public–private interaction, which fragmented the industry sector. Small construction companies, for example, preferred to lobby for work with local governments while large firms preferred communication with higher levels of government (Cuenya, 2005). Unavoidably, this often led to a duplication of actions (Budds, et al., 2005; Rodriguez-Acosta and Rosenbaum, 2005). Notwithstanding this, when subsidies were kept centralized in order to control activities, critics claimed that the powers of the municipalities (who were in a better place to assess local needs) were being curtailed (Chiape de Villa, 1999; Tachópulos Sierra, 2008).

Not only was the process of decentralization facing discontentment, but the effect of legalization of land titles also started to be questioned. This was mainly because even if a family managed to succeed in clearing the bureaucracy of obtaining the title, their inability to fulfill land tax duties meant that they could be dispossessed by the government for non-payment, which was seen as a worse situation than the previous one when they had at least a "perceived" ownership of the land; see, for example, Van Gelder (2009) for the latest discussion on titling theories. And there were also those who stated that even with legal titles, poor families, after undergoing the lengthily process of obtaining the subsidies, were refused credit by the banks (Varley, 1987; Gilbert, 1994; Van Gelder, 2009; Curevo and Jaramillo, 2009). Finally, the subsidies themselves were deemed unsustainable. In Brazil, the 1990s was a decade dominated by high inflation, which did not offer stable conditions for the private sector nor did it allow for government subsidies (Fonseca, 1998; Valença, 1998; Morais, Saad Filho and Coelho, 1999; Valença, 2001; Azevedo, 2007; Valença and Bonantes, 2009). Argentina saw most of the social housing projects delivered by NGOs as the government was unable to create sustainably priced mortgages (Murillo, 2001). In Colombia, the lengthy process of mortgage applications (usually over 10 months) combined with the housing boom created by these very same programs meant that by the time candidates managed to gather all necessary funds (savings, subsidy and credit), the dwelling or plot of land of their choice was out of reach (Gilbert, 1997; Chiape de Villa, 1999).

However, the biggest failure for all the countries was that despite the promises of targeting the very poor, those with no access to formal jobs were unable to benefit from any of the plans. Even in Chile, targeting the very poor proved difficult and the system was heavily subsidized by the government (Gilbert, 1997; Rojas, 2000). Furthermore, as Rojas suggests, much of what was achieved in terms of attracting the private sector was possible due to the country's economic growth (rising employment rate, consumer confidence and rise in interest rates, all beneficial conditions for the private sector). The fragility of the system became evident at the end of the 1990s when unstable economic conditions generated an increase in mortgage interest rates that saw the collapse of many programs and institutions across all the countries but Chile. Unavoidably, when transferring the system to other countries with larger housing deficits and thousands living informally hoping for subsidies, the so called Chilean model faced bigger challenges. It is interesting to point out that towards the turn of the century, Argentina, Brazil and Colombia had to divert subsidy funds to support the administration of an increasing demand (Gilbert, 1997; Chiape de Villa, 1999; Tachópulos Sierra, 2008; Cuervo and Jaramillo, 2009). Still, a decade later, some argue that the financial conditions at the time were not mature enough to attract institutional investors capable of providing private funding for housing programs, let alone provide a solution for the lowest sector of the population, as it was claimed by the World Bank (Soto and Torche, 2004; Zanetta, 2005). As it will be demonstrated in the following section, the failure to complement government subsidies with a share of the more profitable high-end residential market also played a part in the collapse of the system.

4.4 Gated communities during the 1990s

Simultaneously with the opening-up of the markets for private capital to support social housing, the process facilitated private investment in other areas of RE. As will be explained later, the decentralization of planning and lack of coordination across different government levels described above provided a fertile ground for high-end residential investment, which was reported as the most profitable RE business in Latin America during the 1990s (de Duren, 2006).

Gated communities have been defined as residential areas where public spaces have been privatized; they can either be in the city or the suburbs and in affluent or poor neighbourhoods (Blakely and Snyder, 1997). The mode of transport in the gated communities is the car, following the American example of peripheral cities, which was encouraged through the New Urbanism charter. In all the countries under study, urban sprawl accelerated when motorways were expanded during the 1990s, fuelled by private investment. As motorways are mainly private enterprises, most operate on toll systems, which, for those who can afford it, considerably reduces the commute to the city (Thuillier, 2005; Bäbr and Borsdorf, 2005; de Duren, 2006; Borsdorf et al., 2007). Inevitably, shorter commutes increased the demand for dream houses in suburbia.

During the 1990s, as it is today, the families relocating to gated communities are mostly of the same socio-economic background, as the level of service charge for maintenance of communal grounds and facilities (which can be as diverse as swimming pools, golf courses and helipads) acts as a barrier to social mix.

The newest trend is "affordable" condominiums for the middle classes, which have fewer amenities but still enjoy private security, one of the main concerns for many Latin Americans (Coy and Pöhler, 2002; de Duren, 2006). Infrastructure around these complexes usually includes health centres, education facilities and cemeteries, all privately paid by locals. New businesses either follow or precede the opening of a gated community, including large shopping malls and supermarkets (Coy and Pöhler, 2002; Borsdorf et al., 2007). During the 1990s, and given that most municipalities lacked a long term plan, most of these infrastructure projects were executed by private investors – who are mostly responsible for the layout of the periphery of most cities in Latin America.

Briefly and broadly speaking, all the countries share the same pattern: the new community is initiated by a developer or a business consortium which assumes responsibility for the entire project, from gathering investors to buying the land, architectural design, planning application, construction and sales of units to final consumers. After the families move in, services and maintenance of the community's site can be provided by the developing firm, outsourced or undertaken by the residents themselves organized in a cooperative (Coy and Pöhler, 2002; Borsdorf et al., 2007). The urban periphery is the preferred place for most gated communities, sometimes encroaching agricultural lands. This is due to land availability and also given the high return that can be achieved compared to more consolidated areas of the city. Turning rural land into urban, for example, reportedly generates profits of up 550% of the original value (Chiape de Villa, 1999).

Some argue that these types of developments bring benefits to areas that are characterized by informality (Sabatini et al., 2001; Salcedo and Torres, 2004). The argument is that the new spaces of the wealthy are providing workplaces for the poor, for example in construction, security and domestic service. Figures published so far show that this is partly true (Thuillier, 2005). However, the benefits that municipalities receive in terms of developers' contributions and taxes from new residents have been questioned by some (Brain and Sabatini, 2006; Thibert and Osorio, 2014). These authors maintain that new developments tend to be located in poor municipalities with a large informal population and little infrastructure. This means small local authorities already struggling to collect revenues. Therefore these offices see the arrival of a new gated community as a way to generate jobs and tend to compete with other municipalities to attract these developments by offering tax incentives to developers. In Chile, for example, the increase in land prices and the failure of the system to impose stricter regulations on developers means that the State is the main provider of the sites for social housing (Brain and Sabatini, 2006). As these authors suggest, increased construction costs are also part of the problem, and given that the subsidies have a cap, the only solution to keep developing social housing is by building on fiscal land, making the authors speculate the possibility that the State is undervaluing sites in order to complete politically charged projects to win more votes. Considering the argument explained in the previous section, stating that progressive building is lowering construction costs, it seems plausible that Chilean construction companies are indeed benefiting from the subsidized system.

The case of the small municipalities in metropolitan Buenos Aires is also para-digmatic. Here the planning process is lengthy and opaque, as planning delay usually pushes developers to break ground as soon as the land is purchased. This is done with the hope that planning permission will be "solved" at some stage by the local officer (de Duren, 2006; Thuillier, 2005). To add to the problem, once the new wealthy residents move in, their tendency is to contest municipal tax: they consider that all their services are privately paid and so they see no reason to pay municipal taxes.

Faced with similar problems during the late 1990s, local authorities in Colombia were requested to develop a strategy plan (Ley de desarrollo Territorial 388 of 1997, still in force; Alcaldía de Bogota Ley 388, 1997). New municipal funds were also created (Fondos Municipales de Vivienda de Interés Social) with the aim of collecting and managing revenues for social housing programs. While the strategy plan allows municipalities to identify land for social housing, the municipal funds equip them with regulations to pursue development contri-butions from the private sector. This is done either by the demand of land for social housing from the land owner or by a development levy paid by develop-ers (Chiape de Villa, 1999; Tachópulos Sierra, 2008). However, there are con-cerns that the regulation is not fully enforced (Carrión Barrero, 2008) and that municipalities only produce a plan just to comply with the national law; and the reality in Colombia is very much like that of Chile and Argentina – developers rule the land.

The 1990s ended with successive economic crises that affected the larg-est economies in the region. In Colombia, mortgage rates increased from 6% to 20% in the year 2000 (Cuervo and Jarmillo, 2009). The national govern-ment had to introduce a very controversial new tax to help the banks cope with defaults. The Instituto Nacional de Vivienda de Interés Social y Reforma Urbana (National Institute of Social Housing and Urban Reform) (INURBE) went into liquidation by a decree and a new institution was created, the Fondo Nacional de Vivienda –Fonvivienda. This national housing fund was to act as facilitator and channel the resources of a new plan, initiated in 2002, which looked again (similarly to previous plans) at decentralizing activities as a solu-tion (Escobar, 2001; Tachópulos Sierra, 2008). Argentina and Brazil's social housing plans and subsidizing programs suffered similar problems. However, as it will be shown in the following section, the severity of their financial cri-ses in 1999 and 2001 saw drastic political swings that dramatically changed their welfare systems.

5. Building the way out of recession

During the critical years at the turn of the millennium, the construction of social housing reduced considerably. Notwithstanding this, as policies in recession years are mainly aimed at keeping unemployment low, most countries continued to sup-port the construction of housing for the more reliable luxury end of the market. Colombia, for example, established the Ahorro de Fomento de la Construcción

(AFC) system (literally, "savings to support the construction industry"). This is basically a mechanism by which the buyer of a new unit of RE has a tax exemption which in some cases is equivalent to three and four times the subsidy of a social housing unit (Cuervo and Jaramillo, 2009). However, just as the wealthy received tax exemptions in this country, the poor received free houses as the economic conditions improved around 2007, thereby improving their standards of living. The government launched the 100,000 Viviendas Gratis plan in 2012 (Gobierno de Colombia, Haciendo Casas Cambiamos Vidas). The beneficiaries were families that suffered from environmental disasters or during the drug trafficking wars. This new scheme was a last attempt to target the elusive bottom of the pyramid, that is, the very poor families, and followed World Bank recommendations. It is a very different approach to the World Bank's attitude during the 1970s, when free housing provision was seen as dangerous as it could induce larger rural–urban migration, straining the cities even more (Pugh, 1994).

In Brazil, the workers' party introduced subsidy programmes on a large scale through the programme Minha Casa Minha Vida (PMCMV). There are claims that this was a way the government faced the 2008 financial crisis and that it was mainly driven by the construction sector (Macedo, 2010; Cardoso et al., 2010). There are also problems with quality (Kowaltowslki et al., 2006). Some critics have stated that if customers had a choice, that is, were able to buy a non-subsidized home, the whole scheme and the construction companies would be out of business (Formoso et al., 2011).

Mass social housing programs in these four countries run parallel with the rise of the *megaproyecto* for the upper classes. These are larger than previously developed communities as they are intended for more than 50,000 people (Coy and Pöhler, 2002, de Duren, 2006). Furthermore, given the increasing fear of crime, all sectors of the population these days prefer the sense of security offered by gated developments, and so the gated *megaproyectos* now serve the wealthy as well as the poor (Bäbr and Borsdorf, 2005; de Duren, 2006; Borsdorf et al., 2007). The fragmented city is thus spreading across Latin America where gated ghettoes of the upper class are "mixing" with gated social housing projects in the periphery. The monotony of the urban landscape exacerbates when the same consortium of investors and developers (bearing in mind they are also architects and designers) operate across many Latin American countries (Coy and Pöhler, 2002, p. 358). Still, most problematic is the land struggle that this generates: where a new gated community for the wealthy appears, it immediately takes up good land that becomes unaffordable for social housing (Brain and Sabatini, 2006). Land squatting becomes then the only solution for those who seek employment in the new gated community. Informality erupts and spreads around enclaves of wealth, usually taking poor quality land that has no interest for RE investors. The pattern of a sea of informality with patches of wealth and housing estates, described as a fragmented city, is then completed.

Despite this, some changes are emerging, such as the territorial law in Colombia explained in the previous section. Still, most authors agree that Brazil is leading the way in development contributions. Special Social Interest Zones

(ZEIS) are created by a new regulation that requires cities to contribute a certain percentage of social housing for all developments (Prefeitura de São Paulo Decree 44.667, 2004). Currently this regulation has only been applied in São Paulo, but with promising results given that social housing schemes resulting from these programmes intend to target different sectors of the population, encouraging a more social mix (Budds et al., 2005). The other system used in Brazil is the sale of vouchers for development rights in exchange for social housing. In this system, if a developer wants to build above the ratio that the size and location of the plot allows, he or she can negotiate this and obtain vouchers that can be either auctioned in the RE market or used in other sites. In addition, properties owned by those who owe land tax that are unoccupied and are considered suitable for social housing debt can exchange the property for land tax credit. The legislation means that there are no monetary transactions, which increases transparency for the parties, the debtors and the creditors – and the local authority in question acquires a new unit of RE for social housing.

In contrast, Argentina and Chile still fail to target the private sector, leaving the responsibility to deliver serviced land to municipalities. As a consequence, and given the extremely difficult economic circumstances that Argentina currently faces, most municipalities have reduced their social housing targets (Barreto, 2012). Furthermore, some municipalities are so short of resources that they have started to divert most of the subsidies to the payment of municipal employees. A similar situation of diversion of subsidies to cover for other shortcomings was seen in Colombia during the time of the Instituto de Crédito Territorial (ICT), the State-run social housing provider that operated until the 1990s reform (Tachópulos Sierra, 2008) and during the closing days of the INURBE. The deviation of funding from housing to pay for an administrative system that is unable to cope with a large population in need seems to be a recurrent fault throughout the decades in these countries.

To tackle these irregularities, Argentina is currently attempting to reverse the decentralization process and intends to control resources from the State (Scheinshon and Cabrera, 2009). This has generated a new wave of clashes between national government and the opposition, particularly, with the Ciudad Autónoma de Buenos Aires. As these authors suggest, the only hope for social housing projects are NGOs and social activist movements who are able to lobby between the two political factions. By profiting from the lack of government cooperation, organizations like Madres the Plaza de Mayo and Piqueteros are finding a place as social housing developers.

6. Discussion

As demonstrated in the preceding sections, all four countries studied have followed a path of developing gated communities for the rich, starting with country clubs in the 1970s and continuing with the megaproyectos of today. Equally, they all share a modality of developing social housing neighbourhoods with very little infrastructure and services. To complete the landscape, a myriad of precariously constructed homes that have steadily spread since the 1960s occupy unwanted plots or lands that are prone to natural disasters. It is true that Chile is solving the housing deficit but the price paid is an increase in inequality, with ghettoes for the poor as well as for the wealthy. As a result of the spatial polarity, all countries fail to achieve social mix. This is troublesome in many ways. The lack of contact with other groups is harboring fear of the other and fueling the trend in security, walls and gates in residential developments at all scales of society. Equally, this lack of engagement with other sectors of the population manifests itself in public places and services. As has been established, the wealthy tend to pay for private education, health and even the maintenance of common public areas, including the streets. This increases the perception that State provision of public services are less effective and only fit for use by those who have no other resources. The decentralization of the state explained here contributes to this idea of a failing state, tarnishing in the process the image of the municipal authorities who have little power to impose planning laws against speculative development.

The most worrying aspect of this fragmentation of society is the lack of empathy that it generates. As it has been established, the unwillingness to contribute with municipal taxes by a sector of the society that sees no benefit in State-run services challenges the provision of social services by the municipality and local government. Repeatedly, after the economic crisis at the turn of the century and after the 2008 crisis, municipalities have tended to divert housing subsidies towards other ends, such as Argentina, which used subsidies to pay employees, or Colombia, where, due to the large number of applicants for social housing programs, the country had to divert funds to the management of these very programs. There are no reports that Brazilian municipalities are diverting housing subsidies to support the maintenance of social services, but the country's growth rate is diminishing and this can have an effect on the deliverance and maintenance of the ambitious PMCMV. As seen in Chile, there is a "hidden" subsidy provided by the State as the main supplier of development land for social housing schemes. This highlights the fact that the very much praised Chilean subsidized system is only superficially working, given that the State has to contribute the land. The success in reducing the housing deficit in Chile is due mainly to the small number of people in need of a home that the country has when compared to the other countries under study.

There is a systematic failure to complement government subsidies with a share of the more profitable high-end residential RE market in all countries. Developers appear as a very powerful group and, as demonstrated, municipal authorities tend to compete with other municipalities to attract large luxury developments by offering tax incentives to developers. In doing so, municipal authorities not only fail to collect much needed revenues but they induce a hike in land prices in some cases in the periphery by over 500%. Given that the periphery is also the preferred location for social housing, the boost in land value by luxury developments immediately curtails the possibilities of housing the poor. The above mentioned Ley 388 in Colombia and the ZEIS in Brazil are a step in the right direction, but much more is needed in order to spread the economic benefits of urban development that the upper classes are currently enjoying so that the poor can also share.

Another underlying principle ruling housing policies in all these countries is the reliance on the construction industry as an important economic driver (Gilbert, 1997, Murillo, 2001; Cuervo and Jaramillo, 2009; Scheinshon and Cabrera, 2009). Whether building luxury homes or social housing, construction is the industry that serves the government to control unemployment particularly in times of recession. It is not a mere coincidence that the rise of the megaproyecto explained above appeared at the turn of the century, just as the region faced the largest economic crisis in history. It seems that the bigger the recession, the bigger the construction project. In a region with a large unskilled labor force, turning to the construction industry to help the economy seems a logical approach. Still, the economic argument of spending in infrastructure and construction to help improve the economies needs to be underpinned by good governance and a robust planning system whereby redistributive mechanisms are secured to share prosperity with all sectors of the population. As it stands, the current system of underfunded municipalities operating with obscure planning regulations that are easily changed to suit speculative developers will only benefit a privileged few and not the poor who constitute the majority, thereby negating any efforts aimed at achieving economic development. It will not, in the long term, support the construction industry as a sustainable economic driver. The jury is out, for example, on whether the frenetic construction activity and projects delivered in Brazil before the World Cup 2014 will deliver a long term legacy. So far, reports from the media are that it will not.

Chile and Argentina are the ones that are falling behind in urban equity while Colombia and Brazil are making better efforts, although the latter with more moderate clear success. Thus, generally, RE is doing little to achieve or improve economic development in the countries studied. The main reasons are because it is failing to target social inequality and curtailing opportunities at the bottom of the pyramid that have no access to the benefits of urban life. A weak State means that unrestricted private development is ruling the land. And, as established by the experiences of the 1990s, private enterprises in a context of fragile regulations and rule of law have not yielded optimal results in Latin America.

7. Conclusion

This chapter has examined the question of inequality in the context of the residential RE sector in Latin America. The countries under study represent all the different political tendencies, from populist Argentina to centre-right Colombia. It can be argued that the provision of free housing would align with a populist government, but in this case it is centre-right Colombia that is providing something for nothing through the 100,000 Viviendas Gratis program. Still, as recent political campaigns show, all these governments use housing programs as a way to win more voters. Using housing development as a political pawn puts pressure on governments to demonstrate quantifiable results to justify their promises with little care for quality and standards. New alternatives to home ownership must be supported at the macro level to address housing deficits. For example, encouraging leasing and rental practices for all sectors of the population and not just for the poor may avoid stigmatizing the rental option. This should be implemented and framed by a fair and robust legislation of landlord/tenant relations. The rental option should be provided and managed by NGOs, local cooperatives or worker organizations who are free from corruption scandals to encourage a trickle-down transparency, avoiding rent default and vandalism, which are usual problems in current government-led rental schemes.

As shown in this chapter, Brazil has found some innovative solutions for land value capture and to some extent so has Colombia. The Chilean model has served to reduce social housing deficit in that country but, as explained above, investment-led planning seems to be stimulating spatial fragmentation and only benefiting a small section of the population. Paradoxically, with the most populist government of the four countries, Argentina is the one falling behind in redistributive mechanism of land taxation. Latin America is a region that has historically directed development towards the middle and upper classes. If the current conditions do not improve, uncontrolled RE development will only aggravate this tendency. If measures to spread the economic benefits of urban development are not evenly distributed among all sectors of the population, the region will continue to be known as the most unequal in the world.

Note

1 The "Gini index measures the extent to which the distribution of income or consumption expenditure among individuals or households within an economy deviates from a perfectly equal distribution [. . .] [A]Gini index of 0 represents perfect equality, while an index of 100 implies perfect inequality" (World Bank Data Indicators, available at http://data.worldbank.org/indicator/SI.POV.GINI).

References

Andrade, S. and Azevedo, L. A. (1982) Habitação e Poder: Da Fundação da Casa Popular ao Banco Nacional da Habitação. Rio de Janeiro: Zahar.

Arellano, J. (1982) Políticas de Vivienda Popular: Lecciones de la Experiencia Chilena. *Colección Estudios Cieplan*, 9, pp. 41–73.

Arismendi, I. (1983) *Gobernantes Colombianos*. Bogotá, Colombia: Interprint Editorial.

Azevedo, S. (2007). Desafios da habitação popular no Brasil. Políticas recentes e tendências, in: Cardoso, A. L. (Ed.), *Habitação social nas metrópoles brasileiras. Uma avaliação das políticas habitacionais em Belém, Belo Horizonte, Porto Alegre, Recife, Rio de Janeiro e São Paulo no final do século XX* (pp. 13–41). Porto Alegre: ANTAC.

Bäbr, J. and Borsdorf, A. (2005) La Ciudad Latinoamericana, La Construcción de un Modelo. Vigencias y Perspectivas. *Ur[b]es*, 2 (2), pp. 207–221.

Barreto, M. A. (2012) Cambios y Continuidades en La Política de Vivienda Argentina (2003–2007). *Cuadernos de Vivienda y Urbanismo*, 5(9), pp. 12–30.

Barreto, M. and Alcalá, L. (2008) Cambios en las prestaciones urbano-ambientales de la política habitacional Argentina: reflexiones a partir de cuatro programas orientados a población en situación de pobreza del Gran Resistencia, in: Pipa, D., Peyloubet P. and de Salvo, L. (Eds), Ciencia y tecnología para el hábitat popular: Desarrollo tecnológico alternativo para la producción social del hábitat, Buenos Aires: Nobuko (pp. 309–328).

Blakeley, E. and Snyder, M. (1997) *Fortress America: Gated Communities in the United States*. Washington DC: Brookings Institution Press.

Boldarini Arquitetura e Urbanismo (Ed.) (2008) *Urbanização de Favelas: A Experiência de São Paulo, São Paulo*. Prefeitura de São Paulo, Habitação.

Bonduki, N. (2008) Política habitacional e inclusão social no Brasil: Revisão histórica e novas perspectivas no governo Lula. *ARQ.URB Revista eletrónica de Arquitetura e Urbanismo*, 1, pp. 70–104.

Borja, J. and M. Castells (1997) Local y global: La gestión de las ciudades en la era de la información. Madrid: Taurus.

Borsdorf, A., Hidalgo, R. and Sánchez, R. (2007) A New Model of Urban Development in Latin America: The Gated Communities and Fenced Cities in Metropolitan Areas of Santiago de Chile and Valparaíso. *Cities*, 24(5), pp. 365–378.

Bourguignon, F., Ferreira, F. H. G. and Menendez, M. (2007) Inequality of Opportunity in Brazil. *Review of Income and Wealth*, 53(4), pp. 585–618.

Brain, I. and Sabatini, F. (2006) Relación Entre Mercados de Suelo y Política de Vivienda Social. *ProUrbana*, 4, pp. 1–13.

Budds, J., Teixeira, P. and SEHAB (2005) Ensuring the Right to the City: Pro-Poor Housing, Urban Development and Tenure Legalization in Sao Paulo, Brazil. *Environment and Urbanisation*, 17, pp. 89–113.

Cardoso, A. L. and Leal, J. A., (2010) Housing Markets in Brazil: Recent Trends and Governmental Responses to the 2008 Crisis. *International Journal of Housing Policy*, 10(2), pp. 191–208.

Carrión Barrero, G. A. (2008) Debilidades del Nivel Regional en el Ordenamineto Territorial Colombiano, Aproximación desde la Normatividad Política Administrativa y de Usos de Suelo. *Architecture, City and Environment*, 3(7), pp. 145–166.

Castañeda, J. (2006) Latin America's Left Turn. *Foreign Affairs*, 85(3), pp. 28–43.

Ceballos Ramos, O. and Saldariaga Roa, A. (2008a) La Institucionalización de la Acción Estatal Frente al Problema de la Vivienda (1942–1965), in: Ceballos Ramos, O. (Ed.), *Vivienda Social en Colombia: Una Mirada Desde su Legislación 1918–2005*. Bogota: Editorial Pontificia Universidad Javeriana, pp. 51–91.

Ceballos Ramos, O. and Saldariaga Roa, A. (2008b) La Transición en el Manejo Institucional del Problema de la Vivienda 1965–1972, in: Ceballos Ramos, O. (Ed.), *Vivienda Social en Colombia: Una Mirada Desde su Legislación 1918–2005*. Bogota: Editorial Pontificia Universidad Javeriana, pp. 107–136.

Ceballos Ramos, O. and Saldariaga Roa, A. (2008c) La Creación de las Corporaciones de Ahorro y Vivienda 1972–1990), in: Ceballos Ramos, O. (Ed.), *Vivienda Social en Colombia: Una Mirada Desde su Legislación 1918–2005*. Bogota: Editorial Pontificia Universidad Javeriana, pp. 137–178.

Chiape de Villa, M. L. (1999) *La Política de Vivienda de Interés Social en Colombia en Los Noventa*, Santiago de Chile, CEPAL.

Chomsky, N. (2010) *Hopes and Prospects*. London: Hamish Hamilton.

Clift, J. (2003) Hearing the Dogs Bark. Jeremy Clift Interviews Development Guru Hernando de Soto. *IMF Papers, Finance & Development* [online]. Available at http://www.imf.org/external/pubs/ft/fandd/2003/12/pdf/people.pdf [Accessed 10 September 2014].

Collier, D. (1978) Industrial Modernization and Political Change: A Latin American Perspective. *World Politics*, 30, pp. 593–614.

Collins, J., & Lear, J. (1995) *Chile's Free Market Miracle: A Second Look.* Oakland, CA: Institute for Food and Development Policy.

Coy, M. and Pöhler, M. (2002) Gated Communities in Latin American Megacities: Case Studies in Brazil and Argentina. *Environment and Planning B: Planning and Design,* 29, pp. 355–370.

Cuenya, B. (2005) Cambios, Logros y Conflicts en La Política de Vivienda en Argentina Hacia Fines del Siglo XX. *Boletín Ciudades Para un Futuro más Sostenible,* 29/30.

Cuervo, N. and Jaramillo, S. (2009) Dos Décadas de Política de Vivienda en Bogotá Apostando por el Mercado. *Universidad de Los Andes, Documentos CEDE,* 31, pp.1–36.

De Duren, N. (2006) Planning à la Carte: The Location Patterns of Gated Communities around Buenos Aires in a Decentralised Planning Context. *International Journal of Urban and Regional Research,* 30(2), pp. 308–327.

De Soto, H. (2001) *The Mystery of Capital: Why Capitalism Triumphs in the West and Fails Everywhere Else.* London: Black Swan.

Economic Commission for Latin America and the Caribbean (ECLAC) (2010) Time for Equality: Closing Gaps, Opening Trails. 33rd Session of ECLAC, 30 May–1 June, Brasilia.

Economic Commission for Latin America and the Caribbean (ECLAC) (2012) Structural Change for Equality: An Integrated Approach to Development. 34th Session of ECLAC, 27–31 August, San Salvador.

Economic Commission for Latin America and the Caribbean (ECLAC) (2014) Compacts for Equality towards a Sustainable Future. 35th Session of ECLAC, 5–9 May, Lima.

Edwards, S. (2010) *Left Behind. Latin America and the False Promise of Populism.* Chicago: The University of Chicago Press.

Escobar, G. (2001) La Vivienda en Colombia en el Cambio de Siglo, in: Brand, P. and Viviescas, F. (Eds), *Trayectorias Urbanas en la Modernización del Estado en Colombia.* Bogotá: Universidad Nacional de Colombia y Tercer Mundo Editores, pp. 235–265.

Fernandes, E. (2002) The Influence of de Soto's *The Mystery of Capital. Land Lines,* 14, pp. 4–7.

Fernandes, E. (2007) Constructing the "Right to the City" in Brazil. *Social & Legal Studies,* 16(2), pp. 201–219.

Fonseca, M.A.R. (1998). Brazil's Real Plan. *Journal of Latin American Studies,* 30, pp. 619–639.

Formoso, C., Leite, F., and Miron, L. (2011) Client Requirements Management in Social Housing: A Case Study on the Residential Leasing Programme in Brazil. *Journal of Construction in Developing Countries,* 16(2), pp. 47–67.

Frankema, E. H. P. (2006) The Colonial Roots of Latin American Land Inequality in a Global Comparative Perspective: Factor Endowments, Institutions or Political Economy? Paper of the Groningen Growth and Development Centre, University of Groningen [online]. Available at http://www.researchgate.net/publication/229051616_The_Colonial_Roots_ of_Latin_American_Land_Inequality_in_a_Global_Comparative_Perspective_Factor_ endowments_Institutions_or_Political_Economy [Accessed 10 September 2014].

Fukuyama, F. (2008) *Falling Behind: Explaining the Development Gap between Latin America and the United States.* Oxford: Oxford University Press.

Garcia de Freitas, F. and Cunha, P. (2013) Chile: Subsidies, Credit and Housing Deficit. *Cepal Review,* 110, pp. 189–211.

Gilbert, A. (1997) On Subsidies and Homeownership: Colombian Housing Policy during the 1990s. *Third World Planning Review,* 19, pp. 51–70.

Gilbert, A. (2002) Power, Ideology and the Washington Consensus: The Development and Spread of the Chilean Housing Policy. *Housing Studies*, 17(2), pp. 305–324.

Gilbert, A. (2004) Helping the Poor through Housing Subsidies: Lessons from Chile, Colombia and South Africa. *Habitat International*, 28(1), pp. 13–40.

González, I. and Martner, R. (2012) Superando el "Síndrome del Casillero Vacío" Determinantes de la Distribución del Ingreso en América Latina. *Revista CEPAL*, 108, pp. 7–25.

Green, M. and Ortúzar, J. (2002) Willingness to Pay for Social Housing Attributes: A Case Study from Chile. *International Planning Studies*, 7(1) pp. 55–87.

Kowaltowslki, D., Gomes da Silva, V. Pina, S., Labaki, L. Ruschel, R. and Moreira D. (2006) Quality of Life and Sustainability Issues as Seen by the Population of Low-income Housing in the Region of Campinas, Brazil. *Habitat International*, 30(4), pp. 1100–1114.

Kusnetzoff, F. (1987) Urban and Housing Policies under Chile's Military Dictatorship 1973–1985. *Latin American Perspectives*, 53, pp. 157–186.

Lima, J. (2006) Urban Reform and Development Regulation: The Case of Belem, Brazil, in: Zetter, R. and Butina Watson, G. (Eds), *Designing Sustainable Cities in the Developing World.* Hampshire: Ashgate, pp. 120–145.

Macario, S. (1964) Protectionism and Industrialization in Latin America. *Economic Bulletin for Latin America*, 9, pp. 61–101.

Macedo, J. (2010) Methodology Adaptation across Levels of Development: Applying a US Regional Housing Model to Brazil. *Housing Studies*, 25(5), pp. 607–624.

McKenzie, D. and Mookherjee, D. (2003) The Distributive Impact of Privatization in Latin America: Evidence from Four Countries. *Economía*, 3(2), pp. 161–233.

Melo, M. (1992, May). Policymaking, Political Regimes and Business Interests: The Rise and Demise of the Brazilian Housing Finance System. Center for International Studies working paper 2597/C92/4. Massachusetts: MIT.

Ministerio De Vivienda y Urbanismo (MINVU) (2004) *Chile: Un Siglo de Políticas de Vivieda y Barrio.* Santiago: MINVU.

Morais, L., Saad Filho A. and Coelho, W. (1999) Financial Liberalisation, Currency Instability and Crisis in Brazil: Another Plan Bites the Dust. *Capital & Class*, 23, pp. 9–14.

Murillo, F. (2001) Private–Public Partnership, the Compact City and Social Housing: Best Practice for Whom? *Development in Practice*, 11(2–3), pp. 336–343.

Ocampo, R. (2004) Contribuciones INVI al debate Teórico y Práctico Sobre el Proceso Habitacional. *Revista INVI*, 52(19), pp. 19–59.

Parris T. and Kates, R. (2003) Characterizing and Measuring Sustainable Development. *Annual Review of Environment and Resources*, 28, pp. 559–586.

Pugh, C. (1994) Housing Policy Development in Developing Countries: The World Bank and Internationalization, 1972–93. *Cities*, 11(3), pp. 159–180.

Pugh, C. (2001) The Theory and Practice of Housing Sector Development for Developing Countries, 1950–1999. *Housing Studies*, 16(4), pp. 399–423.

Rodriguez-Acosta, C. and Rosenbaum, A. (2005) Local Government and the Governance of Metropolitan Areas in Latin America. *Public Administration and Development*, 25, pp. 295–306.

Rojas, E. (2000) The Long Road to Housing Sector Reform: Lessons from the Chilean Housing Experience. *Housing Studies*, 16(4), pp. 461–483.

Sabattini, F., Cáceres, G. and Cerda, A. (2001) Segregación Residencial en las Principles Ciudades Chilenas: Tendencies de las Tres Últimas Décadas y Posibles Cursos de Acción. *Eure*, 26(79), pp. 21–42.

Sacks, J. (2006) The Challenge of Sustainable Development in Latin America. Lecture Series of the Americas transcript. Organization of American States, March 14, Washington.

Salcedo, R. (2010) The Last Slum: Moving from Illegal Settlements to Subsidized Home Ownership in Chile. *Urban Affairs*, 46, pp. 90–118.

Salcedo, R. and Torres, A. (2004) Gated Communities in Santiago: Wall or Frontier? *International Journal of Regional Research*, 28(1), pp. 27–44.

Samuelson, R. (2001) The Spirit of Capitalism. *Foreign Affairs*, 80(1), pp. 205–211.

Scheinsohn M. and Cabrera, C. (2009) Social Movements and the Production of Housing in Buenos Aires: When Policies are Effective. *Environment and Urbanisation*, 21, pp. 109–125.

Silva Lerda, S. (1997) *Estudio Análisis de la Evolución de la Política Habitacional Chilena*. Santiago, MINVU.

Soto, R. and Torche, A. (2004) Spatial Inequality, Migration and Economic Growth in Chile. *Cuadernos de Economía*, 41, pp. 401–424.

Stiglitz, J. (2003) *Globalization and Its Discontents*. London: Penguin Books.

Stiglitz, J. (2007) *Making Globalization Work*. London: Penguin Books.

Subsecretaría de Urbanismo y Vivienda (SUV) (2007) Lineamientos Estratégicos para la Región Metropolitana de Buenos Aires, Dirección Provincial de Ordenamiento Urbano y Territorial.

Tachópulos Sierra, D. (2008) El Sitema Nacional de Vivienda de Interés Social (1990–2007), in: Ceballos Ramos, O. (Ed.), *Vivienda Social en Colombia: Una Mirada Desde su Legislación 1918–2005*. Bogota: Editorial Pontificia Universidad Javeriana, pp. 181–238.

Texeira de Melo, E. (2014) FGTS Ação para Alteração do Índice de Correção dos Depósitos, *Ferreira & Mello Bulletin* [online]. Available at: http://www.ferreiraemelo.com.br/midias/fgts_pdf.pdf [Accessed 28 May 2014].

Thibert, J. and Osorio, G. A. (2014) Urban Segregation and Metropolitics in Latin America: The Case of Bogota, Colombia. *International Journal of Urban and Regional Research*, 38(4), pp. 1319–1343.

Thuillier, G. (2005) Gated Communities in the Metropolitan Area of Buenos Aires, Argentina: A Challenge for Town Planning. *Housing Studies*, 20(2), pp. 255–271.

Turner, J. (1977) *Housing by People: Toward Autonomy in Building Environments*. London: Pantheon Books.

Valença, M. M. (1992) The Inevitable Crisis of the Brazilian Housing Finance System. *Urban Studies*, 29, pp. 39–56.

Valença, M. M. (1998) The Lost Decade and the Brazilian Government's Response in the Nineties. *The Journal of Developing Areas*, 33(1), pp. 1–52.

Valença, M. M. (2001). Globabitação Sistemas habitacionais no Brasil, Grã Bretanha e Portugal. São Paulo: Terceira Margem.

Valença M. M. and Bonantes, M. F. (2010) The Trajectory of Social Housing Policy in Brazil: From the National Housing Bank to the Ministry of the Cities, *Habitat International*, 34(2), pp. 165–173.

Valladares, L. P. (1978) Working the System: Squatter Response to Resettlement in Rio. *International Journal of Urban and Regional Studies*, 12(1), pp. 12–25.

Van Gelder, J. L. (2009) Legal Tenure Security, Perceived Tenure and Housing Improvement in Buenos Aires: An Attempt towards Integration. *International Journal of Urban and Regional Research*, 33(1), pp. 126–146.

Varley, A. (1987) The Relationship between Tenure Legalization and Housing Improvements: Evidence from Mexico City. *Development and Change*, 18, pp. 463–481.

Varley, A. (2002) Private or Public: Debating the Meaning of Tenure Legalization. *International Journal of Urban and Regional Research*, 26(3), pp. 449–461.

Werlin, H. (1999) The Slum Upgrading Myth. *Urban Studies*, 36(9), pp. 1523–1534.

World Bank (1993) *Housing: Enabling Markets to Work.* Washington DC: World Bank.

World Bank (1994) *Vivienda un Entorno Propicio para el Mercado Habitacional.* Washington DC: World Bank.

World Bank (2013) *Planning Connecting and Financing, Cities – Now: Priorities for City Leaders.* Washington, DC: World Bank.

Zanetta, C. (2005) Seeking Better Policies or Just Giving up Responsibility? The Decentralization of Argentina's National Housing Fund (FONAVI). Third Urban Research Symposium [online]. Available at: http://www.pragueinstitute.org/ GUDMag07Vol3Iss1/Zanetta%20PDF.pdf. [Accessed 28 July 2014].

14 Corporate real estate management in Africa

A case study of Nigeria

Timothy Tunde Oladokun

1. Introduction

Prior to the 1980s, real estate (RE) was not generally regarded as an important asset of an organisation whose core business was not RE. It was regarded as a tool necessarily required for housing a company's core business and its staff. Thus, to business executives, corporate RE (CRE) was just a necessary overhead; a cost of conducting business, which was to be managed tactically and reactively (Luokko, 2004). The result is that the CRE manager has historically struggled to provide service in an environment in which cost management was a main focus. Little attention was given to the assets of the company, which take huge capital to procure and the decision to procure them is always taken at the topmost level of the organisation, often at the board level and also as a strategic decision. It is thus of concern that not much attention is paid to RE, suggesting that RE has been neglected in corporate management (Ilsjan, 2007).

Traditionally, most corporations have owned RE that they use in their business activities (Petison, 2007). Existing literature has, however, brought to attention that CRE is being under-managed. For instance, RICS (2002) reported that UK businesses throw away £18 billion a year through inefficient use of RE, which could have improved gross trading profits by up to 13% and contributed to economic development. Hwa (2003) found great loss of the value contribution of RE due to the fact that many companies have little ideas of their RE costs and the extent to which their assets could be used to increase productivity and contribute to economic growth. This is in spite of the fact that no corporate body or organisation can function without RE since it is RE that provides space for its operations.

In African countries, like in many parts of the world, capital commitment to CRE seems to be on the increase within the last two decades. Activities and involvements of corporate organisations are becoming more visible in the RE market. During this period, corporate organisations such as banks, insurance companies, oil companies and a host of other organisations are seen to be acquiring RE (space) for their business activities (Onamade and Adejugbe, 2014). According to these authors, emphasis seems to have been shifted from renting to owner-occupation in order to use RE as a means of brand identity. In spite of this, African countries such as Nigeria have witnessed a lot of corporate downsizing and collapse with its negative effect on the economy. For instance, 64 of the 89 banks in the country could not satisfy the 2004 recapitalisation policy of the government and were declared distressed since they could not raise their financial capital base to 25 billion as required by the law (Sanusi, 2010). This is regardless of the fact that the organisations have substantial RE holdings that would have been used to liberate them from financial illiquidity. The results of the collapse are joblessness and a high rate of unemployment, thereby increasing crime rates, especially in the cities, which all ultimately negatively affect economic development.

In recent times, the changing global business environment has attracted business managers' attention to the significance of CRE as an operational asset and redirected their focus to it as a valuable resource. Since the first discovery of the importance of CRE in 1983 by Zeckhauser and Silverman (1983), namely that a major portion of all assets (approximately 25% to 41%) of corporate organisations in America were invested in RE, other studies have confirmed that the occupancy costs of corporate space represent some 10% to 20% of operating expenses and remains 41% to 50% of corporate net operating income (Manning and Roulac, 1999; Veale, 1989).

In developed countries, the grave consequences of the neglect of CRE have been documented. Authors such as Sharp (2013), Englert (2001), Seiler et al. (2001), Rodriguez and Sirman (1996) and Nourse (1993) have challenged the practice of inactive management of CRE. Adendorf and Nkado (1996) expressed a similar opinion, and in a report by Cornet Global (2005) based on Ernst and Young's survey, 52% of all organisations were found to be either doing nothing or did not know what to do regarding their RE portfolios. Veale (1989) noted with dismay that many organisations' attitude of not clearly and consistently evaluating the performance of their RE by treating it as an overhead cost has the tendency to hinder national economic growth.

With a historic reputation as just an unavoidable "cost of production" (Laws, 2007), the current environment of increasing corporate governance, regulation and financial transparency necessitates that in best practice occupier organisations, RE receive a strategic focus and be firmly embedded into corporate decision-making (Bouris, 2005). Strategic use of RE increases organisational efficiency and improves profitability (Ilsjan, 2007; Bouris, 2005; McDonagh, 2002; Gibson 2002; Breitenstein et al., 1998; Nourse and Roulac, 1993), which in turn can enhance economic development.

Corporate organisations that are not in the RE business have, either by taking leases or direct ownership, acquired RE to accommodate their core business activities and/or to house their staff. The drive towards having a competitive edge, which has made, for example, Nigerian organisations such as banks strive to own RE in all parts of the country, also seems to have made them ignore its attendant cost of acquisition. As a result, a bulk of the shareholders' funds are being invested in RE. The inability to proactively manage RE often makes recouping its cost take a longer period at the expense of the core business of such organisations. Oftentimes, this has the tendency of reducing the overall profit of the organisation with resultant negative effects on the economy.

In the business world, global competition, technological innovations, business knowledge and the restructuring of local economies have been identified as some of the reasons compelling corporate organisations world-wide to transform their RE portfolios in order to increase market share and enhance shareholders' value (Roulac and Heaney, 2003). The result is that, as corporations become more international or global in their approach to business, the role of CRE is becoming more crucial (Wills, 2008). Companies therefore have to try to create and maintain their competitive advantage and one way to achieve this is by transforming their RE portfolios in order to increase market share and enhance shareholders' value (Roulac, 2001). As African countries embrace the concept of their economies being driven by the private sector, foreign investors will want to be convinced of the efficient utilisation of the RE of these organisations as a potential resource for increased productivity; efficient organisations will likely employ more hands, which reduces unemployment, thereby improving the standards of living of those employed. The increased economic activities generated by these organisations will bring about development and growth in the ailing African economies.

In Africa, however, CRE executives seem to still identify and treat RE management as a non-core activity with very fragmented functioning across business units. It can then be said that CRE in African countries is undervalued within a company's business assets, which require little or no training to manage. In addition, while the capital commitment to RE appears to be high on the balance sheet of corporate organisations relative to other major costs, there seems to be few or no record of the contributions of RE to achieving the objective of the organisation so as to justify the huge investment in RE, which is always the second largest cost of organisations after staff salary (Oladokun, 2011). This is because while many corporate organisations (especially the banks) fail to exploit the opportunities to enhance their financial status by effective CREM, many of them are currently facing difficulties regarding the required operational funds for the running of their core businesses (Sanusi, 2010). Nevertheless, the erstwhile practice of reactive management of RE as an investment asset has changed drastically and the strategic view of CRE requires organisations to treat RE in an active manner, since its cost comes second to payroll at 20%–40% of business value (Veale, 1989).

The above have posed serious problems and challenges to corporate executives and shareholders in the quest to wriggle out of the liquidity crisis that is rocking many of the corporations as a result of their failure to tap the benefits of effective CREM to leverage their business operations. Therefore the question remains as to whether or not organisations in African countries like Nigeria manage their CRE in such a manner to ensure that it contributes significantly to achieving corporate objectives. The need for companies to become conscious of RE holdings and their importance in enhancing productivity as well as engendering economic development is beginning to command attention.

This chapter therefore seeks to examine the practice of CREM in the context of Africa using Nigeria as a case study and its implications for economic development. This is necessary in view of the recent aggressive opening of branches and acquisition of units of RE by African corporate organisations. With the huge investment in the acquisition of units of RE by these corporate organisations, one wonders whether or not the managers are conscious of an appropriate balance in the allocation of resources between RE and other assets of the organisation. A similar concern is whether or not corporate managers manage their RE in a manner that contributes to economic development. The chapter is structured as follows. Immediately after this introduction is a description of the research methodology used. The third section gives an overview of CREM whilst section four deals with data presentation analysis and discussion. Conclusions are summarised in the last section.

2. Research methodology

The quantitative research methodology was deemed appropriate and adopted for this study. Using companies that are listed with the Nigerian Stock Exchange, a survey was conducted in 2012 where structured questionnaires were administered face-to-face to RE managers to ensure a prompt response. The focus on these organisations is because they are likely to occupy substantial RE, and thus will provide sufficient insight into the practices of CREM by these companies.

The sampling frame comprised 50 CRE managers from 21 banks, 5 GSM communication companies and 24 insurance companies that are listed with the Stock and Exchange Commission in Nigeria as contained in the registered list of the Nigerian Stock Exchange Commissioner in 2012. Out of this sampling frame, 38 managers (representing 14 banks, 21 insurance and 3 GSM managers) who were willing to participate in the survey were selected for the questionnaires to be administered to them. Thus, the sample size represented 78% of the sampling frame.

The questionnaires were in three sections. Section one asked questions pertaining to the respondents and their organisations. Questions included the status of the respondents in the organisations as well as educational and professional qualifications of the respondents. In section two, questions about the organisations' involvement in CREM and CREM strategies adopted in the study area were posed. Respondents were asked to indicate their years of experience and their respective designations. This section also asked questions on the volume of their RE portfolio and their geographical spread across the country. The last section asked questions pertaining to the status of CREM in Nigeria. The data obtained were analysed using descriptive statistics of percentages and cluster analysis while findings were displayed with the use of tables and charts.

3. An overview of CREM

The exploratory work of Zeckhauser and Silverman (1983) and the subsequent work of Hite et al. (1984) appraised many American corporations and discovered a gross neglect of between 25% and 40% of their total assets, which they viewed could leave them open to takeovers, profit losses and lower stock price performance. The findings aroused interest in the importance of CRE to business organisations.

In 1987, Nourse and Kingery studied how American corporations that are not in the RE business undertake profitable disposal of surplus RE. The study, which found that American corporations lose the opportunity to profitably dispose of surplus RE, did not take cognisance of African countries. The study, however, concluded that US corporations have not evolved profitable management of their RE holdings. Gale and Case (1989) studied the state of CREM practices and found that RE was largely being perceived as a factor of production and not as a resource for increased profitability.

Rutherford and Stone (1989) limited their investigations to examining factors influencing CRE unit formation. The findings that wholly owned subsidiaries were likely to be motivated by development and profits while centralized and decentralized RE departments were likely to be motivated by contracting efficiency and cost control needs to be investigated in African countries like Nigeria.

Nourse and Roulac (1993) examined the relationship between RE strategies and CRE decisions and discovered that CRE decisions are effective if they support the corporation's overall business objectives. Manning and Roulac (1996) identified the expected roles of a CRE manager as necessary to contribute to the wealth of a company and its shareholders and suggested a central organisation and management as well as the training of a significant proportion of CRE staff to work closely with the operating business units, their support staff and local business unit issues. Schaefers (1999) evaluated factors influencing CREM in Germany and found gross under-management of CRE in spite of the great value such assets carry in the corporations' balance sheet. The study, however, found gradual recognition of the importance of CRE functions by the organisations even though it requires a more formal and systemic approach. An investigation of how decisions relating to CRE are taken in African countries is yet to be documented.

The study of McDonagh and Hayward (2000) was limited to an investigation of 457 organisations in New Zealand. Its main findings are that valuations, brokerage and building services are the most usual services being outsourced while strategic planning is the most common job being kept in-house. In his study, Roulac (2001) demonstrated how strategic CREM can impact on a business strategy. The result, which described CRE as a multidimensional asset that can add to corporate profitability and shareholder value in many ways, identified seven distinct ways by which CRE can bring competitive an advantage to a corporation. These include creating and retaining customers, attracting and retaining outstanding people, contributing to effective business processes, promoting enterprise values and culture, stimulating innovation and learning, impacting core competency and enhancing shareholder wealth.

Woollam (2003) categorised CRE into core, tactical and surplus. Asson (2002) examined the practice of sale-and-leaseback in corporate organisations and found that what a RE partnership does is to take the germ of a good idea of sale-and-leaseback and take it to a more refined and sophisticated level. McDonagh (2002) proposed a methodology to combine previously identified CREM performance input variables into a single relative measure of CREM performance using factor analysis. The study adopted the model to study the performance of CREM in organisations in New Zealand and established that a small group of variables tend to occur together to provide a strong indicator of performance. Though Bryan (2003) limited his study to an examination of the opportunities for CRE decision makers for different types of lease finance, the study found that the cost of sale-and-leaseback financing has decreased lately quite significantly as the amount of capital available for sale-and-leaseback transaction has increased rapidly. In an examination of the practice of CRE in Singapore, Liow and Tay (2006) employed statistical data-driven analytical techniques to study the direct and indirect effects of performance factors on CRE. The study pointed out that CRE is under-managed in Singapore. In addition, the study revealed that the CRE planning and the existence of CRE units have direct impact on CRE performance. Applying the result of these studies to African countries might, however, require caution, hence this study.

McDonagh (2006) carried out a time series analysis of the development of CREM practice in New Zealand. The author identified the substantial number of qualified CREM personnel and the development as substantial and continuous improvement in CRE practice, but found that the percentage of the organisations with a separate RE unit and the allocation of RE costs remained stable over time. Kaluthanthri (2009) examined the CREM practices of the banking sector in Sri Lanka. The author administered questionnaires on commercial banks with a minimum of 50 branches. The study found that, in general, CRE is under-managed in the commercial banking sector in Sri Lanka and it is in a passive mode. The study further found that three variables, namely the CREM organising structure, the knowledge and skills of CRE managers and employees and the role of CREM in the bank, are in a passive mode, while CREM policies, function and activities are in a selective mode, with the fifth variable, CREM practice, being in an active mode.

Oi (2010) investigated the CREM of organisations in Japan. The study found that the importance of CREM has been recently recognised and that some Japanese companies have integrated CREM into their corporate strategy. The author categorised CREM in Japan as cost reduction, workplace strategy and portfolio optimisation. The study provided statistics in respect of each category of the real estate assets of Japanese companies and their adopted strategies. The study also provided information and explained the manual developed by the Ministry of Land, Infrastructure, Transport and Tourism (MLIT) as policies for appropriate CRE management. Hartmann et al. (2010) identified different organisational models concerning both the functions and responsibilities assigned to CRE professionals in European and North American companies. The result of an empirical survey of 74 major European and 38 North American companies left out the opinion of the business executives. Nevertheless, the study found five typical models describing the allocation of responsibility of RE functions within a company and identified the performance of those responsibilities. The study concluded that there is no one "best practice" CREM model in a specific situation, as often stated, but instead various promising organisational models seem to exist.

Nunnington and Haynes (2011) examined the complex decision making process involved in corporate relocation and the validity of a tool designed to improve the objectivity and strategic management of this process and to change the focus of the decision upon the strategic management objectives rather than the RE deal. The author identified the progression of the decision making process and components of that process and evaluated a tool designed to improve the process. The study concluded that the size of the organisation can have a significant impact on the building evaluation and decision making process and that smaller firms with fewer resources are more likely to make the relocation decision based on "gut feeling" rather than detailed evaluation. The study of Ristaniemi and Lindholm (2011) was limited to the examination of the added value elements of CREM in industrial companies in Finland.

The work of Christersson and Rothe (2012), drawing from existing literature, examined the impacts that relocation has on the relocating organisation itself and identified the economic, social and environmental impacts of office occupier relocation. Whereas the study did not consider the attitude of business executives, the findings are that relocation has various impacts including relocation costs, disruption, employee reactions to change, altered lease attributes and a changed environmental footprint. Furthermore, the changes in productivity, employee satisfaction, employee turnover, organisational dynamics, ways of working, commuting, accessibility for external stakeholders, and organisational culture and image are all possible impacts of organisational relocation.

The examination of the risks for CRE departments arising from global trends was the focus of the study of Sharp (2013). The study identified risks such as portfolio right-sizing in mature US and European markets versus growth opportunities in emerging markets. The author concluded that the global financial crisis has increased the importance of CRE to CEOs as a tangible lever for enhancing revenue growth – even as cost-cutting has resulted in historic lows in the departmental budgets and a slashed pool of in-house talent. Amid continuing global economic challenges, CRE executives now face new C-suite demands to dramatically affect the culture, productivity and performance of their companies.

Palm (2013) examined different strategic pathways for structuring the commercial CREM organization as well as the alignment of their business models with the environment. Based on an analysis of 15 interviews with top-level managers in the Swedish commercial RE sector, the author identified two pathways when making strategic plans for a company. The first is to choose whether the organisation should have its own frontline personnel or outsource this function. The second is to decide how the leasing task should be treated: should it be treated as a RE manager's task or should it be a function of its own in the organisation? The conclusion of the study is that the organisations studied can be structured using both pathways and the firm can still be successful.

In summary, a look at the studies reviewed above will leave no one in doubt that there is evidence to suggest that there are studies that have focused on the importance of CREM in organisations whose business is not RE. There is also evidence to show that the empirical studies focused on the practice of CREM in the advanced countries. However, a critical examination of the results of the above studies missed a fundamental issue regarding addressing CREM in an emerging economy like Nigeria.

Many studies, for instance, Kaluthanthri (2009), Schaefers (1999), Avis et al. (1993) and Joroff et al. (1993) in Sri Lanka have addressed the practice of CREM in a country that does not share similar economic and political characteristics with African countries like Nigeria. Meanwhile, as more corporate organisations aggressively open branches and acquire RE space, a major issue that confronts both the managers of these businesses and their shareholders is the issue of effective allocation of resources between all the company's assets. A study of CREM should also be of interest to them.

In African countries, Oladokun et al. (2009) explored the relevance and usefulness of CREM to business organisations in developing countries like Nigeria. The study undertook an extensive literature review to argue for the need to embrace the strategic use of CRE and highlighted new directions for CREM practice. Oladokun (2010)'s work "Towards Value-Creating Corporate Real Estate Asset Management in Emerging Economies" aimed at increasing the awareness of developing countries, corporations and academics concerning the significance of CRE holdings in the corporate assets portfolio. Building on extensive review and analysis of related studies in advanced countries, the author identified the value adding attributes of RE as a strategic resource of an organisation. The paper argued that the loss of the value contribution of RE in African countries like Nigeria can be avoided if organisations can begin to adopt the practice of performance measurement of their CRE. Apart from the fact that the paper was not empirical, the other major shortcoming is the lack of evidence on the measurement of the impact of CRE on a business organisation.

Oladokun (2011) examined the skill requirements for the practice of CREM in Nigeria. With a focus on 270 practising RE firms in Lagos State, 145 final-year students of Estate Management in Obafemi Awolowo University, Ile Ife, as well as corporate real estate executive officers of the 24 recapitalised commercial banks, 21 insurance and 5 GSM communication companies in Nigeria, the study found that, in rank order, the skills requirements for CREM were in financial performance, investment in corporate strategy, productivity, space efficiency management and the management of customers and employees. Oladokun (2012) investigated corporate site selection and acquisition process in a Nigerian global system for a mobile (GSM) communication company. The study integrated a literature review and a case study and found that the process of site acquisition follows the global trend of outsourcing of non-core services by organisations to consultants who have the technical ability to handle such assignments. The case study also revealed an arrangement of successful collaboration between multinational companies and local/indigenous firms.

The above studies are, however, exploratory works which investigated the preparation of companies for CREM practice. Most empirical studies come from the English-speaking world and Northern Europe; these countries do not share or perhaps do not have the same business culture and economic challenges like emerging countries such as Nigeria. The writer is not aware of any empirical study in the country on measuring the impact of CRE and if one ever existed it is rather rare. This clearly shows a substantial gap in CREM performance literature and reinforces the need for a study on CREM in Nigeria. This study intends to fill the vacuum by drawing on the experimental approaches adopted in previous research on the subject in countries like US, UK, Malaysia, Finland, Singapore, New Zealand and Germany.

4. Data presentation, analysis and discussion

4.1. Characteristics of respondents regarding company profile

The study adopted the classifications of listed firms with the Nigerian Stock Exchange and provided two categories that best described the respondents' organisations. Table 14.1 summarises the characteristics of the organisations.

A majority (70%) of the companies are private companies while the remaining (30%) are public companies. This could best explain the success of the privatisation and commercialisation policies of the federal government to allow the private sector to run the economy. Regarding the core business of the organisations, 42.9% are into core commercial banking services and 35.7% are into insurance, while the remaining 21.4% are GSM communication providers. The average number of people employed by the respondents' organisations is over 500 with a majority of the firms operating from more than 100 sites and locations in the

Table 14.1 Characteristics of organisations

	Respondents' organisations	*Percentage (%)*
Description	Public company	30
	Private company	70
Core business	Banking	42.9
	Insurance	35.7
No. of employees	Communications company	21.4
	Average	>500
No. of sites in Nigeria	< 50	21.4
	51–100	–
	Over 100	78.6
Respondents' education	Graduate degree	46.8
	Master's degree	37.9
	Others	15.3
Year of experience	Mean	12 years
Qualifications	Probationer	30
	ANIVS	61
	FNIVS	9

country. The organisations can therefore be regarded as substantially big with many of them having branches outside the country.

Most (84.7%) of the respondents have Bachelor of Science (BSc) as their minimum qualification and had an average length of 12 years' experience. The responses show that 61% are associate members of the Nigerian Institution of Estate Surveyors and Valuers (NIESV). A substantial 30% are probationers under NIESV while 8% are fellows (FNIVS). The above results show that most of the respondents are experienced and mature enough to operate at the management level required of CRE managers. They can also be relied upon to give adequate data for the study.

Table 14.2 contains responses on the characteristics of the RE units of the organisations. Responses to the questions on the job title of the head of the CRE unit revealed divergent views. Titles indicating a clear RE focus including the description "Property", "Facilities" or "Asset" Manager were the most popular. For the sampled organisations, the job title of the RE unit head varies considerably between organisations. The findings are that while a majority (70%) of the organisations referred to the head of the RE unit as a Property Manager and some (6%) referred to him as a Project Manager, he is called Asset/Facilities Manager in some other organisations (15%). Further findings indicate that 37% of the heads of the RE units in the sample function as Property Manager/Facilities Managers. Less than half (32%) operate as Team Leader/Supervisor while about 31% operate as Unit Manager/National Manager.

In line with the findings in the literature, CRE unit reporting level is key to effective CREM performance. The study found that more than 72% of the RE

Table 14.2 Characteristics of RE units

	Respondent organisation	*Percentage (%)*
Job title	Property manager	70.0
	Asset/facility manager	15.0
	Project manager	6.0
	Roll-out implementation manager	9.0
Status of CREM head	Board	0.00
	CEO or equivalent	0.00
	GM or equivalent	0.00
	Unit manager/national manager	31.0
	Team leader/supervisor	32.0
	Property manager/facilities manager	37.0
	Others	0.00
Size of real estate unit	Minimum	10 staff
	Maximum	155 staff
CRE reporting level	To CEO/MD	72.0
	To financial controller	28.0

units report to the Chief Executive Officer or the Managing Director while the remaining 28% report to the Financial Controller. There is a wide range in the size of the RE units. The smallest unit has 10 staff while the biggest unit has 155 staff.

The results of the RE portfolio profile of the sampled organisations are contained in Table 14.3. On average, the sampled organisations have substantial RE holdings. The largest holdings come from the banking sector. The communications companies have the second largest holdings followed by the insurance companies. Their RE portfolio includes residential, commercial (offices) and cell sites. On the average, the banks and the communications companies are actual owners of more than 67% of their RE portfolios while the insurance companies own on average 50% of their RE holdings. The reason for the higher percentage of the former is understandable. The Universal Banking Policy compelled all banks to locate in all major urban and rural centres in the country while the quest for national coverage requires all communications companies to have their towers even in the remotest part of the country.

In order therefore to benefit from the huge investment involved in this act, the organisations opted for outright purchase and/or acquisition of sites (RE). This will enable a long use of the RE for a long time and avoid interruptions that occasionally characterise lease option. The seeming low RE holding of the insurance companies is explained by the recent transfer of a large portion of the previously joint holdings to the recently formed Pension Administrators, which has hitherto been part of these organisations. This act was carried out in line with the requirements of the Pension Reform Act of 2007.

On the average, the value of the owned portion of the CRE of the sampled organisations are 20 billion, 10 billion and 2 billion naira for banks, communication companies and insurance companies respectively while the rental value of the leased portion of their CRE range from N8,500,000 to N250,000,000 per annum. The findings confirm earlier researchers' submissions that the CRE of a business organisation is substantial enough to warrant or attract management attention.

Table 14.3 RE portfolio profile of sampled organisations

	RE portfolio	*Percentage (%)*
No. of buildings in portfolio	<50	28.6
	51–100	7.1
	Over 100	65
Owned portion of portfolio	Banking average	73
	Insurance average	50.23
	Communications company	67.25
Value of owned portion of portfolio	Banking average	N20 billion
	Insurance average	N2 billion
	Communications company	N10 billion
Annual rental	<N100,000	25.6
	N100,000–N400,000	22.7

4.2 Status of CREM

This section shows the "activity level" in the area of CRE management in the respondents' companies. The activity level is used to depict the level to which the organisations have succeeded in using CRE as a contributor to their growth and development. On the basis of the activity level, internally homogeneous and externally homogenous clusters are identified showing the analysis of sixteen managerial and organisational characteristics, which are in theory and practice well known as critical success factors in the operations of a (pro-)active CRE management system (Schaefers, 1999; Teoh, 1993; Nourse and Roulac, 1993; Joroff et al., 1993; Pitman and Parker, 1989; Gale and Case, 1989; Avis et al., 1989).

Measured in terms of how important the organisations perceive these factors, as well as how they perform in respect of each of the factors, the findings revealed diversity. Responses to the question to respondents asking them to rate how important each factor is (importance) and how well their companies performed with regards to each factor (performance) were rated on a Likert scale from 1 to 5. The results are contained in Figure 14.1, which shows the factors by their performance score. In Figure 14.1, the vertical and horizontal axes represent performance scores and variables (factors) respectively.

With the help of cluster analysis, three types of organisations were identified that differ significantly with respect to their CREM systems. The distribution of respondent organisations among these three categories as shown in Figure 14.1 is as follows:

☐ Active	30.8%	3.5 to 4.5 level (Series 3)
☐ Selective	61.5%	2.5 to 3.4 level (Series 2)
☐ Passive	7.7%	1.5 to 2.4 level (Series 1)

The "active" companies – 30.8% of the organisations – are those that are conscious of the various key factors of CREM at a very high level. All characteristics are scored on a 3.5 to 4.5 level. On the other hand, a large 61.5% are the "selective" companies with a lower level performance and the "passive" companies are those with the lowest realisation level. In all, a total of 69.2% of the organisations have not been using CRE to improve the performance of their organisations. The loss of the value addition of CRE to the financial performance of organisations could explain the inability of many of the organisations to survive the financial crisis that led to the collapse of many giant organisations in the country.

A further interesting revelation in the results is the large differences between importance and performance scores by the different clusters (see Figure 14.1), which is an indication of mismatch of resources and needs and thus the organisations may require improvement in that area.

With respect to the "active" organisations, Figure 14.2 – with the vertical and horizontal axes representing performance scores and variables (factors) respectively – shows a small difference between importance and performance scores. In fact, the finding indicates an even level of importance and performance in terms of the organisational treatment of the management functions under study. At the

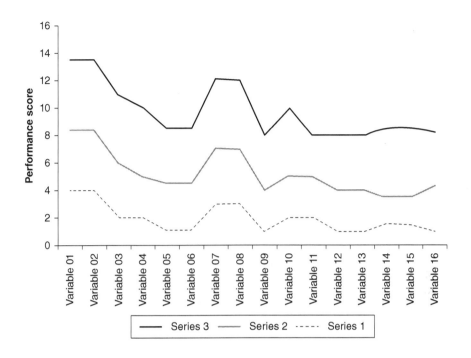

Key to Figure 14.1 variables
01 = Detailed and up-to-date information on RE
02 = Centralised keeping of RE data by REM
03 = Integration of both RE and corporate information systems
04 = Detailed and formal strategic planning for facilities and REM
05 = Bottom-up integration of strategic planning for RE and business units
06 = Top-down integration of corporate objectives and strategies in RE planning
07 = Central location of RE unit in overall organisational structure
08 = Access to top management
09 = Operation of RE unit as separate and distinct responsibility centre
10 = Positive attitude by top management towards RE
11 = Centralised RE authority and responsibility
12 = Internal renting system for RE space
13 = Well-defined and regular RE performance measurement
14 = Well-defined and regular strategic RE control
15 = Transparency of RE costs
16 = Professionally trained and qualified human resources in RE

Figure 14.1 Status of CREM by different clusters

same time, there is "over-performance" with respect to centralised RE authority and responsibility and internal renting system. The result suggests that active
organisations obviously utilise CRE to reduce operation costs, especially occupancy costs, thereby conserving resources for enhancing the efficiency of other
resources of the organisations. Efficiency of corporate organisation generates economic activities and thereby enhances national economic growth.

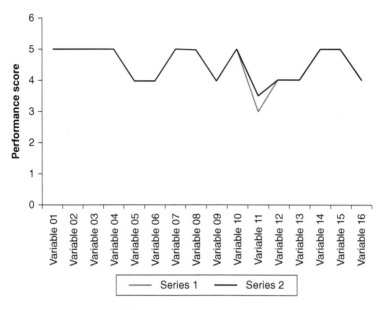

Figure 14.2 Status of CREM practice: active

In contrast, those organisations that have realised selective and passive modes of CREM exhibit some large discrepancies between the importance and performance level of the different success factors (see Figures 14.3 and 14.4 where for each figure, the vertical and horizontal axes represent performance scores and variables (factors) respectively). Even though these organisations substantially recognise the importance of CRE in improving their productivity, they have not been largely utilising it, thereby losing its contribution to economic development. Clear discrepancies arise primarily with respect to detailed and up-to-date information on RE, centralised keeping of RE data by the RE management unit, the integration of both RE and corporate information systems and detailed and formal strategic planning for facilities and the RE management area. Moreover, organisations in those groups have typically neglected transparency of RE costs. Although most of these organisations regard the key factors to be of critical or moderate importance, the operational system clearly lags behind the importance assessment.

In summary, two-thirds of the respondents' organisations do not have an adequate styled management system for the RE holdings. There is clear evidence that RE, though a key asset for non-RE organisations, is under-managed by the majority of the sampled organisations. Only one-third fulfil the prerequisites to seriously consider RE as an asset to be actively managed to support the organisation and to improve productivity. The study therefore indicates that there is much room for improvement with respect to the performance of the various factors by the sampled organisations. Nigerian organisations, and for that matter other African

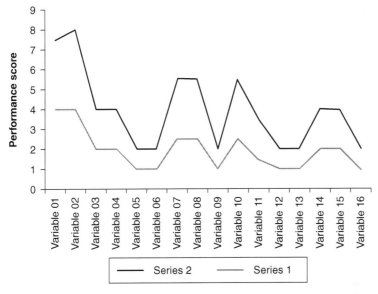

Figure 14.3 Status of CREM practice: passive

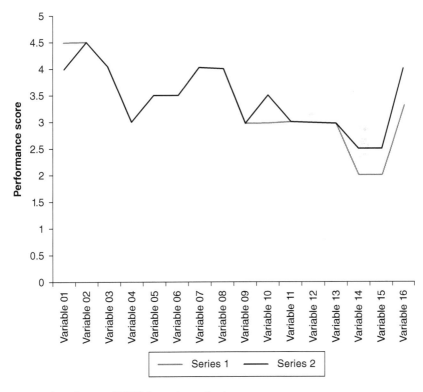

Figure 14.4 Status of CREM practice: selective

organisations, must awaken to the need to exploit the added value of their RE to increase their financial standing as well as to enhance economic development.

5. Conclusion

The study has provided a good insight into the practice of CREM in an African country like Nigeria. It has thus strived to bridge the gap arising from the scarcity of empirical studies on the issue of CREM and added to the limited body of knowledge in RE education and practice in Nigeria. It is the conclusion of this study that CREM is at its infancy in the country as most organisations have failed to realise its benefits to make the African economy vibrant and developed. The practice therefore requires proper refocusing and nurturing to enable it be developed towards offering a value contribution to the economic development of Nigeria. The study provides a useful resource both for educators and for the professional body retraining professionals for contemporary RE practice. It also serves as a good tool that can be used to stimulate the economy for better and productive development.

Emerging from the research is the need for business executives to properly align with the trends in the importance and performance of the key factors for effective CREM. There is an obvious need for the industry to attribute more importance to CRE as an organisational resource that could be utilised to enhance productivity and increase profitability. The industry must, with effective CREM, increase productive activities in African countries, thereby transforming the countries from being import-dependent to productive economies. This will untie the Nigerian – and, for that matter, African – potential which has hitherto been neglected. There is also the need for business organisations to move into the mainstream of CREM in order for its relevance to be established within the context of the "whole" firm, and to properly train CRE managers in the art of contemporary and strategic RE management in order to be able to run the RE units as profit centres.

References

Adendorff, M. J. and Nkado, R. N. (1996) Increasing Corporate Value through Strategic Real Estate Management. *Real Estate Review*, 26(3), pp. 69–72.

Asson, T. (2002) Real Estate Partnerships: A New Approach to Corporate Real Estate Outsourcing. *Journal of Corporate Real Estate*, 4(4), pp. 327–333.

Avis, M., Gibson, V. and Watts, J. (1989) Managing Operational Property Assets: A Report of Research of Organisations in England and Wales. Reading, UK: Department of Management and Development, University of Reading.

Bouris, G. (2005) Value Creation and the Contribution of Corporate Real Estate: Mapping Enterprise Value and Driving Positive Financial Impact. *Corporate Real Estate Leader*, July.

Breitensten, O, May, A. and Eschenbaum, F. (1998) *The Component of Corporate Real Estate Management*. Berlin, Heidelberg & New York: Springer Verlag.

Bryan, R. T. (2003) Managing Virtual Project Teams. A research project submitted to the Faculty of San Francisco State University in partial fulfilment of the requirements for the degree of Master of Business Administration [online]. Available at: Online.Sfsu. Edu/Ceb/B895/Btrautsch895.Pdf.

Christersson, M. and Rothe, P. (2012) Impacts of Organizational Relocation: A Conceptual Framework. *Journal of Corporate Real Estate*, 14 (4), pp. 226–243.

CoreNet Global (2005), Real Estate in the News: Corporate Real Estate Leader, *Corporate Real Estate Network*, 4(1), p. 12.

Gale, J. and Case, F. (1989) A Study of Corporate Real Estate Resource Management. *Journal of Real Estate Research*, 4(3), pp. 23–34.

Gibson, A.V. (2002) Risk Management and the Corporate Real Estate Portfolio. Paper presented at American Real Estate Society Annual Meeting of 1987.

Hartmann, S., Linneman, P., Pfnu¨ r, A., Moy, D. and Siperstein, M. (2010) Responsibility For and Performance of Corporate Real Estate Functions. *Journal of Corporate Real Estate*, 12(1), pp. 7–25.

Hite, G.L., Owers, J.E. and Rogers R.C. (1984) The Separation of Real Estate Operations by Spinoff. *AREAU Journal*, 12(3), pp. 318–332.

Hwa, K. T. (2003) The Reorganisation and Restructuring of Corporate Real Estate. Paper presented at the 9th Pacific RIM Real Estate Society Annual Conference, January 20–23, Brisbane, Queensland, Australia [online]. Available at: www.prres.net/Papers/ Hwa [Accessed 27 October 2009].

Ilsjan, V. (2007) Corporate Real Estate Management (CREM) in Estonia. *Journal of Corporate Real Estate*, 9(4), pp. 257–277.

Joroff, M., Lonargand, M., Lambert, S. and Becker, F. (1993) Strategic Management of the Fifth Resource: Corporate Real Estate. Report of Phase One: Corporate Real Estate 2000. Atlanta Georgia: Industrial Development Research Foundation,

Kaluthanthri, P. C. (2009) Corporate Real Estate Management Practices in Sri Lanka: Experience of the Commercial Banking Sector. *Sri Lankan Journal of Real Estate*, 2, pp. 61–83.

Loukko, A. (2004) Competitive Advantage from Operational Corporate Real Estate Disposal. *International Journal of Strategic Property Management*, 8(1), pp. 11–24.

Manning, C. and Roulac, S. E. (1999) Corporate Real Estate Research within the Academy. *Journal of Real Estate Research*, 17(3), pp. 265–279.

McDonagh, J. (2002) Measuring Corporate Real Estate Management Performance. Paper presented at the 7th Annual Pacific Rim Real Estate Society Conference, 21–23 January, Christchurch, New Zealand.

McDonagh, J. and Hayward, T. (2000) The Outsourcing of Corporate Real Estate Asset Management in New Zealand. *Journal of Corporate Real Estate*, 2(4), pp. 23–32.

Nourse, H. O. (1994) Measuring Business Real Property Performance. *Journal of Real Estate Research*, Fall, pp. 431–444.

Nourse, H. and Kingery, D. (1987) Survey of Approaches to Disposing of Surplus Corporate Real Estate. Journal of Real Estate Research, 2(1), pp. 51–59.

Nourse, H. and Roulac, E. (1993) Linking Real Estate Decisions to Corporate Strategy. *Journal of Real Estate Research*, 8(4), pp. 475–494.

Nunnington, N., and Haynes, B. (2011) Examining the Building Selection Decision-Making Process within Corporate Relocations: To Design and Evaluate a Client Focused Tool to Support Objective Decision Making. *Journal of Corporate Real Estate*, 13(2), pp. 109–121.

Oi, T. (2010) Corporate Real Estate Management in Japan. Paper presented at the 16th Annual Pacific Rim Real Estate Society Conference, 24–27 January, Wellington, New Zealand.

Oladokun, T. T. (2010) Towards Value-Creating Corporate Real Estate Assets Management in Emerging Economies. *Journal of Property Investment & Finance,* 28(5), pp. 354–364.

Oladokun, T. T. (2011) Corporate Site Selection and Acquisition in a Nigerian GSM Communication Company. *Journal Corporate Real Estate,* 13(4), pp. 247–260.

Oladokun, T. T. (2012) An Evaluation of the Training Needs of Nigerian Estate Surveyors for Corporate Real Estate Management Practice. *Property Management,* 30(1), pp. 86–100.

Oladokun, T. T., Aluko, B.T. and Odebode, A. A. (2009) Corporate Real Estate Management: A Need for Paradigm Shift in Nigeria. *The Estate Surveyor and Valuer, Journal of the Nigerian Institution of Estate Surveyors and Valuers,* 32(1), pp. 67–73.

Onamade, B. and Adejugbe, A. (2014) Establishing Corporate Foundations in Nigeria: Some Legal and Compliance Considerations [online]. Available at: http://www.stra chanpartners.com/ESTABLISHING%20CORPORATE%20FOUNDATIONS%20 IN%20NIGERIA SOME%20LEGAL%20AND% 20COMPLIANCE% 20 CONSIDER ATIONS %2014 0414.pdf [Accessed 20 January 2014].

Palm, P. (2013) Strategies in Real Estate Management: Two Strategic Pathways. *Property Management,* 31(4), pp. 311–325.

Petison, P. K. (2007) A Review of Lease versus Own Decision in CRE: The Way Forward for Developing Countries Such as Ghana. An MSc. thesis submitted to the Department of Real Estate Construction Management, Division of Building and Real Estate Economics, Royal Institute of Technology, Ghana.

RICS (2002) Property in Business, A Waste of Space? [online]. Available at: http://www. rics.org/site/download_feed.aspx?fileID=2418&fileExtension=PDF [Accessed 25 September 2012].

Ristaniemi, E. and Lindholm, A. (2011) Added Value of Corporate Real Estate Management in Industrial Premises [online]. Available at: www.cres.scix.net/data/ works/att/eres2011_134.content.pdf [Accessed 3 January 2012].

Rodriguez, M. and Sirman, C. F. (1996) Managing Corporate Real Estate: Evidence from the Capital Market. *Journal of Real Estate Literature,* 4(1), pp. 13–36.

Roulac, S. (2001) Corporate Property Strategy Is Integral to Corporate Business Strategy. *Journal of Real Estate Research,* 22(1/2), pp. 59–80.

Roulac, S. and Heaney, G. (2003) Corporate Real Estate in Ireland: A Current Perspective on Corporate Strategic Decision Making. *Journal Property Finance and Investment,* 20(1), pp. 31–44.

Rutherford, R. and Stone, R. (1989) Corporate Real Estate Unit Formation: Rationale, Industry and Type of Unit, *Journal of Real Estate Research,* 4(3), pp. 121–130.

Sanusi, L. S. (2010) The Nigerian Banking Industry: What Went Wrong and the Way Forward. A Convocation lecture delivered at the Convocation Square to mark the Annual Convocation Ceremony of Bayero University, 26 February, Bayero University, Kano [online]. Available at: http://www.cenbank.org/OUT/SPEECHES/2010/THE% 20NIGERIAN%20BANKING%20INDUSTRY%20WHAT%20WENT%20 WRONG%20AND%20THE%20WAY%20FORWARD_FINAL_260210.PDF [Accessed 23 September 2013].

Schaefers, W. (1999) Corporate Real Estate Management; Evidence from Germany Companies. *Journal of Real Estate Research,* 17(3), pp. 301–320.

Seiler, J. M. Chatrath, A. and Webb, R. J. (2001) Real Asset Ownership and the Risk and Return to Shareholders. *Journal of Real Estate Research,* 22(1/2), pp. 199–212.

Sharp, D. (2013) Risks Ahead: The Transformation of the Corporate Real Estate Function. *Journal of Corporate Real Estate,* 15(3/4), pp. 231–243.

Tay, L. and Liow, K.-H. (2006) Corporate Real Estate Management in Singapore. *International Journal of Strategic Property Management*, 10, pp. 93–111.

Veale, P. (1989) Managing Corporate Real Estate: Current Executive Attitude and Prospects for Emergent Management Discipline, *Journal of Real Estate Research*, 4(3), pp. 1–21.

Wills, P. C. (2008) Corporate Real Estate Practice in Australia. *Journal of Corporate Real Estate*, 10(1), pp. 40–53.

Woollam, C. (2003), Flexibility at Any Price? Challenging a Costly Convention on Leases. *Journal of Corporate Real Estate*, 6(1), pp. 73–82.

Zeckhauser, S., and Silverman, R. (1983) Rediscover Your Company's Real Estate. *Harvard Business Review*, Jan/Feb, pp. 111–117.

15 Real estate markets and valuation practice in Central and Eastern Europe

Slovenia, Hungary, Poland and Lithuania

Vida Maliene, Isabel Atkinson, Maruška Šubic Kovač, Andrea Pödör, Judit Nyiri Mizseiné, Robert Dixon-Gough, Józef Hernik, Maria Pazdan, Rimvydas Gaudėšius and Virginija Gurskienė

1. Introduction

The value of real estate (RE) extends far beyond the land, buildings and fixtures that define it. As well as a reliable financial investment, RE is valued as a place of residence, work, business and leisure, which can provide security and produce an income. Consequently there are many stakeholders (including sellers, buyers, developers, creditors, agents, investors, valuers and the State) with varying demands and requirements with respect to the RE. To satisfy their needs (for example, to invest in RE, buy, sell, develop and mortgage RE), there are various procedures and controlling measures, such as RE law, market analysis and valuation practice, which have to be considered. The RE market is a unique measure that contributes to the economy locally, regionally, nationally and internationally. By definition „a market is a situation in which buyers and sellers of [a] particular commodity or services are in sufficient numbers to create the opportunity of comparing prices and quality, thereby enabling the forces of supply and demand to operate" (Wootton, Jones Lang, Estate Gazette Limited and South Bank Polytechnic, *The Glossary of Property Terms*, 1989). Whilst this definition is correct, it does not fully describe the RE market, as this is a market which presents greater complexity due to its diverse nature.

The RE market is intangible and exists all over towns, cities and the countryside. Understanding the complexity of the RE market is quite challenging as the nature of this market is imperfect; units of RE themselves are heterogeneous and difficult to compare. The RE market is also influenced by the behaviour, psychology and taste of stakeholders, and many other factors including:

- demand and supply of RE type;
- the regional, national and global economy;
- government policy including taxation;

- mortgage or business credits availability;
- transparency of transactions;
- availability of market information;
- geographical characteristics;
- changing fashion;
- communication and accessibility; and
- location, size, age, quality and specification of RE.

The RE market can typically be arranged into sub-sectors geographically: local, regional, national and international. The range of markets can be subdivided further by considering the nature of the RE investment: primary, secondary and tertiary. These sectors can cover RE whether it is regional, local or national, and as such, these decisions make the market even more complex and fragmented. The market has no formal organisation and can be further divided into occupational, investment and development markets. The characteristics of these markets vary as different factors affect performance, demand and supply in each sector. The RE market can also be subdivided into commercial (with further sub-sectors such as retail, industrial and leisure), residential, corporate and agricultural sectors.

Currently, in the globalisation era, when trade and businesses have been established across the globe, there is an even bigger challenge to define boundaries of markets and recognise all features influencing RE markets. Today RE markets overlap and have become dependent upon one another. In order to understand the current state of the RE market, including valuation practice in any country, it is necessary to consider past economics and social events that have occurred and fundamentally shaped the state of today's economy.

Up until 1990, Slovenia, Hungary, Poland and Lithuania remained under the command of the communist regime. Consequently RE was owned by the State in a majority of cases. Independence, however, re-introduced the concept of private ownership and subsequently re-activated the economies and RE markets of each respective country. As RE for purchase and investment became available, demand for RE valuation grew, along with the need for up-to-date legislation on planning, RE law and taxation.

In 2004, Hungary, Poland and Lithuania joined the European Union (EU). The Office of the Committee for European Integration (2008) observed that between May 2006 and May 2008 Poland witnessed accelerated development and an economic growth rate in excess of 6%. Unemployment also fell and household disposable incomes rose, resulting in an annual consumption growth of 5.2%. A significant factor influencing these figures was recongised to be the economic migration of Polish workers to the older European member states (Office of the Committee for European Integration, 2008). In its 10 years of EU membership, Lithuania has also witnessed significant economic stimulation, as well as the increased living standards of citizens (Ministry of Foreign Affairs of the Republic of Lithuania, 2014). Similarly, in Hungary, for the period of 2007 to 2013, expectations for employment, development and the economy as a whole were high, thanks to European investment accessed through EU membership (Hungarian Chamber of Commerce and Industry, undated). Slovenia's accession to the EU, however, was described to have a"positive, but negligible" affect on macro-economic measures (Lavrac and Majcen, 2006).

Whilst newly invigorated capitalist RE markets, along with accession to the EU, demonstrated the benefits to be had from globally interdependent markets, in 2008 the global finanical crisis highlighted the disadvantages and risks of operating an open market. The global recession was triggered by the downturn of the sub-prime sector of the US housing market (Verdick and Islam, 2010). The US was reported to have fallen into recession in December 2007 (Verdick and Islam, 2010), however it spilled over into Europe and the rest of the world following falling global stock markets and banking system crises. Rose and Spiegel (2012) observe how industrialised countries, as well as those with emerging economies, were greatly affected. Unstable economies bring unemployment, reduced job security and tightened lending. These factors deter or even prevent potential buyers from investing in RE. Consequently, residential RE markets suffer. Similarly, as economies contract, business and commerce can suffer from decreased trade and difficulties in terms of obtaining finance. This subsequently has repercussions on RE markets.

This chapter provides an overview of Slovenian, Hungarian, Polish and Lithuanian RE markets, including the analysis of main players/actors, RE legal framework and RE valuation practice. In 1989 the countries in Central and Eastern Europe (CEE) began a remarkable transition from centrally planned economies towards market economies when the Berlin Wall fell and the Iron Curtain lifted (Hartvigsen, 2014). Whilst Slovenia, Hungary, Poland and Lithuania joined the EU at the same time, on 1st May 2004, their economies, including RE markets, have developed in some distinctive ways over the years.

2. The Slovenian RE market

2.1 Introduction

After the declaration of independence during the 1990s, the Slovenian RE market developed relatively slowly in comparison to other countries which have also joined the EU (Lithuania, Latvia, Czech Republic and Poland). In the late 1990s, one could observe the first steps towards increased activity in the RE market. In terms of legislation, the RE market began developing after 1998 when the demand for RE, especially apartments, began to grow in accordance with the growing purchasing power and population migrations driven by adaptation to job availability. However, according to the Geodetical Administration of the Republic of Slovenia (GARS, 2014), the supply lagged behind the growing demand. This was due to unimplemented systemic solutions which affected the availability of RE in the market (unclear ownership situation, legal procedures, absence of a market-based RE tax, the State as an over-large owner of primarily agricultural and forest land).

Compared with other EU Member States, the Republic of Slovenia is a relatively small country with a surface area amounting to 20,273 km^2, and is heterogeneous with regard to the natural demographic and economic characteristics. National level statistical data is shown below. These figures are significant for understanding the performance of the RE market in Slovenia.

Figure 15.1 shows that the number of inhabitants in Slovenia has been increasing since 1999. However, since 2008 the situation in the economic sector has not been favourable: Figure 15.2 illustrates that GDP has been on a downward trend and unemployment has been increasing. These factors have consequently had an impact on the RE market. With regard to natural characteristics, the country is predominantly wooded and most inhabitants live in detached houses and in their own homes/apartments.

At a local level, the public administration functions are performed in 58 general territorial administrative units in accordance with the principle of de-concentration. Local self-governance is also ensured as the inhabitants of Slovenia exercise self-government functions in municipalities. Currently there are 211 municipalities in the Republic of Slovenia.

Regarding the settlement structure, it should be noted that there are about 6,000 settlements, of which only approximately 90 have more than 2,000 inhabitants. The capital and national centre of Slovenia, the city of Ljubljana, on the other hand, has more than 200,000 inhabitants. Slovenia is also characterised by the uneven socio-economic development of different parts of the country, which is reflected in the RE market. Consequently there are big differences in the development and price levels between regional markets and even between the local markets within them. Whilst in urban areas the RE market is developed, market activity remains low in certain pockets.

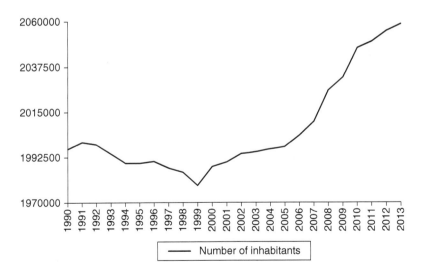

Figure 15.1 Population of the Republic of Slovenia (1990–2013)

Source: Own study based on Statistical Office of the Republic of Slovenia (SORS) data (1990–2013).

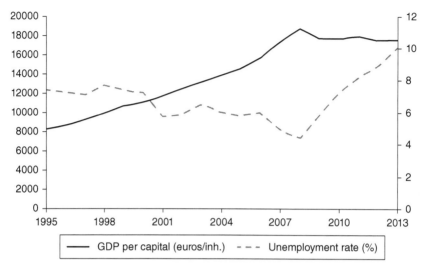

Figure 15.2 GDP per capita (euros/inh) and unemployment rate (%) (1995–2013)
Source: Own study based on SORS data (1995–2013).

Table 15.1 Statistical data on certain characteristics of Slovenia

Categories of land cover (ha) (2007)

Total	Wooded areas	All agricultural Areas	Bare soils	Water	Built-up areas
2.027.300 (100%)	1.338.654 (66.03%)	562.753 (27.76%)	31.764 (1.57%)	13.503 (0.67%)	57.158 (2.82%)

Distribution of population by dwelling type (2012)

Total	Flat	Detached house	Semi-detached house	Other
100%	28.9%	66.6%	4.1%	0.4%

Population by tenure status (2012)

Tenant at market price	Tenant at reduced price or free	Owner-occupied (with mortgage)	Owner-occupied (no outstanding mortgage)
5.5%	18.3%	8.4%	67.8%

Source: Own study based on SORS data and EUROSTAT (2007, 2012).

2.2 National RE market structure and performance by sector

Financially, the economic crisis reverberated in the RE market of Slovenia. Whilst in the global RE market the level of transactions and RE price growth trend continues, Figure 15.3 illustrates that, since 2008, the level of transactions

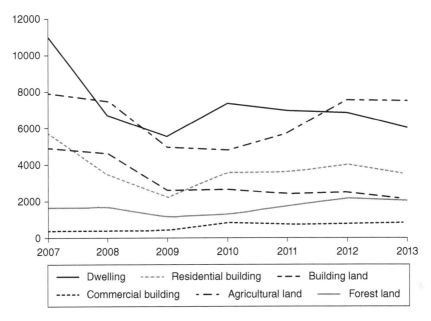

Figure 15.3 Number of transactions by type of RE (2007–2013)
Source: Own study based on SORS data (2007–2013).

and RE prices in Slovenia have tended to show a downward trend, and this has been linked with financial and economic instability.

Whilst supply of used apartments is relatively stable, the supply of new apartments continues to decrease; this is somewhat due to the purchases made by the Apartment Fund of the Republic of Slovenia. It is anticipated that in the years to come there will be more RE in the market from the so-called "stranded investments", which are managed by the Bank Assets Management Company (BAMC), established in 2013. Following the collapse of the major construction companies that had been key investors in the RE market, there has been an observed increase in the amount of smaller construction companies who are building and renovating minor residential houses for known buyers. Banks are extremely cautious when lending to households and economic entities. This, along with the decreased purchasing power, has ramifications on the aforementioned conditions in the RE market.

2.3 The main players/actors in the RE market

With regard to the number of transactions, the main actors in the RE market are individuals and households. Following the change of the socio-economic system in Slovenia in 1990, residential RE went through great changes: the economic function of a residential apartment strongly supplanted its social function. An apartment became, to a greater extent than previously, an object of capital investment. The State, which used to provide apartments to its citizens, now only makes

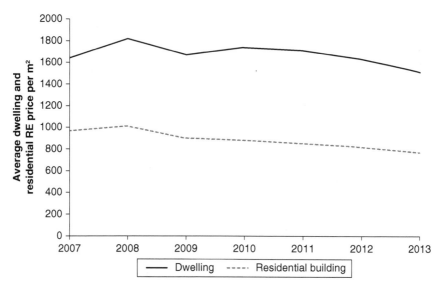

Figure 15.4 Average dwelling and residential RE price per m² (euros/m²) (2007–2013)
Source: Own study based on SORS data (2007–2013).

it possible for them to acquire their own apartments. Individuals mostly purchase apartments for their own household's needs and from their own financial means, with the help of bank loans.

Over past 20 years, the State, or more specifically the joint-stock Motorway Company of the Republic of Slovenia, was to a large extent involved in highway network construction in Slovenia and somewhat less in housing construction. The main actor of the RE market in this field is the Housing Fund of the Republic of Slovenia Public Fund, established in 1991. The Fund, serving the interest of the State, covers the whole territory, and in accordance with its business policy, finances and implements the national housing programme and promotes house building, renewal and the maintenance of flats.

Following the collapse of the former methods of construction finance (through the mandatory collection of financial means for such purposes from all employees) in 1991, the circumstances surrounding the financing of housing construction were rather unfavourable. Domestic/national banks were recapitalised and they were rather unprepared for the new methods of operation under these new circumstances. In the first half of the 1990s, enquiries for loans were relatively low (Cirman, 2010) and interest rates were relatively high. After inflation was brought under control and when mortgage loans were introduced in 1997, circumstances gradually improved. Following 2000, more and more foreign banks came to Slovenia; offering loans, not only to buyers, but also to investors of new apartments, thus increasing the stock of apartments in the market.

Many RE agencies were founded in the early 1990s when the field of RE brokerage was not legally regulated. Today the activities of RE agencies are regulated in the Real Estate Agencies Act 2003 (Official Gazette of the Republic of Slovenia, hereinafter UL RS 42/03). This Act distinguishes between the RE agency service providers and the RE agent profession. Within such agencies, services are carried out by regulated RE agents.

The passing of the Real Estate Agencies Act in 2003 saw the profession legally regulated for the first time. The activity of RE agents is relatively new and thus many infringements occur – for example: the inadmissible duplication of RE offered for sale on the market; compelling third persons to sign the RE agency contract; the inappropriate division of payment for agencies; advertising without an agency contract; and shifting the entire agency payment to a third person. The quality of operations performed by RE agents in Slovenia is also affected by the unregulated conditions in the field of RE records, for example, inadequate data in the Land Register; RE Register; the Cadastre of Buildings; and the Land Cadastre.

In the past, construction companies frequently acted on the RE market as the implementers of construction as well as investors in construction. As evident from Figure 15.5, the economic crisis is reflected in the value of construction put in place.

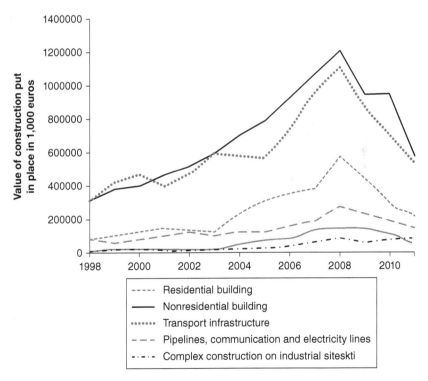

Figure 15.5 Value of construction put in place by type of structure (1998–2011), in 1,000 euros

Source: Own study based on SORS data (1998–2011).

Foreign investors into RE are not common in Slovenia. The Court of Auditors found in its Audit Report (2013) that the Ministry of Economic Development and Technology of the Republic of Slovenia, during most of the period that the Audit Report refers to, failed to stimulate foreign direct investment in an efficient manner. Additionally, the Report observed that the Public Agency of the Republic of Slovenia could have more efficiently implemented the measures of prompting foreign direct investments as laid down by the Promotion of Foreign Direct Investment and Internationalisation of Enterprises Act 2006. This, in turn, could have stimulated entrepreneurship, innovation, development, investment and tourism. Also, the World Investment Report shows that in the past year, Slovenia was the least attractive country within the EU for foreign direct investment. The Slovenian economist Damijan (2013) states that as a foreign investor in Slovenia, one needs to have a "strong desire and nerves of steel" as, in addition to other expenses, the cost of labour is amongst the highest in the world; the tax legislation is amended on a monthly basis, as well as the budget where applicable; the business environment is fully unstable and the labour legislation inflexible; most public tenders are tailored; one does not get anywhere without corruption; the judiciary does not function or the proceedings take a decade or more; prices of plots of land are too high; and the taxes are becoming an unbearable encumbrance.

2.4 Challenges of the existing RE legal framework

2.4.1 Legal aspects of RE investment

On 1 January 2003, after a transitional period of several years, Slovenia adopted the new Spatial Planning Act 2002. At the same time, the new Construction Act 2002 was passed, which instead of two permits [the location (planning) permit and building permit], introduced a single permit – the new building permit. In 2007 the new Spatial Planning Act was introduced. The Act manages topics relating to spatial planning and building development. The individual planning acts/documents at the different levels are shown in Table 15.2.

For areas containing individual towns, settlements of urban importance, or other development centres which will become towns or settlements of urban importance, an urban plan may be adopted. In 2010 an Act regarding the siting of spatial arrangements of national significance in physical space was passed. This Act newly regulated the national level of spatial planning in Slovenia. The two spatial planning systems presented here play an important role in the investment of construction processes.

Land for building is generally obtainable on the basis of a legal transaction or deed (purchase/sales contract, exchange contract, or donation contract), inheritance, the law (of usucaption/adverse possession/acquisitive prescription), or of a decision issued by a national authority (for example, judicial decisions on the division of RE in joint ownership, administrative decisions on the reallocation and consolidation of agricultural land, an administrative decision on re-parcelling or an administrative decision on denationalisation). The State and local communities

Table 15.2 Administrative division of Slovenia and types of planning documents

Slovenian administrative division	The institution responsible for planning	Planning documents (studies)
National level	Ministry of Infrastructure and Spatial Planning of the Republic of Slovenia	1 National strategic spatial plan 2 National spatial plan
Regional level	Slovenia has not yet formally established the regional administrative level	3 Inter-municipal spatial planning document as a "regional spatial plan"
Local level	Local community; municipal council, mayor	4 Municipal spatial plan with the strategic and operational part 5 Municipal detailed spatial plan 6 Urban plan

Source: Own study based on Spatial Planning Act 2007.

may acquire property rights over RE provided that all the necessary prerequisites have been met. They may also do so on the basis of a final and binding administrative decision on expropriation.

Encroachment upon the right of property within the public interest is carried out according to the Spatial Planning Act 2002. Land for building may be obtained under public interest on the basis of a legitimate pre-emptive right and on the basis of expropriation. For the first time in 70 years, the law is developing with regard to land policy instruments and the reallocation and consolidation of land for building; however, to date, it has not been put into effect.

The construction of buildings and networks, with the exception of commercial public infrastructure, is permitted on land with developed facilities. Within a municipal spatial plan, a municipality determines the types of public utility infrastructure, which has to be constructed by individual spatial planning units. The supply of utility services to land is carried out on the basis of a programme for the supply of utility services. The costs of construction of the public utility infrastructure envisaged in a contract are charged to the investor. By being charged the costs, it is considered that the investor has paid, in kind, the public utility charges for the construction of the infrastructure that they have installed. A public utility charge is a payment for part of the costs of the construction of public utility infrastructure, which the debtor of the costs (hereinafter referred to as the debtor) pays to the municipality. The amount of the public utilities charge does not include costs for the maintenance of the infrastructure.

Procedures applicable to the granting of building permits have been laid down in the Construction Act 2002. Despite the several amendments to the Act, the procedures applicable to the granting of building permits have remained relatively lengthy.

2.4.2 RE Taxation System

The more significant taxes linked to RE include:

- inheritance and gifts tax (Inheritance and Gift Taxation Act 2006);
- RE turnover tax (Real Estate Turnover Tax Act 2006);
- value added tax (Value Added Tax Act 2006);
- capital gains tax on RE sale (Personal Income Tax 2006);
- rental income tax (Personal Income Tax 2012);
- agricultural land development tax (Agricultural Land Act 1996);
- tax on RE of greater value (Fiscal Balance Act 2012);
- planning gain tax (Fiscal Balance Act 2012); and
- RE tax (Real Estate Tax Act 2013).

In the recent years, Slovenia has instituted several new *ad valorem* taxes on RE, including the new RE tax (Real Estate Tax Act 2013). Two months after its adoption, in February 2014 the government decided, on account of the disputability of the higher tax rates or tax values, to equalise the tax rate for residential and non-residential RE, whilst the category of "land for building construction" (envisaged land use) is to be abolished and the land will be valued and taxed according to its actual use.

2.4.3 Legal aspects of RE valuation

In Slovenia there are no legal regulations in place to regulate individual RE valuations. The Real Estate Agencies Act 2003 only temporarily (in the period 2003–2006) laid down the individual RE valuation and predicted the issuing of individual RE valuation methodology; however, the relevant Article was abrogated at a later date. Due to the inertia of the administrative RE valuation from the previous socio-economic system, the aforementioned Act explicitly abrogates the administrative RE valuation.

Depending on the Valuers' Society, the valuers may, or must, in their operations use the international valuation standards of the International Valuation Standards Council (http://www.ivsc.org/), and the Slovenian Business and Financial Standards (UL RS 56/2001, as amended), which include the real estate valuation standards, the hierarchy of standards, the recommended literary sources, and similar. Specialised laws (important in particular in the 1990s) regulate the characteristics which need to be taken into account by valuers. These laws include the regulations governing the privatisation of RE, denationalisation of RE, the expropriation of RE and limitation of property rights on RE.

For the purposes of RE mass valuation, a specific law was passed titled the Real Estate Mass Valuation Act 2006, which jointly with the Real Estate Tax Act 2013 constitutes the legal framework for RE taxation in the Republic of Slovenia. The Act states that the mass appraisal system shall be used to determine the generalised value of RE. The Act provides the legal framework but does not prescribe in detail the procedures and data contents; this is to avoid frequent amendments caused by the system's adaptation in response to changes in appraisal models and in the RE market.

At present, a major problem concerning mass RE valuation is inadequate data. Consequently, RE valuers are not utilising the generalised market values of the mass valuation models as a basis for appraisal.

2.5 Existing governmental and/or financial institutions' support schemes for RE investment

The Housing Fund of the Republic of Slovenia Public Fund is one of the most significant governmental institutions on the RE market. The activities of the Housing Fund include the provision of loans for the purchasing, construction or reconstruction of apartments or residential houses of natural persons, and the setting-up and implementation of specific saving schemes within the so-called National Housing Saving Scheme.

Since 2000, the Fund has been operating as a RE fund which has constructed, autonomously or by co-investment with the communities, a major quantity of residential apartments. These apartments are then sold, in accordance with the public tender conditions, under more favourable conditions than those on the market. Alternatively, some apartments have been allocated to the beneficiaries as rented non-profit apartments.

In 1999, the Fund set up the National Housing Saving Scheme, a saving-crediting system for natural persons in collaboration with volunteer business banks. The Scheme constitutes a systemic basis for the promotion of long-term savings and favourable crediting (lending of funds). On conclusion of the first cycle, it became evident that the offer of loans by the (business) banks increasingly resembled the loans according to this Scheme; however, the funds saved did not suffice for resolving the housing problem.

Slovenia is among the countries with an extremely high proportion of proprietary apartments. For this reason, the tenancy market for such apartments should be stimulated. In addition to the provision of better legal security for the tenant and the lessor, the Housing Fund should play an important role in this particular area by providing a greater quantity of non-profit apartments.

The need for non-profit apartments is great and difficulties lie in the provision of funds for their construction or purchase. The national budget does not supply funds to this end, and the non-profit rents do not suffice even for the coverage of the legally-defined costs. A non-profit rent is paid for non-profit apartments. This rent is defined by an administrative method (a points-scoring method) which does not take into account the location of apartments. The research carried by Zwoelf (2006) shows that market-determined rents in the centre of Ljubljana are as much as five times higher than the non-profit rents defined by these methods. In such cases, this is more of a social rent than a non-profit rent.

2.6 RE valuation practice

In the field of individual Slovenian RE valuation, there is no legally defined system for the valuation of RE in terms of value definition, RE valuation methods, information bases for valuation (costs of construction of typical RE and adjustment factors) and specific Slovenian standards of RE valuation, which would regulate the activity of individual RE valuation in Slovenia and improve the quality of valuation reports.

RE valuers, as previously mentioned, make use of the International Valuation Standards and methods utilised by the American Society of Appraisers (ASA). The start of major changes that introduced market-related methods to RE valuation took place in 1992. At that time, experts of the American Society of Appraisers, in conjunction with the Agency of the Republic of Slovenia for Promotion of Economic and Company Reconstructions, organised a training seminar on RE as part of the company privatisation process. The seminar mainly dealt with RE valuation methods used by the ASA experts, which are still used in all valuations in Slovenia today.

The main sources of data for RE value assessment, in addition to the more specific ones, include:

- the RE Market Record (accessible free of charge via the web portal of the Geodesic Administration of the Republic of Slovenia; separate login for the registered users, and for the other users);
- the RE Register (including the records of all RE in Slovenia – data has been obtained from the existing public records, the Land Cadastre, Cadastre of Buildings, Central Register of Inhabitants, and other records) and supplemented with the data of the RE census and accessible free of charge via the web portal of the Geodesic Administration of the Republic of Slovenia);
- the collection of Spatial Acts (collective Spatial Acts are accessible free of payment at the common or communal web portal);
- the Design Engineers' Investment Assessments collection (which contains over 400 value-rated and already constructed buildings whose end price is divided by post-calculation into percentages, construction works, craftsmen's works, and installations, and is accessible on payment at the web portal of the PEG-Construction Portal/PEG-Gradbeni portal); and
- the Land Register (a collection of data on the legally valid rights and facts on RE, accessible free of payment at the web portal of the Law Courts).

Based on this data, two payable applications have been created: the TRGOSKOP application and later, on account of dissatisfaction of the valuers, the CENILEC application.

For the purposes of the law courts in Slovenia, there are the judicially sworn-in valuers of RE and for the needs of financial reporting there are the so-called authorised appraisers of RE value. Under the Courts Act 1994, the power of nomination of sworn-in valuers and experts in civil engineering (within this also RE sworn-in valuers) was transferred from the primary courts to the Ministry of Justice. On the initiative of this Ministry, the Association of Sworn-In Valuers and Experts in Civil Engineering of Slovenia was established in 1997.

During the 1990s, there arose a need for so-called authorised appraisers of RE value for financial reporting purposes. In 1994, the Slovenian Institute of Auditors was established as a representative body of qualified licensed auctioneers and valuers. It issues licences to RE valuers and prescribes special training programmes for new applicants. Most of these valuers are sworn-in valuers.

2.7 Conclusion

Based on the new Constitution of 1991 (UL RS 33/1991), the Republic of Slovenia set up the new foundations for the capitalist market economy and thereby also the foundations of the field of RE. In the transition period, the inertia of the old system still governed, in particular in the fields of data collection on the functioning of the RE market, data on RE, RE valuation, and data on the RE agents. Legal regulations in this field were adopted more than 10 years after the passing of the new Constitution. During this period the privatisation and restitution processes were taking place. In 2004, Slovenia acceded to the European Union and in 2007 it became part of the Eurozone. Following a period of expansion, in 2008 Slovenia fell into recession. The economic conditions had ramifications on the Slovenian RE market including a relatively extensive offer of RE for sale, a large number of unfinished construction investments, empty RE, and a decline in RE prices.

3. The Hungarian RE market

3.1 Introduction

Hungary is a landlocked country in Central Europe. It is situated in the Carpathian Basin and is bordered by Slovakia to the north, Ukraine and Romania to the east, Serbia and Croatia to the south, Slovenia to the southwest and Austria to the west. The country's capital and largest city is Budapest.

Hungary is generally divided into four geographical regions: the Alföld (Large Plain), the Northern Highland, Dunántúl (Transdanubia) and the Kisalföld (Small Plain). A high proportion of the territory, about two-thirds of the total surface, is suitable for agricultural production. The continental climate and the good soils enable the production of a broad range of crops. Other comparative advantages include the long-term tradition of plant production and animal husbandry, and the commitment of the professional workforce and population to agricultural activity. Production is also supported by a developed agricultural research, education and extension network.

Administratively, Hungary is divided into 19 counties, with the capital, Budapest, independent of any county government. The counties and the capital are the 20 NUTS third-level units of Hungary. Since 1996, the counties and Budapest have been grouped into seven regions for statistical and development purposes (see Figure 15.6). These seven regions constitute NUTS' second-level units of Hungary. They are Central Hungary; Central Transdanubia; Northern Great Plain; Northern Hungary; Southern Transdanubia; Southern Great Plain; and Western Transdanubia. The counties are then further subdivided into districts. There are also 23 towns with the same rights as counties.

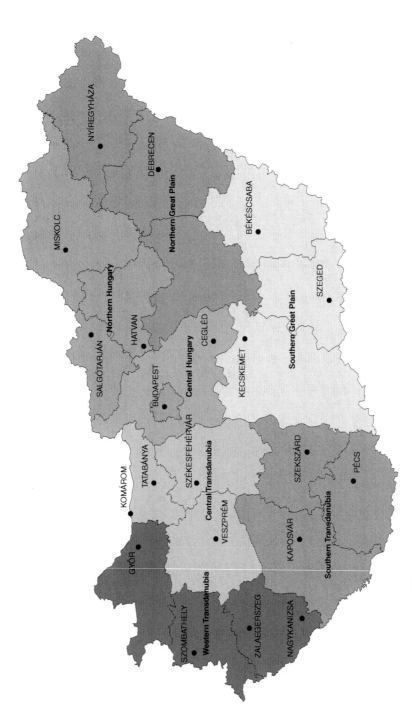

Figure 15.6 Regions of Hungary

Source: Own study based on Hungarian Central Statistical Office data (2014).

In 2014, Hungary's population was recorded as 9,879,000, with a population density of 107 inhabitants per square kilometre. More than one quarter of the population live in the Budapest metropolitan area, with 69.5% in cities and towns overall. Roughly on a par with most other European countries, the total fertility rate in 2013 was estimated at 1.41 children born per woman, which is lower than the replacement rate of 2.1. Additionally, life expectancy in 2012 was observed as 71.55 years for men and 78.38 year for women. These figures have continued to increase since the fall of Communism.

As Figure 15.7 indicates, the number of inhabitants of Hungary is decreasing in line with the population trends of other European countries. However, in Hungary this decrease is not mitigated by immigration, as Hungary is not considered an attractive destination for migrants.

The economy of Hungary is a medium-sized, upper-middle-income, structurally, politically and institutionally open economy. The market liberalisation started in the early 1990s as part of the transition from a socialist economy to a market economy. Additionally, Hungary has been a member of the Organisation for Economic Co-Operation and Development (OECD) since 1995, a member of the World Trade Organisation (WTO) since 1996, and a member of the European Union since 2004 (OECD Economic Survey, 2014).

The private sector accounts for more than 80% of the Hungarian gross domestic product (GDP) and foreign ownership in Hungarian firms is common. Hungary's main industries include metallurgy, construction materials, processed foods, textiles, chemicals (especially pharmaceuticals) and motor vehicles.

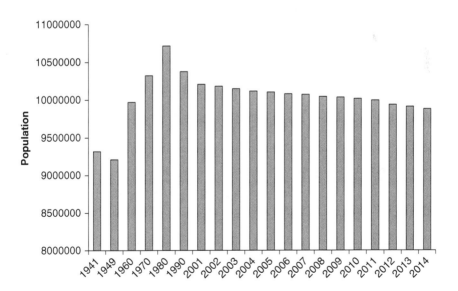

Figure 15.7 The decrease in the population of Hungary (1941–2014)

Source: Own study based on Hungarian Central Statistical Office data (1941–2014).

Agricultural cultivation is also a widespread tradition across the country; consequently the land market has a significant influence on the state of the Hungarian economy.

Due to the favourable conditions in agriculture, land ownership and the farming structure have always been important and politically charged issues in Hungary. This is illustrated by a history of three land reforms in the past half-century, their only common feature being the underlying political motivation at the expense of economic efficiency. These land reforms created huge problems for a rural population dependent on agriculture, and it is generally agreed that agriculture and land ownership have failed to develop functionally in the past decades.

As a result of the first land reform of 1945–1948, the agrarian structure became characterised by the dual structure of many small-scale farms co-existing with a few, relatively large, State farms, which covered 15% of the arable land. Despite the dictatorial collectivisation following the Second World War, by the 1980s a specific, quite efficient, production structure developed.

3.2 Recent issues of RE

During the 1990s, in parallel with political changes, the transformation of Hungarian agriculture began. After the collapse of command economy structures and State socialism, a return to private ownership and farming through a third agricultural land reform was pursued as a potential remedy to the problems. This agricultural land reform, however similar to other former socialist countries, revealed numerous contradictions. The reform has caused a significant change in the agricultural structure, the outcomes of which have affected more than half of the total area of the country and over 2.5 million new owners were created. Compensation and land privatisation have resulted in inadequate land size and shape, and an overly fragmented distribution of plots belonging to one owner, failing to support viable family farming and competitiveness. As a result of land privatisation, the previous farming on-large-scale was replaced by farming based on private ownership, which is today characterised by inadequate RE field size for sustainable and competitive family farming. Plot ownership patterns differ significantly from the land use structure of viable agricultural plots. However, the introduction of a well-established land consolidation procedure, supported by National Land Fund integrating the rural development approach of some EU Member States, has significantly contributed to a better quality of life in rural areas.

The recent economic situation can be connected to the former privatisation processes. Parallel to the privatisation of the agricultural lands, a similar, but not quite analogous, process occurred in the RE market. This procedure produced an abundance of privately owned RE without any financial background for maintenance; only a few units of RE remained in the hands of local government. The current economic crisis has also had ramifications for the Hungarian RE market as the following section will outline.

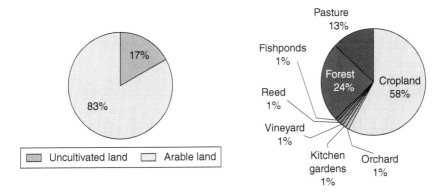

Figure 15.8 Percentage of different types of arable land

Source: Own study based on Hungarian Central Statistical Office data (2013).

3.3 Current situation and outlook of the Hungarian RE market

GKI Economic Research Co. conduct RE surveys twice a year based on the opinion of market players. GKI published two indices, one for Budapest, and another for the country as a whole. These relate to residential, office space, retail space and warehouse markets, and provide an overall view of the entire RE sector. In particular, the indices reflect the relationship between supply and demand, as well as the future plans and expectations of market actors.

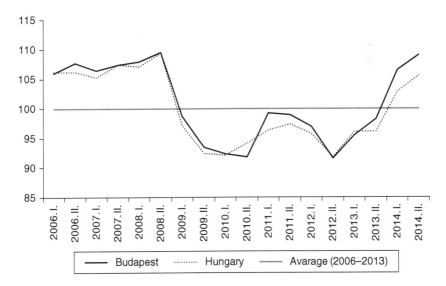

Figure 15.9 RE indices produced by GKI (2006–2014)

Source: Own study based on GKI data (2006–2014).

According to the research by GKI (2014), recently there have been no new office developments and occupancy rates have remained high. This lack of new development has caused office space market indices (which reflect the expectation of RE developers, realtors and companies) to plummet. The expectations of respondents showed some optimism, but hopes for a market recovery have disappeared. Furthermore, expectations for the capital market are significantly worse than those expected for rural areas. The retail RE market, on the other hand, has stabilised at a high level of turnover rate due to strong retail competition.

In 2011 retail trade sales were observed to stagnate, and went on to decrease during 2012 and 2013 due to the regression of consumption of the Hungarian households. In 2012 retail space market indices (which reflect the expectation of RE developers, realtors and companies) were also observed to be shrinking.

In the residential RE market, house prices have fluctuated since the global recession of 2008. According to Hungarian Central Statistical Office (KSH) house prices in Budapest have dropped sharply, since:

- in 2008, house prices fell by 1.7%;
- in 2009, house prices fell 11.34%;
- in 2010, house prices rose by 0.11%;
- in 2011, house prices fell 3%; and
- in 2012, house prices fell 5.57%.

In addition to falling prices, longer selling times have also been observed across the board. Table 15.3 shows that in the given period, the number of sold homes fell dramatically. Whilst this trend continues, expectations of the residential and office space markets have considerably improved. Expectations for the retail market, on the other hand, have not improved, reflecting the uncertainty of future government measures.

If Table 15.3 and Figure 15.10 are analysed, a significant correlation can be detected between the prices of RE and the number of newly built homes.

In 2014 the above-mentioned indices published by GKI show a significant increase, and optimism was reported among the actors of the RE market. In 2014, 8,358 new homes were built, which was an increase of 15% on the previous year. Whilst the number of new dwellings increased in Central Hungary and in Transdanubia, the negative trend has not changed in the Eastern part of Hungary. The number of dwelling construction permits has also increased year-on-year by 28%, to 9,633 (Hungarian Central Statistical Office).

3.4 Challenges of the existing RE legal framework

A new Hungarian Civil Code regulates the new order of contracts and inheritance, which are closely connected to the RE market.

3.4.1 Legal procedure for purchasing RE in Hungary

On 15th March 2014 the new Hungarian Civil Code (new HCC) entered into force with the intention of strengthening legal certainty. The new HCC is based on the

Table 15.3 Number of sold homes (2007–2013)

Number of sold homes and construction for selling (1,000 pieces)

Year/quarter of a year	total number of sold flats	new homes built	homes for resale	construction for market purposes
2007	191,20			17,90
2008	154,10	140,00	14,10	17,40
2009	91,10	82,90	8,30	16,90
2010	90,30	85,50	4,80	10,70
2011	87,70	83,90	3,90	4,80
2012	86,00	83,30	2,60	3,50
2013, first three quarters	55,90	54,60	1,3	1,7

Source: Hungarian Central Statistical Office (KSH) (2007–2013).

autonomy of the individual and is typical of its dispositive regulation; that is, any-thing which is not forbidden by law is allowed. Among the provisions of RE law, the new HCC introduces an online registry of the pledges on movables, allow-ing online registration and deletion. The general rules of contract law have also been changed, as well as the rules of certain types of contracts. The main purpose was to implement existing judicial practice into the new HCC. Additionally, the rules relating to family law have been implemented based on the new HCC. The rights of partners and spouses are converging, mainly in favour of the protection

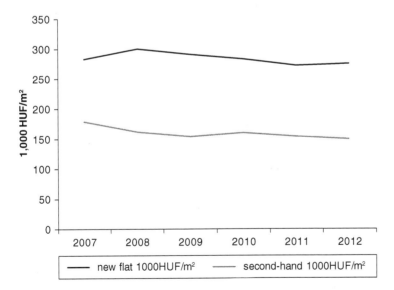

Figure 15.10 Prices of RE in Hungary
Source: Ingatlannet (2014).

of children. The regulation of the matrimonial property regime has also been extended; by the use of a matrimonial property agreement spouses may decide, themselves, how to arrange their property relations. The provisions on parental supervision also changed, both in their scope and content. The new HCC focuses on the responsibility of parents, and gives a key role to the agreements made by them. The rules of adoption and guardianship have also become more realistic.

3.4.2 The process of buying RE in Hungary

Buying RE in Hungary is said to be relatively easy. Nowadays foreign nationals can also buy RE without limitations (not counting agricultural lands, which do have some restrictions). Firstly, a solicitor is needed to go through the whole process for a commission of 1% of the sales price. After the RE is chosen the owner's status is checked at the Land Registry, a 10–20% deposit then has to be paid and there should be a signed purchase agreement which ensures the owner cannot sell it to anyone else. The solicitor will then file the contract of sale at the Land Registry and the new owner's name will appear as a "side note" on the ownership certificate, as the next owner. This is enough to prevent the previous owner from selling it or putting any mortgage on it. Once the total purchase price is paid, the possession is fully registered. The RE can only be sold when the registration on the side note is confirmed. RE can be bought by a private individual or company.

The general legal framework governing the types of RE ownership is set out in the Act on Arable Land, the Act on Real Estate Registration (Act CXLI of 1997) and the Civil Code of Hungary (Act V of 2013). From 1 May 2014, the Act on Arable Land has ceased to exist, and the rules on arable land are now regulated by the Act on Trade of Agricultural and Forestry Lands Act CXXII of 2013. According to the rules, Hungarian private individuals and legal entities may acquire ownership of RE freely; however, a few restrictions and prohibitions exist.

The new Act on Arable Land CXXII of 2013 allows citizens of EU member states to purchase agricultural land in Hungary. According to the Act, Hungarian and EU citizens can buy land in Hungary, but if the area exceeds one hectare, buyers have to prove that they are genuine farmers. This will be identified by a degree in forestry or agriculture, or at least three years' experience of an agrarian activity in Hungary prior to the purchase. Only natural persons are allowed to purchase agricultural land, but some exceptions still exist. Farmers, or close relatives of farmers, are able to acquire most of the land: up to 300 hectares. Farmers can lease up to 1,200 hectares of land for farming purposes. If a tenant farmer also owns land, then the amount of land available for lease is reduced by the amount of land that he owns.

3.5 The main players/actors in the RE market

The actors and the services of the Hungarian RE market are typical. They include the seller, buyer and renter as well as the services provided by the likes of RE agencies, mortgage providers and solicitors.

In Hungary the number of privately owned apartments (typically residential apartments) is extremely high. This has an unfavourable impact on the RE market. This is due to the higher cost of rent in comparison to loans in the period prior to the financial crisis.

The global financial crisis has had a considerable impact on the Hungarian economy and RE market. Developers must cope with the restricted availability of funding. However, the few developers who have their own capital can take advantage of the under-utilisation of construction firms, and the resulting lower prices. In Hungary the volume of investment is still significantly below the levels prior to the global financial crisis. The market consists typically of international cross-border transactions or transactions in connection with the restructuring of the RE portfolio of a company. The prospects for the RE market in Hungary remain generally linked to the macro-economic outlook of the country. The market is still stagnant, and investors are waiting for an upward shift in the Hungarian economic environment.

Most of the transactions made by physical persons (as opposed to legal persons) and the majority of sold units of RE are second-hand flats. In the first nine months of 2014, the housing market data showed an upturn in sales; the growth in the number of homes sold was more than 30%. Predominantly second-hand homes changed hands: the share of new homes accounted for only 2.5% of the housing market (Hungarian Central Statistical Office, 2015).

Before the economic crisis, newly built houses were mainly financed by loans offered by foreign banks in foreign currencies. Following the crisis, these loans were observed to cause social problems as individuals we unable to meet their repayments; consequently, the government stepped in to implement the following actions:

- the State, which previously provided apartments to citizens, is now investing in apartment developments, or the purchase of flats, to rent to those who have lost their apartments due to the foreign loans;
- there are a number of so-called debtor rescue practices such as "the foreign currency loans" which convert debt to HUF loans at the actual exchange rate; and
- the government is assisting the foreign currency debtors to reclaim a total HUF1,000 billion back from the banks and reduce the monthly instalments of troubled borrowers.

These actions have implications for the loan market, and it is likely that this will include a reduction in bank lending.

The government, along with the ministries, has recently initiated some general construction. Whilst house building is typically initiated by developers, the government has allocated funding for families who wish to invest in house building. There are a number of different initiatives, however, these are predominantly intended to support young families with children who wish to buy or build a new flat.

New Government Regulation 314/2012 (XI.8) on settlement development concepts, integrated settlement development strategy and means of spatial planning have been accepted. The government has also set up a separate fund for its Integrated Urban Development Strategy (IUDS). In 2009 the preparation of an IUDS became a legal requirement for every city. For cities of county rank, the government provides the following assistance at a national level: a call for proposals for ensuring financing; methodological guidelines; and individual mentoring. In line with spatial planning initiatives, many cities receive funds for rehabilitation and the renewal of certain areas.

The activities of RE agencies are regulated by the Housing Act as amended. The latest amendment of the Housing Act was passed through Parliament (by Act No. CXXXII of 2005) on 14 November 2005, and the amendments became effective on 31 March 2006. The new Act contains provisions concerning the preconditions of the activities involved with RE agency and appraisal, and the trade of RE. These activities, which are specified by the Act, may be pursued exclusively by business organisations where at least one member, or one employee, engages personally in that activity, and who has the requisite qualifications prescribed in statutory regulations. The person who is engaged in this activity has to satisfy the following criteria: the individual has adequate qualifications; does not have a criminal record; and has no debt. RE agents are required to register themselves, and provide the urban local government and competent court of registration with their registration certificates.

3.6 RE taxation system

Taxes payable in Hungary when purchasing a unit of RE are as follows.

Stamp Duty is payable on the purchase of RE. Rates range from 2% to 10% depending on the value of the RE. A purchaser is exempt from Stamp Duty if the value of the RE is under €60,000. Tapered relief applies for RE valued between €60,000 and €120,000.

VAT: the selling of newly built RE is always subject to VAT. The VAT payable on the sale of second-hand units of RE depends on the VAT status of the RE seller. The Hungarian VAT rate was raised to 25% from July 1st 2009.

Luxury Tax (Wealth Tax) applies on residential and holiday dwellings.

Local Property Taxes (rates) are payable and vary per location. The average cost of local tax is €3.95 per square metre per year; building tax is 900HUF per square metre or 3% of the market value of the RE; and land tax is 500HUF per square metre or 3% of market value. In all cases the final decision rests with the local tax authority.

The Capital Gains Tax (CGT) rate is 27%. CGT deductions will apply depending on the length of time for which the residential RE was owned, as follows:

- If the RE to be sold was owned for five or more years then there is no obligation to pay CGT;
- If the RE to be sold was owned for between four and five years then the CGT will be paid for the amount corresponding to 30% of the capital gain;

- If the RE to be sold was owned for between three and four years then the CGT will be paid for the amount corresponding to 60% of the capital gain;
- If the RE to be sold was owned for between two and four years then the CGT will be paid for the amount corresponding to 90% of the gain; and
- If the RE to be sold is owned for less than two years then 100% of the gain will be chargeable.

A seller is exempt from CGT if he or she re-invests the gain back into a room or an apartment in a retirement home in Hungary or any European Member State.

Inheritance Tax (IHT) is payable by non-resident beneficiaries on certain transferred assets. The tax rate depends on the relationship between the beneficiary and the donor as well as on the value of the RE. The rates vary from 2.5% to 40%. Any inheritance below 20 million HUF, where the inheritor is the child, parent or spouse, will be exempt from stamp duty. This exemption also applies to an inheriting niece or nephew without a parent as a legal guardian.

Taxation is very much dependent on the land registry system. Moreover, this administrative system is relatively well-established and modernised compared to the European standard.

3.7 RE administration in Hungary

In Hungary, there is an integrated Cadastre and Land Registration agency providing one single supervising function at the Ministry of Agriculture and Regional Development. The system is under continuous redevelopment, but it is based on the following organisational and structural elements. The role of the Department of Lands and Mapping is twofold: it serves as National Mapping Agency for large-scale base mapping (cadastral and topographical maps), and also as national Land Administration (land registration, land evaluation, land use and land consolidation). It operates and supervises the Land Office Network consisting of 19 county-level and 116 district Land Offices, the Capital Land Office, and the Institute of Geodesy, Cartography and Remote Sensing. This latter runs the application-oriented remote sensing and geographic information system/land information system (GIS/LIS) research, technology developments and operational services (Márkus and Nyiri, 2005).

3.8 RE appraisal in Hungary

The practical penetration of the concept of RE appraisal began in Hungary between the two world wars. In the first decades of the twentieth century significant RE investment created an economic environment in which the RE appraisal played an important role. Kotsis' work regarding the appraisal of RE was published in 1927. In 1942 his work was also published by the Institute for Engineering Education, which already contained most of the RE appraisal studies – on a high level – necessary today.

After the war, the progress of the professional field slowed down. The appraisal usually arose as a task performed by duty offices and tax authorities for official purposes. The traditions of classic (independent, individual) value assessment were perpetuated by legal experts. These experts gained a huge database of experience. The legal experts' work built on the analysis of market facts, as well as, in many cases, the analysis of price and cost.

A mass demand for RE appraisals arose again from the late 1980s. As the Hungarian economy was developing and transforming, the transformation of RE appraisals for various purposes, including appraisals for loan security, played an important role.

Whilst the ownership reorganisation brought about by social transitions (privatisation, compensation) did not happen according to the rules of the open market, it focused on the necessary breakdown of State RE operating with an improper effectiveness. Additionally, it focused on the social administration of justice. Since the end of mass privatisation, economic policy has seen increasing demand for real values measured in the open market behind the economical movements. The economic policy provides this requirement with legal and institutional background.

A new element of the Hungarian market is the appearance of institutional investors. A great number of deals have been made since the late 1900s (mainly in the capital and in some larger cities). They have been significantly contributing to the stability of the RE market and to the development of a structured, mature market. The institutional investment is typically a market transaction in which the real market value provides a base for the investor's decision; the return-based appraisal gives support to the investors.

A key element in the RE market is the valuation method or process. In Hungary, as in other EU countries, the method which is permanently in use is elaborated by TEGoVA, the European Group of Valuers' Associations. The only exclusion is the valuation of the arable lands where a scientifically developed method is used. In the case of Hungary, this is very important – as explained in detail in the following section.

3.9 Valuation methods

Different approaches to patent valuation are used by companies and organisations. Generally, according to Hajnal (1999), Berdár et al. (2005) and TEGoVA's European Valuation Standards (2012), these approaches are divided into two categories: the quantitative and qualitative valuation. Whilst the quantitative approach relies on numerical and measurable data with the purpose of calculating the economic value of the intellectual property, the qualitative approach is focused on the analysis of the characteristics and potential uses of the intellectual property, such as the legal, technological, marketing or strategic aspects of the patented technologies. Qualitative valuation deals also with assessing the risks and opportunities associated with the intellectual property of the company.

Several methodologies are used in the quantitative approach, but generally they can be grouped in four methods (Rakvács, 1999; Soós, 2002, Fórizs, 2000):

- the cost-based method;
- the market-based method;

- the income-based method; and
- the option-based method.

The cost-based method is based on the principle that there is a direct relation between the costs expended in the development of the intellectual property and its economic value. Two different techniques are used to measure costs:

1 the reproduction cost method: estimations are performed by gathering all costs associated with the purchase or development of a replica of the patent under valuation; and

2 the replacement cost method: estimations are performed on the basis of the costs that would be spent to obtain an equivalent patent asset with similar use or function.

In both methods, present prices are taken into account; that is, the expenditures as of the valuation date and not the historical costs when these actually happened. For assessing costs, two cost sources should be included: direct expenditures, such as costs with materials, labour and management; and opportunity costs, relating to the lost profits due to delays in market entrance or investment opportunities lost with the aim of developing the assets.

The market-based valuation method relies on the estimation of value based on similar market transactions (for example, similar licensing agreements) of comparable patent rights. Given that often the asset under valuation is unique, the comparison is performed in terms of utility, technological specificity and property, having also in consideration the perception of the asset by the market. Data on comparable or similar transactions may be accessed in the following sources:

1 company annual reports;

2 specialised royalty rate databases and publications; and

3 in-court decisions concerning damages.

The income-based method is based on the principle that the value of an asset is intrinsic to the expected income flows it generates. After the income is estimated, the result is discounted by an appropriate factor with the objective of adjusting it to the present circumstances and therefore determining the net present value of the intellectual property. There are different methods of calculation of the future cash flows, such as:

1 the discounted cash flow method: this method aims to estimate future cash flows, which are projected and discounted by applying an appropriate discount factor. The main source of information to estimate the cash flows is generally the business plan of the company that exploits or intends to exploit the asset; and

2 the relief-from-royalty method: in this method the value of the asset is considered as the value of the royalty payments from which the company is relieved

due to its ownership of the asset. Hence, the appropriate royalty rate must be determined, allowing the estimation of the future royalty income stream. A discount rate is applied to determine the present value of the asset.

3   Different from the other methods, the option-based method takes into consideration the options and opportunities related to the investment. It relies on option pricing models (for example, Black-Scholes) for stock options to achieve a valuation of a given intellectual property. In these cases, patents may be valued using the techniques developed for financial options, as applied via a real options framework. The key parallel is that a patent provides its owner with the right to exclude others from using the underlying invention, so both patents and stock options represent a right to exploit an asset in the future and to exclude others from using it. The patent (option) will have value to the buyer (owner) only to the extent that the expected price in the future exceeds the opportunity cost of earning just as much by a riskless alternative. Thus patent rights can be thought of as corresponding to a call option and may be valued correspondingly.

3.10 Current methods of arable land assessment

As mention previously, the market of arable land plays a special and important role in the Hungarian RE market. This is explored in greater detail in this part of the study based upon the works of Sipos and Szücs (1992) and Mizseiné Nyiri (2008).

Since the arable land is not amortised during production and cannot be reproduced, no cost-based assessment (depreciated replacement cost method) can be applied for the establishment of its market value. So the value of the arable land can be established through two methods:

1   by analysing the comparative data of the market; and
2   with a method based on the calculation of the yield.

3.10.1 Assessment of arable land by market comparison

In the case of market comparison, the assessment is made by the extension of the prices of the completed and known transactions of sale. The market value is the price for which the RE can be sold within the frame of a private, legal contract and at the date of the assessment, assuming the following:

- the seller is willing to sell;
- due time is available for the arrangement of the sale;
- the value does not change in the period of negotiation;
- the sale is made with proper publicity; and
- an offer deviating significantly from the average, due to special interest, is accepted.

The steps of the assessment are as follows:

- *Selection of the basic set.* The geographical location of the RE in the basic set is similar to the examined RE while the land use should be the same as that of the examined RE. The extreme values of the basic set significantly deviating from the average have to be ignored during the analysis.
- *Selection of the RE suitable for the comparison.* Only identical forms of values and only identical rights (for example, unencumbered ownership and rental right) may be compared. Correction factors have to be applied between the different forms of values and rights.
- *Determination of the specific basic value.* The specific basic value has to be projected, usually to the hectare.
- *Analysis of the value-modifying factors.* Only significant factors influencing the value can be taken into consideration, which is not typical of the basic set of the comparative data.
- *Modification of the specific basic value, calculation of the specific value.* From among the value-modifying factors, those significantly influencing the value have to be quantified (in per cent, or nominally in Hungarian forint) and in accordance with it, the base value can be adjusted. The so-gained value will be the specific value of the RE.
- *Calculation of the final (market) value.* The RE market value is gained by the product of the specific value and the size of the RE.

The limit of the application of the method means there is no intensive land turnover at present, and the data of the completed transactions are not published either, so in many cases it means a difficulty in taking up the suitable basic set.

3.10.2 Land assessment with yield calculation (income and expenditure method)

The classical economic theories establish the land price in a relatively simple way as the quotient of the land allowance (income part attributable to the land) and the capitalisation rate of interest. If the owner (seller) wants to sell his land at such a price and deposits this amount into a bank, he will receive an annual income comparable to the alternative utilisation of leasing out the land.

The application of this relatively simple formula causes several problems. The basic question is how much of the allowance can be attributed to the land as a factor of production. In order to resolve the problem, the literature advises several methods (in accordance with the residual value, hoped-for net income and substitute cost). A description of the methods can be found in the literature referenced at the end of this chapter.

The Hungarian Agricultural Economic Research Institute elaborated a practical method in which the income is determined on the basis of the present land assessment system ("AK") and the value of the rent is also involved in the calculation. This is the Golden Crown. The cadastral golden crown net income is a complex index number which shows the quality of the land in accordance with the estimation of the expenditures and yield results at the end of the nineteenth century and at the beginning of the twentieth century. Its compulsory use was ordered in the registration of the RE by the Act of VII/1875. It is still available for all land uses and land parts nationally; it is included in the national RE registration.

3.11 The role of geographic information systems (GIS) in market analysis

RE investment means a process in which investors take part in proportion to their investments. It is essential to reduce the risk and enhance the quality of the investment. GIS has a very important role in the decision support phase of this process, and also in the feasibility study. Displaying information in maps often reveals trends, patterns and opportunities that may not be detected in tabular data alone.

The location or the position of the RE is one of the most important factors determining its value. It is not merely a geographical term; in fact it is rather an economic term, as it is summing up the totality of those environmental conditions and services that jointly influence the quality, comfort, use value and market value of the given RE. Location is not just about finding any site but finding the best site. GIS technology can deliver the results that investors require. By analysing data connecting to locations – demographics, aerial photographs, traffic networks, shopping centre usage, merchandise potential data, and competitive influences – the investors can find ideal locations for RE. GIS helps the RE industry to analyse, report, map and model the merits of one site or location over another (Pödör and Nyiri, 2010).

The sensitivity to location in the residential RE sector is key. Sensitivity linearly increases with the specific value. A block of freehold flats of medium standard can be built in many places in a large city as more locations can fulfil the conditions of such an investment; however, the construction of a luxury villa may be only appropriate in a special environment and under special conditions if it is to be sold efficiently.

GIS allows residential RE developers to analyse a lot of different information, such as parcel, zoning, tax, census, flood risk and demographic data. This enables the creation of accurate business models that establish the economic potential of different sites or land units.

In Hungary, as an emerging market, there are definite differences in real estate value connected to location and also to economic productivity (see Figure 15.11), although there are some examples of when big investors like Mercedes can change the economic background of the location. In the case of Kecskemét, RE development was also changing with the emerging market needs because of investment by Mercedes in the city. Nevertheless, overall indices showed that there was a definite decrease in the RE market in the region of Kecskemét.

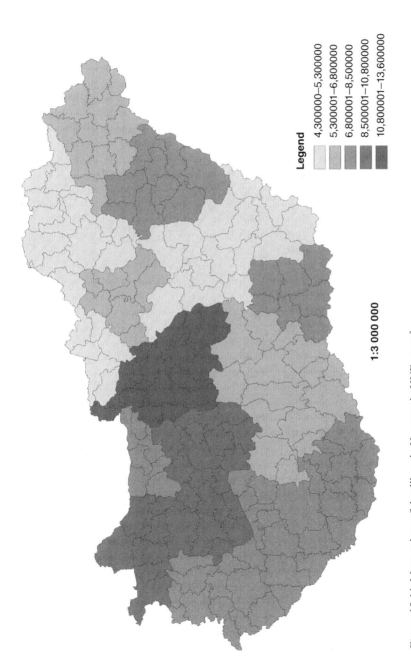

Legend

4,300000–5,300000
5,300001–6,800000
6,800001–8,500000
8,500001–10,800000
10,800001–13,600000

1:3 000 000

Figure 15.11 Mean prices of dwellings in Hungary in HUF per m².

Source: Own study based on Hungarian Central Statistical Office data (2014).

3.12 Conclusion

Following independence in 1991, the Hungarian RE market returned to a state of private ownership and farming. Agriculture has continued to play an important role in the economic and physical make-up of the country. Since the fall of communism and accession to the European Union, Hungary has seen a continued decrease in its population. This, together with the global economic crisis of 2008, has placed further pressure on the country's RE market. Commercial and residential RE values have fallen and the market has become stagnant. Furthermore, the country is not deemed an attractive place to buy for foreign investors, and these investors are also prohibited from purchasing agricultural land. However, despite these factors, expectations for residential and office space markets are optimistic. The implications of the global recession have had ramifications on home owners across the country, with individuals unable to meet their loan repayments. Subsequently the State has intervened through a number of measures in order to help those affected. However, these actions have implications for the loan market and it is likely that future lending will be affected, therefore potentially creating greater uncertainty in the Hungarian RE market.

4. The Polish RE market

4.1 Introduction

Poland is the ninth largest country in Europe and, since joining the EU in 2004, is the seventh largest in the EU. With a GDP of 4.3% (2011), it is ranked as the eleventh most wealthy country in the EU. However, with its highly-skilled workforce and relatively low remuneration, the potential of its abundant natural resources (for example, shale oil and gas) is significant.

Since transformation in 1989, the country has undergone significant changes in its social structure, governance and the development of its infrastructure. With a population of 38,200,200, it is the sixth most populous state in the EU. Still a mainly rural country, the majority of the population lived in the urban areas prior to 1989 (approximately 61%) and this has not changed significantly. The major cities and their populations are Warszawa (1.717 million), Kraków (0.758 million plus 0.2 million students), Trójmiasto [Gdańsk, Gdynia and Sopot] (0.742 million), Łódź (0.740 million), Wrocław (0.633 million), Poznań (0.553 million), and Katowice (0.308 million). Poland is divided into 16 voivodeships and 314 districts (powiaty). The smallest unit of administrative division in Poland is a community, of which there are 2,478 in Poland. The land use structure in Poland is presented in Table 15.4.

4.2 The Polish RE market

During the last decade of the twentieth century, urban space in Polish towns and cities has become a commodity and, for the private investor, it is one of the key

Table 15.4 Statistical data on land use in Poland by hectares (2013)

Agricultural land	Forest and covered with trees	Waters	Developed and urbanised areas	Ecological areas	Fallow lands	Other areas
18,770,139 (60.03 %)	9,633,820 (30.81 %)	647,378 (2.07 %)	1,612,791 (5.16 %)	35,565 (0.11 %)	476,147 (1.52 %)	92,128 (0.29 %)

Source: Główny Urząd Statystyczny (2013).

actors of those spaces (Kotus, 2006). New housing and shopping centres have been developed leading to denser populations in suburban areas, leading to dramatic changes in the spatial and social structures of Polish urban areas. In Poland, the city authorities and investors encourage the fullest commercial use of the urban spaces, locating commercial facilities and good quality housing in the city's open areas.

The relationship between investors and city authorities have had direct consequential results for the development of urban space with many public places in some of the largest Polish cities gradually changing their function. Kotus (2006) observes that in the central parts of many cities, office buildings, hotels, entertainment centres, and shopping centres have appeared; the latter becoming the open-spaces of the twenty-first century. Generally, most urban investments are intended to make the city space more attractive and are largely oriented towards visitors coming to town on business; tourists; suburban residents spending only a few hours in town; and young city residents with comfortable incomes. Thus, public places in most Polish cities have acquired a new, commercial meaning and have been increasingly appropriated by the categories of investment that include objects setting a new lifestyle for the younger groups of city dwellers, for example, entertainment centres accessible to specified social groups endowed with enough financial means to patronise them, and financial centres, hotels, offices, and congress centres that are of little practical use to most urban residents.

In addition to these pressures, there has been a growing trend, since 1989, for communities adjacent to the urban areas to encourage the growth of suburbia through the designation of areas of land suitable for housing. In urban areas, many former heavy industries have closed or relocated and their locations have been taken by retail and services industries, whilst the construction of modern office blocks has taken place around the periphery of most major towns (Dixon-Gough et al., 2013). Within the town, many former houses that had multiple occupancy tenancy prior to 1989 have either been converted to low occupancy apartments, high-end businesses (often in the service sector) or hotels.

One of the greatest impacts on the urban areas of all major cities is the newly constructed office space, the key investors being international companies. The major beneficiaries according to Michniak (2011) are Warszawa (3.5 million m^2), Kraków (0.48 million m^2), Wrocław 0.390 million m^2), Trójmiasto (Gdańsk, Gdynia, Sopot) (0.303 million m^2), Katowice (0.263 million m^2), Poznań

(0.246 million m^2), and Łódź (0.215 million m^2). These developments have made significant changes to the appearance of many outer areas, the working conditions of their employees, and the traffic flows of those cities.

In a topological accessibility survey of a typical urban area within a Polish city, determined on the basis of selected neighbourhood and public places, it may be observed that the zone of highest accessibility in an analysed set includes the public places, especially newly located ones, whilst the housing estates, both old and new, are situated outside it (Kotus, 2006). Other objects located in the best-access zone include office buildings, trade (in the form of new shopping centres) and warehouse areas, indicating that despite advances in personal mobility, locations competitive in terms of accessibility are those near the inner-city rings.

The location of large-scale retailing facilities – as close to the potential customer as possible – seems to be a unique feature of Polish trade. Although the number of car owners has recently increased, there are still many who still do not own cars. Furthermore, the state of repair and the quality of many urban roads are still unsatisfactory in most Polish urban areas. Hence the tendency to site retail facilities as close to the city centre with its urban clusters as possible.

Further significant changes have been made to the retail market, which may be divided into two sectors: (1) the DIY market, which is often linked to other outlets in home improvement, such as white goods, furnishings, and garden centres, located in large retail parks with good car parking facilities, and (2) the general retail sector – the individual shops and commercial premises on the ground floors of buildings located at the busy streets of the city, forming traditional strings of trade, and the galleries or shopping centres built close to the core of the cities, often on former industrial sites. These are still a relatively new innovation in Polish cities, the major retail outlets being the traditional retail outlets at street level, often with living accommodation above them.

As mentioned above, the trading areas such as the galleries still tend to be situated in some of the most accessible parts of Polish towns and cities, often close to public transport routes. The major difference between the trading and the DIY markets is probably in the nature of the products sold. This latter market is orientated towards the car-driving elements of the community, since the products sold cannot easily be carried on public transport: hence the need for the integrated planning for a good road network and associated infrastructure with these developments.

Also linked into this infrastructural development is the major category of development since 1989, which has been the proliferation of distribution centres, often on the periphery of the large cities, which have developed in two directions. Firstly, there are large logistic centres constructed for the developers of the international market – both west to Europe and east to Ukraine and Russia. Secondly, there are the small-scale centres consisting of a small to medium size warehouse or manufacturing facility developed by local investors and rented out to local businesses.

Against the backdrop of Europe, the Polish housing situation is alarming. On many indicators the country is ranked last place against other European Union

Member States (Polski Związek Firm Deweloperskich, 2012). In recent years, despite the relatively large number of houses constructed in the peri-urban areas, there has been no actual change in the number of dwellings in relation to the number of households. This is largely the result of the consolidation of inner city apartments and the change of use of many others (see above). This does not take into consideration the gradual improvement of the general housing stock. Much of the housing stock around the periphery of most Polish urban areas still consists of communist-style apartment developments, many of which have been significantly improved. These factors are exemplified in Figures 15.12 and 15.13 where, given the market trends across Western and Central Europe, it might be anticipated that the price of housing stock (per m^2) might have shown indications of a rise.

The price of housing in Poland has declined, the rate of decrease being greater in Warszawa than in Kraków due to the greater choice of both primary and existing stock. Largely due to the static number of house units in Kraków, the decline in value has been less marked. This variation may be extended across Poland as identified in Figure 15.14.

For Poland to provide its citizens with minimum levels of housing at a European level, against a backdrop of declining availability of credit, there should be an immediate development of a market for rental housing that requires more flexible rules regarding leases whilst also providing incentives for investors. To date, because of the excessive levels of risk associated with residential RE, the greatest level of investment has been in commercial buildings (often aided by foreign investment).

There is, however, a proven system for building societies, whose contribution to construction is now minimal but which could potentially be encouraged to increase their level of intervention through Government support. Based on statistical data and estimates of the Central Statistical Office (Główny Urząd Statystyczny), at the end of 2011 the number of households in Poland stood at approximately 14.65 million (Główny Urząd Statystyczny, 2011). The total number of apartments is approximately 13.6 million, which indicates that the number of individual houses is only 0.93. This result is identical to that achieved in 2002, which shows a disappointing level of change that is at odds with many of the developments that have taken place in peri-urban areas over the past decade. However, a significant market recovery was observed on the RE market following Polish accession to the EU. Opened borders led to the sale of RE and a sudden increase in construction. Since 2004, developers have begun to invest more in housing. Table 15.5 shows that the trend continued until 2008. After 2008, the housing market decreased together with the decreasing number of building permits issued, largely due to the global economic crisis that started in the US RE market.

The ineffective loan policy experienced in the USA, and to a lesser extent in the UK, was also evident to a lesser extent in Poland. Although banks became more liberal in granting loans between 2004 and 2008, the Polish currency (PLN) was strong in comparison to EURO, USD, and CHF, which resulted in a growing demand for RE and an inflation of house prices. Loans in foreign currency were also very popular, although it is now difficult to pay back such loans.

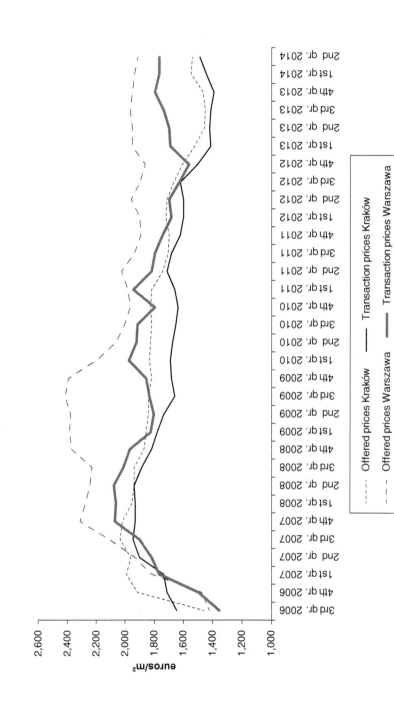

Figure 15.12 Offered and transaction prices per m² of apartments on the primary market

Source: Own study based on National Bank of Poland data (2006–2014).

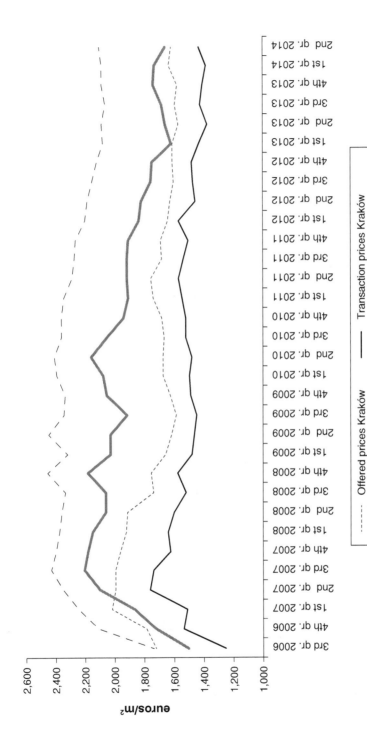

Figure 15.13 Offered and transaction prices per m² of apartments on the existing stock market

Source: Own study based on National Bank of Poland data (2006–2014).

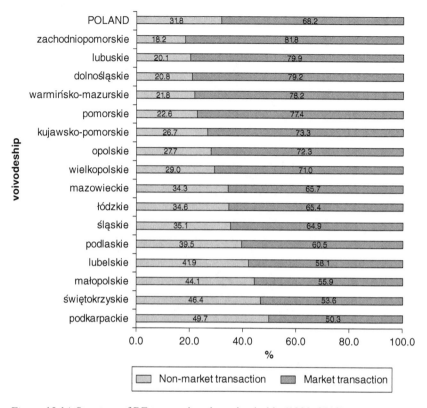

Figure 15.14 Structure of RE transactions by voivodeship (1990–2012)

Source: Own study based on data from the Instytut Rozwoju Miast (2013) 23 lata polskiego rynku nieruchomości – Monitoring za lata 1990–2012 (23 Years of the Polish Real Estate Market 1990–2012). Instytut Rozwoju Miast (Institute of Urban Development), Kraków.

Note: Non-market transactions are transactions without money: endowments, inheritances, transfers of farms in return for social security benefits.

This situation continued until 2009 when the global stock markets crashed, exchange rates increased and banks decided to tighten their loan policies. As a consequence, loans were more difficult to obtain, and the current supply in the RE market is higher than demand, which results in decreasing prices.

There are also other significant factors relating to the housing situation in Poland. For example, according to the Overcrowding of Housing Index calculated by Eurostat, Poland has the third lowest ranking in the EU. Conversely, statistics show that in 2010 some 81.5% of the population had title to their RE against the EU average of 70.7%; this is no doubt related to the cost of RE rental in Poland, which is amongst the highest in the EU, with an average of 30% of expenditure on the cost of housing (Raport, 2012). A growth in the availability of rented dwellings would aid the general structure of housing since it would facilitate both the

Table 15.5 Housing developments (2005–2013) in Poland

Year	Commenced housing investments (quantity of apartments)	Housing investments with building permits (quantity of apartments)	Apartments put into use (quantity)
2005	105,836	123,863	114,066
2006	137,962	168,378	115,353
2007	185,117	247,671	133,698
2008	174,686	230,146	165,189
2009	142,901	178,801	160,002
2010	158,064	174,929	135,835
2011	162,200	184,101	130,954
2012	141,798	165,092	152,904
2013	127,392	138,681	145,136
2014	148,122	156,752	143,235

Source: Główny Urząd Statystyczny (2005–2013).

mobility of labour and couples to have families through the provision of relatively cheap rental housing opportunities. One can see a huge disparity between investment made in offices, warehouses, and trades, respectively, which is in contrast to the actual lack of investment business in apartments for rent. This is illustrated below in Fig. 15.15, which shows a clear gap between the respective investments in offices, trading buildings, and warehouse/storage complexes. Despite the reduced level of transactions since 2006, trading buildings are clearly one of the success stories for Poland.

The images shown as Figures 15.16–15.18 are of office buildings in Kraków and are typical of the modern office space currently being constructed in Poland. Figure 15.16 is the Lubicz Office Centre, an 'A' class office block that satisfies the highest international standards. The building is situated in the city centre at the intersection of Lubicz Street and Rakowicka Street and is located close to the railway station (with rail links to the nearby international airport). Figure 15.17 shows the Edison, Newton, and Galileo blocks of an office complex belonging to GTC Corporation. They provide approximately 10,000m^2 of office space and are modern buildings offering a high standard of office accommodation. They are located in the very fast growing district of Kraków in close proximity to the city centre.

Figure 15.18 shows the Lubicz Brewery complex, the history of which dates back to 1840. In that year, the first brewery buildings were erected for Rudolf Jenny. At that time it was the largest brewery in the city. Despite wars and hard times, the brewery continued to function very well, although it was impossible to extend the brewery as it was built in the city centre and constricted by other buildings. Consequently in the 1970s the financial condition of the brewery declined and, in 2001, production ceased. The original buildings in the centre of the city have been renovated one after another by an investor. In the vicinity of old

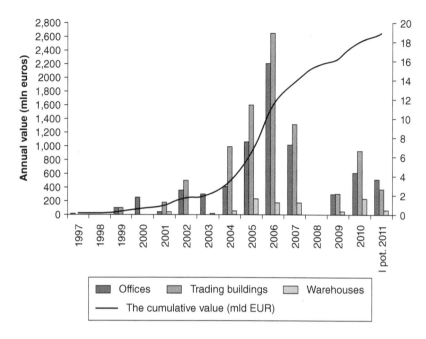

Figure 15.15 Value of investment transactions in Poland (1997–2011)

Source: Knight Frank (2011) Broszura – Rynek nieruchomości w Polsce.

Figure 15.16 Lubicz Office Centre, Kraków

Photo: M. Pazdan.

Figure 15.17 Office buildings Galileo, Newton and Edison, Kraków
Photo: M. Pazdan.

Figure 15.18 Kraków Lubicz Brewery – first phase of investment
Photo: M. Pazdan.

buildings, new buildings have been constructed that mimic the design of the original buildings. Based on the Local Spatial Plan, all of the buildings are designed to fulfil the three main functions: housing, offices, and services.

In rural areas, the situation has not changed significantly in terms of the area of arable, forest, orchards, meadows, pasture, and other land uses. Since 1990, there has been a decline in the profitability of small family farms, and there is an ageing workforce as many young people move from rural to urban areas. The main trend has been the decline in arable farming and the increase in forests and land for other uses. This is coupled with a decline in the value of land, particularly in those areas that are more difficult to farm efficiently using modern, mechanised techniques. In many rural communities, well away from the peri-urban areas of cities, there have been growing trends of land abandonment, encroachment of arable lands by woodland, and the purchase of plots of land for large weekend houses (Hernik et al., 2013).

4.3 The main players/actors in the RE market

For an effective transformation of the political system, the RE market had to change from one that was regulated by numerous prohibitions, with the major players and actors being State institutions such as housing cooperatives. It was only in 1990 that the ban on the sale and purchase of land and buildings was cancelled, followed, in 1994, by the sale of housing units; prior to that date there was a ban on having more than one flat, and low rents were regulated. Subsequently, during 1994–1995, six specialised groups of stakeholders emerged, consisting of investors, lenders, developers, tenants and leaseholders, intermediaries, and the maintenance market (Kucharska-Stasiak, 2006).

The basic operation of the RE market belongs to investors and lenders. Both groups bear the risk associated with investment projects occurring in the first years after the transition. Currently, there is a high risk associated with investments at the financial participation of the EU. The intermediaries group consists of specialists, which include expert advisors on RE, architects, lawyers, insurance agents, RE managers and planners (Kucharska-Stasiak, 2006).

The general process of the sales formalities of RE transactions is regulated by the Civil Code and includes the Property Act 1997, the Property Ownership Act 1994, and the Agricultural System Act 2003. The sales process is based upon the conventional process of an agreement between a willing seller and a willing buyer, the contract being as to the price and the object being sold, which must be defined in details recorded in the mortgage and the land register.

4.4 RE valuation practice

This process is driven by values. All RE has a perceived value and in Poland this is the role of the licensed appraisers who estimate the value of RE by making a thorough appraisal. They are obliged to carry out an appraisal according to the standards of the professional regulations and law. Currently the only standard

Table 15.6 Number of licensed appraisers

	Data valid for 31.12.2005	Data valid for 30.04.2014	Data valid for 31.07.2014
Mazowieckie voivodeship	702	999	1,102
Małopolskie voivodeship	325	464	494
Poland	4,396	5,568	5,970

Sources: Instytut Rozwoju Miast (2013) 23 lata polskiego rynku nieruchomości – Monitoring za lata 1990–2012 (23 Years of the Polish RE Market 1990–2012). Instytut Rozwoju Miast (Institute of Urban Development), Kraków; and Ministerstwo Infrastrukture i Rozwoju (2014).

of the profession binding on the appraisers is a standard called the "real estate appraisal for insuring liability". In addition, during the appraisal, adherence to the Polish General Rules for Real Estate Appraisal are recommended. These rules were prepared by the Polish Federation of Associations of Real Estate Valuation Experts (Polska Federacja Stowarzyszeń Rzeczoznawców Majątkowych) as rules for best practice. The number of licensed appraisers has increased significantly since 2005, as may be seen in Table 15.6.

The training of licensed appraisers involves passing a difficult State examination preceded by practice. Subsequently, the valuation is performed on the basis of appropriate approaches, methods and techniques of RE valuation. In addition, there are numerous standards. This ensures that the appraiser knows exactly the procedure for determining the valuation.

Recently, however, there has been an increasing number of RE valuation experts, whilst conversely there are fewer RE transactions, due to the economic situation in Europe. Consequently, the salaries of appraisers are decreasing. The danger is that this spiral could result in the lowering of the quality of the work done and in the rigorous professional responsibility.

Four approaches to RE valuation are generally recommended, each approach having its own methods and techniques (see Table 15.7).

The sales comparison approach is used if the transaction prices of similar units of RE, conditions of transactions, and features influencing the value of RE are known. Prices are corrected on the basis of features that differentiate similar units of RE from the subject RE and on the basis of the changes in the level of prices in time.

Knowledge of income obtained, or the potential to obtain from rents and other incomes from the subject RE or similar RE, is necessary in order to use a cost approach. In the cost approach, the cost of acquiring the land and the cost of reproduction of its components are determined separately. The cost approach is based on determining the value of the RE, assuming that this value represents the costs of the reproduction of the RE reduced by the value due to wear and tear. The mixed approach is a mix of the income and sales comparison approaches; an appraiser analyses similar RE and the incomes that they generate (Act on Real Estate Management 1997).

Table 15.7 Approaches to RE valuation

Approaches	Methods	Techniques
Sales comparison approach	Comparison of pairs Method of correcting the average price Statistical analysis of the market	
Income approach	Investment method Profit method	• Technique of simple capitalization • Technique of discounted cash flow
Cost approach	Cost of reproduction Cost of replacement	• Detailed technique • Technique • Technique of integrated components • Technique of factors
Mixed approach	Residual method Estimated land value method Liquidation value method	

Sources: Regulation of the Council of the Ministers from 21 September 2004 on the valuation of the property and the preparation of the appraisal, and Regulation of the Council of Ministers from 14 July 2004 amending the above.

An appraiser chooses the appropriate approach, methods and techniques of RE valuation, taking into consideration the aim of the valuation; the type and location of the RE; the designation of the RE in the Local Spatial Plan; the condition of the RE; and the available data on prices, incomes, and features of similar RE.

The sales comparison approach is the most common approach used by appraisers. Based on market analysis, one may say that 90% of valuations are based on this approach. For the last twelve years, appraisers have typically limited methods to the comparison of pairs and correction of the average price: statistical analysis of the market has been beyond the interests of an average appraiser. However, currently statistical analysis of the market is becoming popular among appraisers as it allows for the best justification of the assumptions made during the valuation process. Naturally, the method has its supporters and opponents but, until now, there was no substantive discussion in which both sides could present their opinions on the method (Kosmulska, 2009). The income approach is usually used for commercial RE as it is the best option to valuate RE that could yield a profit.

4.5 Challenges of the existing RE legal framework

The major challenges after 1990 involved adapting the land registry to a market economy (in PRL land registers were deliberately neglected by the State and marginalised); the regulation of the legal status of the property (mostly the consolidation of the ownership of the property); the possibility of crediting the purchase of dwellings (including the possibility of establishing land registers for cooperative housing companies); the enfranchisement of State entities (based on the conversion (by a committee) to land and buildings in the law of ownership of the buildings, and the perpetual usage of land, which permitted, following the privatisation of the company, the right to land and buildings); changes in economic, residential, and real estate regulation acquired from the Church by the State (returns or compensation); the new regulation of expropriation in accordance with the Constitution and the market economy; and elimination of the State-owned farms and the subsequent privatisation of agricultural land. Before Polish accession to the EU, the new laws were subsequently amended, changing the right of perpetual usage in the right of ownership and abolishing restrictions on the acquisition of agricultural property by entities from the EU and changes in the real estate acquisition by foreigners.

After 1990, the former administrative measures and penalties, which had effectively limited the investment in real estate, were cancelled. These changes related, primarily, to the unrestricted acqusition of real estate by citizens and companies; the creation of many new banks (including foreign) for investment and lending; new construction laws to adapt to a market economy; and new regulations relating to the acquisition of property under public investment (changes in expropriation and compensation).

In contrast, the problem of spatial planning has not yet been satisfactorily solved, which continues to limit any investment. At the beginning of 2000, many previous local development plans expired and there was no mandatory adoption of new ones. As a result, many Polish regions do not have such plans. The new Act on spatial planning and development was introduced in 2004. Table 15.8 shows the individual planning documents at the different levels. In 2013 only 26.8% of Poland was covered by Local Spatial Plans. In some communities, there was complete coverage by Local Spatial Plans whilst others had only partial coverage. The situation is similar in the cities and urban areas. The coverage is highest in

Table 15.8 Administrative divisions of Poland and types of spatial planning documents

Administrative divisions of Poland	Planning documents	Obligatory/optional
Country level	Strategy of spatial development for Poland	Obligatory
Voivodeship level	Spatal development plan for voivodeship	Obligatory
District level	None	–
Community level	Study of the condition and the directions of the spatial development	Obligatory, no legal impact
	Local spatial plan	Optional
	Decision about building and land development	Obligatory when spatial plan doesn't exist

Gdańsk (68%) and lower in Kraków (48.2%) and Warszawa (31.6%), with the lowest being Łódź, with only 6.2% coverage (Główny Urząd Statystyczny, 2013). The process of creating a plan is time consuming and expensive, which is the main reason for the lack of plans. Investments conducted with only planning permits, without the existence of a Spatial Plan are risky. According to the ranking in "Doing Business" in 2015, Poland is in the 32nd position (out of 189 countries) in the ranking of the "most business friendly economies" (World Bank 2015). Sadly Poland is in 137th position when dealing with construction permits, as compared with Lithuania (15th), Slovenia (90th) and Hungary (103rd).

This situation was resolved by the so-called Special Act which is issued to specific investment projects, such as the construction of motorways and express-ways, or the construction of stadiums for Euro 2012 (the European Football Championship organised by Poland and Ukraine).

4.6 RE taxation

Currently in Poland the following taxes that affect property are in force: RE tax; agrarian tax; and forestry tax. RE tax applies to parcels of land smaller than 1 hectare, and to buildings, apartments, and commercial lots. According to the Act on Forests, forests have to be larger than 10 hectares, covered with forest vegetation, or temporarily uncovered but designated for forestry, and registered in a monuments record, or be a natural reserve or national park.

Agrarian tax is paid by the owners of farms having arable land greater than 1 hectare, or by the owners of land classified in the land register as arable land, apart from land used for non-agricultural activities. Agrarian tax is currently very controversial since it is based upon the area of land and not upon the farmer's income. Changes are currently being discussed.

Furthermore, there is still no tax based on the value of RE, but tax is calculated only from the surface area. In 1997, a law was passed on RE which introduced the possibility of charging tax based on the value of land. However the Act also stated that there would be a deadline to introduce this tax; so far there is no such regulation.

One could argue that whilst the Polish RE market changed significantly after 1990, the tax system is still reminiscent of the communist era. It is reasonable to question whether a capital RE market and communist tax system make for an effective combination.

By 1990, the valuation of RE was one of the professional rights of surveyors. After transformation, those provisions were repealed, and State licensed experts carried out the valuation of RE. The Real Estate Management Act 1997 defines the requirements to obtain the powers of RE valuation. In addition, this Act contains rules for the valuation of RE, which includes appropriate approaches, methods and techniques for estimating the value of the RE. The expert can only determine the value of the RE. It should be emphasised that the new so-called deregulation of professions is covered by licences. At present, this includes RE brokers and managers. However, deregulation did not include RE valuation.

4.7 Existing Governmental and/or financial institutions' support schemes for RE investment

The current Government support schemes and financial institutions for RE development come down primarily to housing: direct support for the acquisition of housing or support for utilities (infrastructure), which relates to multi-family buildings.

Major support is directed at housing and initially this is provided through the recognition that the provision of housing is a principal objective of Government, which is reflected in the Polish Constitution. In accordance with Article 75 paragraph 1 of the Constitution (Basic Act 1997), public authorities are obliged to pursue policies conducive in satisfying the housing needs of citizens, in particular: (a) combating homelessness; (b) supporting the development of social housing; and (c) supporting activities and efforts to obtain their own homes. For many years, the basic form of support for housing was based on tax instruments. The main form of financial support was through the development of a system of tax reliefs and exemptions to help meet housing needs. This was relatively ineffective, and from 2006 there was a complete withdrawal of the State from housing allowances, which formed part of a broader reform of public finances. These changes reflected the increasingly available mortgage loans and aid in the repayment of credit debt (interest relief).

In addition, the housing support policy for the period 2011–2020 has identified a number of objectives including the elimination of quantitative deficits in the social housing sector; the introduction of effective forms of affordable housing supply in the segment of apartments for rent and building ownership; the elimination of key risks associated with the development of private housing flats for rent; the reduction in the cost of housing construction and rationalisation of spatial-functional new housing stock by creating the appropriate resource armed land for housing; the elimination of the gap repair; the reduction of energy demand in the residential sector; and the rationalisation of the principles of managing public housing stock and resources.

4.8 Conclusion

Since 1989, the entire spectrum of RE in Poland has been transformed from virtually nothing to a process that is gradually maturing. The entire process has not been without difficulties – both economically and developmental. However, it is showing a great deal of resilience at a time when much of Europe has been in recession. Business recession started in the USA, but quickly spread to Europe. In response to the crisis, the loan policies were tightened, which resulted in decreased demand for RE and deflating prices. This may be considered a threat to an emerging market; however, it may also be considered a blessing since the development and price of RE dropped after 2007 to what might be recognised as more realistic values and developments. This is encouraging both from an economic and professional perspective since there is now the opportunity to develop within the bounds of the nation, voivodeship, and the location in a more refined and realistic manner.

5 The Lithuanian RE market

5.1 Introduction

The Republic of Lithuania is a European State located on the south-eastern coast. It has borders with Latvia, Belarus, Poland and Russia (Kaliningrad Oblast). The area of LR is 65,300 square kilometres (forests 33.3%; water 1.35%) with a population, according to the Lithuania Statistics Department, of 2,944,459 as of 1 January 2014. The capital of Lithuania is Vilnius. According to the 2011 population and housing census data, the occupation of residents is focused on industry (17.8%), wholesale and retail trading (17.4%), education (11.1%) and economic activities (Lietuvos Statistikos Departamentas [Lithuania Statistics Department], 2011).

The results of RE market activeness have immense influence on State economics. Moreover, State political and various social factors greatly impact the main parties of this market (for example, buyers and sellers). Immovable property is usually assigned to another owner who knows how to utilise the property in a more effective way and is able to put more investment into it. The greater the number of these types of transactions, the greater the benefit for the national economy.

The political independence of Lithuania was declared on 16 February 1918 and was restored on 11 March 1990, after the occupation of the Soviet Union. The immovable property of the citizens was nationalised during the occupation. In order to rebuild the rights to this property, a law restoring the property rights of citizens (Lietuvos, 1997) was established. This process consequently activated the RE market. However, some citizens did not require their restored property and instead transferred their property rights to other persons. Not all land owners saw themselves as farmers, and in other cases the inherited property was distant from the recipient's permanent place of abode.

Entry to the European Union, and a number of its support programs (investment) for agricultural activity, further activated the RE market. As the economic status of citizens improved, investment in the development of agricultural and other activities by buying agricultural land increased. Additionally, the purchase of land near big cities increased in anticipation of the urbanisation expected from citizens moving from flats into private houses. Due to these rapid changes inside the country, the price of RE exploded, due in part to administrational procedures associated with change of use. However, the country suffered from the global economic crisis in 2008, which adjusted the urban development plans of investors. Consequently, the land purchased near cities became uncultivated and depreciated in value due to the tax burden accrued by unsold land.

5.2 National RE market structure

Under the Civil Code of the Republic of Lithuania (Lietuvos, 2000) an immovable object is a plot of land and related objects which cannot be moved from one place to another without changing their purpose and reducing their value. Lithuania

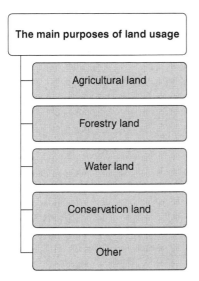

Figure 15.19 Main purposes of land usage

Source: Own study based on Lithuanian Act for Land 2004.

(Lietuvos, 2004) indicates that the entire private, State and municipal land com-
poses the Lithuania land fund, and according to the basic objective, the land fund
is divided into five main objective land usage purposes (Scheme 1) (see Figure
15.19). In 2003 the Ministry of the Environment approved a Technical Regulation
for Buildings which dealt with classification of buildings by their use (Scheme 2)
(Lietuvos, 2003), which is shown in Figure 15.20.

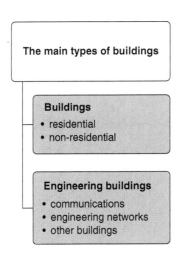

Figure 15.20 Main types of buildings

Source: Own study based on Technical Regulations for Buildings 2003.

From the statistical data stored and publicised by the State Enterprise Centre of Registers (Valstybės įmonė, 2014), one can clearly see the pace of various procedures in the RE market based on the sectors of interest (for example, land or building usage under purposes or proprietary forms (see Tables 15.9 and 15.10).

Data of different needs, since the middle of 2003, is available in the State Enterprise Centre of Registers. Between 1 January 2004 and 1 January 2014, the number of officially registered immovable objects increased by 1,258,829 units. Figure 15.21 shows how the number of registered immovable objects is gradually increasing. A sudden leap is visible during 2004 and 2006; however, this rate did not continue after the technique of data calculation changed in 2007.

5.3 The main players/actors in the RE market

The Lithuanian RE market functions due to transactions and related administration. The key players in the RE market are sellers, buyers, RE development specialists, brokers, investors and assessors.

RE agents mediate between the persons selling and buying RE. Collaboration between the buyer, seller and RE agent creates more opportunities in the market. These services do increase the cost of transactions and RE by about 3% or more. RE agents are expanding their services to include all validation documents regarding land and

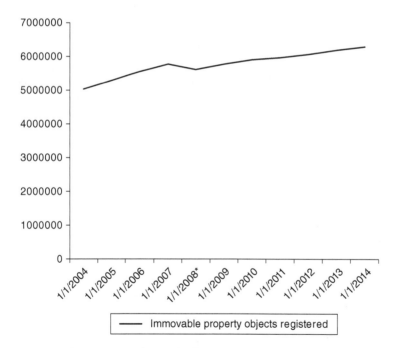

Figure 15.21 RE registration (2004–2014)

Source: Compliled from the data of Centre of Registers (2004–2014).

Note: *In the second half of 2007 the data calculation technique was changed.

Table 15.9 Buildings registered in the RE Registry based on their purposes and proprietary rights for selected periods

	01-01-2008					01-01-2014				
	Registered buildings	Of which				Registered buildings	Of which			
		Private RE (natural and legal persons)	State RE	Municipal RE	Other*		Private RE (natural and legal persons)	State RE	Municipal RE	Other*
Buildings (total)	2,284,614	2,061,154	15,381	19,241	188,838	2,470,072	230,1457	17,108	21,673	129,834
Residential	476,155	418,408	996	1,617	55,134	523,299	467,394	1,072	1,689	53,144
Non-residential	1,808,459	1,642,746	14,385	17,624	133,704	1,946,773	1,834,063	16,036	19,984	76,690
Engineering buildings (total)	475,849	265,224	4,723	9,271	196,631	562,122	426,244	12,775	20,676	102,427
Communications	3,775	1,321	742	1,354	358	12,312	1,795	3,866	6,355	296
Engineering networks	43,073	34,001	1,233	3,490	4,349	76,608	63,401	3,965	6,424	2,818
Other buildings	429,001	229,902	2,748	4,427	191,924	473,202	361,048	12,775	20,676	102,427
In Republic, total	2,760,463	2,326,378	20,104	28,512	385,469	3,032,194	2,727,701	29,883	42,349	232,261

Source: Compiled from the Data of Centre of Registers (2008, 2014).
* General RE of natural and legal persons, State and natural and legal persons, municipalities and natural and legal persons, State and municipalities, State and natural and legal persons, municipalities and natural and legal persons or no registration of proprietary rights.

Table 15.10 Land fund composition of Lithuania based on its purpose and proprietary rights for selected periods

	01-01-2002			01-01-2008			01-01-2014		
	Total (ha)	Of which		Total (ha)	Of which		Total (ha)	Of which	
		Private (ha)	State (ha)**		Private (ha)	State (ha)**		Private (ha)	State (ha)**
Agricultural land	3,956,200	2,090,500	1,865,700	3,956,069	2,949,033	1,007,036	3,941,802	3,388,490	553,312
Forestry land	1,963,600	403,300	1,560,300	1,970,607	629,851	1,340,756	1,977,356	680,166	1,297,190
Water land	184,000	—	184,000	176,458	1,122	175,336	175,394	2,638	172,756
Conservation land	39,700	—	39,700	43,056	104	42,952	43,599	128	43,471
Other	376,100	43,400	332,700	378,968	91,619	287,349	386,715	117,983	268,732
In Republic, total (ha)	6,519,600*	2,537,200	3,982,400	6,525,158*	3,671,729	2,853,429	6,524,866*	4,189,405	2,335,461

Source: Compliled from the data of Centre of Registers (2002, 2014).

* Excluding free State land

** Including land plots belonging to municipalities

buildings. This way, the pressure of paperwork and bureaucratic processes is eased from the RE owner, and the process of the RE sale or privatisation is sped up.

In Lithuania, construction is carried out in accordance with the territorial planning documents. The most important document in this process is the General Plan of Lithuania and of every different region. These documents provide the basis for detailed plans regarding ongoing construction. The administrative procedure involved in the change of use of land significantly increases the price of the said land. Since 2014, following amendments to the Planning Act, preparation of these detailed plans has decreased due to the shorter administrative process involved in the change of use of land. This new version of the law attempts to transform chaotic planning into complex spatial planning.

For individual housing development in Lithuania, individuals and investors are expected to initiate construction. The Government will only initiate construction which relates to the State or public needs. The State, however, will provide discounts for the acquisition of accommodation dependent on a family's economic and social situation. Support is also available to young farmers who purchase agricultural land.

5.4 Challenges of the exisiting RE legal framework

The RE market is regulated by a number of laws. The main law which influences all procedures and proprietary rights is the Civil Code 2000 (amended in 2011). According to this Code, a proprietary right is a right to manage, operate and dispose of proprietary objects at discretion without violating the regulations and rights and interest of other persons. Normally, RE is acquired through transactions or inheritance.

Although the Lithuanian Constitution (1992) declares that a proprietary right is untouchable and guarded by law, in some cases private property might be taken for public needs in accordance with a court order and fair compensation. By virtue of their proprietary right, private land owners are entitled only to the upper layer of the plot and the buildings on it. In certain cases the owner is also entitled to the minerals. A number of restrictions apply in Lithuania in order to acquire agricultural land plots (land area is the main restriction). There are, however, plans to introduce new restrictions within the land acquisition regulations, whereby citizens of Lithuania will be able to sell their agricultural land to foreigners.

5.5 RE tax

Owners who manage their immovable property under a proprietary right must pay taxes. The main legal Act regarding the application of this tax is the Law on Real Estate Tax (Lietuvos, 2005).

Since 1 January 2012, all natural persons (permanent and temporary residents of Lithuania) pay RE tax for the entire immovable property of which they are entitled or that they they plan to acquire under a proprietary right. There are, however, some exceptions such as:

- if a natural person is assigned to use an immovable property (or a part of it) with no specified term or for a period longer than one month; and
- buildings (premises) with a total value of less than 1 million Litas that fulfil the following purposes: living; gardening; garage; farm; greenhouse; steading; utility steading; science; religion or recreation; fishing; and engineering.

If the total value of RE is greater than 1 million Litas, the amount exceeding 1 million Litas is taxed at 1%.

Tax for land rent or exploitation is determined according to the rate assessed by an appropriate municipality board, from 0.3% to 1% of immovable property taxing value (market value established by wholesale evaluation). Municipalities may also determine certain concessions or tax rises (greater taxes are paid for an abandoned land plot).

5.6 RE valuation

RE value is extremely important in the Lithuanian RE market. The law on RE and business assessment (Lietuvos, 1999) permits RE valuation for a variety of purposes including transfer of ownership; obtaining RE insurance; RE taxation and declaration; rent; charge; merger; when RE is taken for public needs; while applying bankruptcy procedures; when judicial decisions are being prosecuted; when RE is acknowledged as derelict; and at the request of the owner. The work of RE assessors influences ensuing decisions. Consequently specialists in this field are classified according to different requirements in the hierarchy (assistant assessor, assessor, senior assessor, surveyor). All assessors must follow the code of professional ethics.

Since RE is assessed for a variety of purposes, there are many kinds of values (market, income, reconstructive and residual). Similarly to many other countries, in Lithuania there are three main methods of assessment – comparable value; reconstructive value; and value of operating income – as well as combinations thereof.

The method of comparable value involves establishing the market value of the property in comparison to the prices of actual transactions of analogous objects and regarding minor differences of the assessed RE and its analogue. This method is generally applied to assess plots of land and flats, as these objects comprise the biggest part of the market, and the more similar the object, the more precise the established market price.

The method of reconstructive value calculates the price of the restoration of objects, present physical state, performance and efficiency according to the work technologies and market prices applied at the moment of assessment. This method is generally used when assessing new or renovated objects.

The method of operating income value is used to calculate possible income from the immovable object. It is used when assessing profit-bearing assets.

There are a number of cases when RE is assessed by State institutions for their own needs. Different cases require different methods to be applied; therefore, the values received can often vary. Whilst assessing the RE, State institutions may use market facts, yet may not refer to them.

The methods of land assessment are as follows:

- the determination of the nominal index-linked price (commonly used during the reform or cadastral measures while using land qualitative rates);
- the value and equivalence methodology of State redeemable land, forest and water bodies (usable during reform in order to ascertain equivalence of nationalised and reversionary RE);
- the determination of average market value according to land value maps; and
- individual assessment.

Buildings are generally assessed by the State institutions based on the following methods:

- determination of average market value according to value maps;
- according to the average market values approved by a Government-established commission of experts for the assessment of the RE to be obligatorily registered; and
- individual assessment.

Mass assessment of RE is a type of RE assessment (a special model is applied) when the value of a particular RE is not determined. Collected data (for example, cadastral and geographic) about the subject RE is analysed to determine borders of value zones including territories with objects of similar qualities. This method is becoming more popular in the tax system and among natural persons who are keen to know the amount of money they may receive for their RE. As it does not determine the value of specific RE, it is necessary to note that this value and the value after an individual assessment may differ by up to 20%.

RE market analysis may be performed for different needs, for instance investment analysis, market research (profitability and competition), potential, and general condition. Investment analysis researches objects' potential for investment pay-off.

Objects such as residential flats built since 2000 are present in the most marketable locations in Lithuania. Table 15.11 shows prices by city. The amount of money invested and the income received from leasehold flats is taken into consideration. Moreover, the seasonally varying demand of the RE in to certain locations is introduced due to unexpected and possible risk factors. It is also necessary to exclude loss due to extra cost which may be not financed if the tenant is not present (for example, heating during the winter season). Figure 15.22 maps the cities and resorts used during this analysis.

It is difficult to say whether greater housing choice has had an impact on prices in Lithuania. The global economic crisis is likely to have had a much greater impact. However, the greater number of immovable objects has increased the number of investors. Observations of the amount of property built and sold suggest that the economic crisis did not stop the construction process, it just reduced it. In Lithuania, the construction of residential RE is currently more common that commercial RE. Consequently there is expected to be a surplus of residential development in the near future.

Figure 15.22 Location of the objects used during the analysis in Lithuania
Source: Own study.

The purchase or renting of RE is greatly dependent on the finances, personal beliefs and values of the buyer or tenant. There was a time when individuals swarmed to banks in order to obtain long-term loans for house purchases. Today, however, the economic instability and geopolitical environment are causing more people to choose rental accommodation.

5.7 Conclusion

Following independence in 1991, a law restoring property rights to citizens was established, consequently activating the RE market. As the economic status of citizens improved, investment in the development of agriculture and other activities, by buying agricultural land, increased. This, together with the purchase of periphery land surrounding towns and cities for urbanisation, caused RE prices to soar. However, the global economic crisis of 2008 has had negative implications for the Lithuanian RE market, particularly with regard to construction and investment into land for development.

The RE market in Lithuania is, however, accelerating. There have been a few active cases regarding international decisions (financial support) after the restoration of independence. Similar activity is anticipated in 2015 when citizens of foreign countries will be able to buy agricultural land. Additionally, the number of registered immovable objects is constantly increasing in the centre of registers.

Table 15.11 Comparative analysis of prices by city

	Number of comparable objects in the market	Average price of saleable object in Euros (1 room (approx. 30m²))	Average price of tenemental object in Euros per month (1 room (approx. 30m²))	Number of accommodated months (seasonality)	Loss due to vacant flat, in Euros (extra cost)	Income per year in Euros	Period in years for the investment to pay off
Vilnius	120	56,000	380	10	60	3,700	16
Kaunas	20	41,000	280	10	60	2,800	14
Klaipeda	20	46,000	230	10	60	2,300	20
Palanga	50	38,000	870	5	580	3,800	10
Druskininkai	10	43,000	430	9	870	3,000	14

Source: Compiled from data at www.domoplius.lt and www.aruodas.lt.

This shows that the qunatity of RE that will be for sale in the future is increasing along with the transfer of ownership rights. Furthermore, market analysis has shown that it is now advantageous to invest in newly built flats in Palanga since it is possible to expect a fast return on investment.

6. Conclusion: Slovenia, Hungary, Poland and Lithuania

In the 1990s, in parallel with political changes, the transformation of economies began in Slovenia, Hungary, Poland and Lithuania. After the collapse of command economy structures and State socialism, a return to private property ownership was pursued as a potential remedy to the coming challenges. In the process of economic transition of all four countries, two stages of development can be identified. The first stage covers the early years of economic transition from independence until 1995–1996, in which the bigger differences in approach and priorities have been observed. The second stage started in approximately 1997–1998, when the reforms became clearly EU-accession oriented, and were thus driven and guided by the set of intermediate and final objectives bilaterally agreed with the EU institutions which permanently monitored the process and provided very important financial and technical assistance for the timely accomplishment of the needed reform goals. The fundamental economic reforms related to liberalisation of prices; trade and foreign exchange; macro-economic stabilisation through control of inflation; restoration of private property; reform of the banking and financial sectors; and the attraction of foreign capital and investments. Albeit the stock of credit relative to GDP started from a very low base in all these countries, the ratio rose sizeably over the period 2000 to 2007, on average by around 15% (Bunda and Ca'Zorzi, 2009).

The average percentage of share of mortgages in GDP over the period 2001–2004 were as follows: Hungary (6%), Slovenia (-3.8%), Poland (-3%), and Lithuania (-4.5%) (Palacin and Shelburne, 2005). In 2003, Hungarian, Slovenian and Polish markets were more attractive to foreign investors than the Lithuanian market; however, since 2007, Lithuania has become significantly more popular than Poland, Hungary and Slovenia from the perspective of international investors (World Bank, 2010). Solid foundations for the free market economy, including the RE market sectors, the RE legal framework and valuation practices based on open market analysis, have been built in each case study country. The newly established legal systems, which include the RE laws of each country, have elastic similarities in their nature and structures both with one another and with most other European countries, as the routes come from the common heritage of the Napoleonic civil law legal system.

RE valuation practices vary from country to country, but worldwide-recognised common valuation techniques based on open market valuation have been successfully adopted in each country. In terms of RE administration systems, each country has an established land register and cadastre system, which successfully contributes towards RE market transparency.

Over the last ten years, RE market performance by sector has been very much dependent on each country's unique Government policy on taxation and investment. In the first stage of economic development, the residential RE market sector has been established quite quickly and effectively in the biggest cities of each country, which has heavily contributed to economic growth. The commercial RE market demonstrated acceleration at the same time. The retail market became the centre of attention by foreign investors, especially in Hungary, Poland and Lithuania. Unfortunately, quite well-performing RE markets were hit by the global RE recession in 2008, significantly affecting the residential RE market sector. Since then, the commercial RE market, including all sectors, has failed to recover. The agricultural market sector has been controlled by policies protecting land ownership from private international investors in each country and has not been developed to the same level as the residential and commercial sectors.

The 2008 recession has affected each RE market in slightly different ways. As Lithuania's market suffered, the value of land purchased for urbanisation depreciated, leading to a fall in construction. Similarly, in Slovenia, many large-scale construction companies collapsed, halting new development. More recently, however, there has been a growth in minor developments by small-scale companies. In Hungary the RE market continues to decline; however, expectations of the residential and office space markets have improved. RE values have also declined in Poland, along with the number of building permits issued. However, trading buildings have been greatly resilient to the economic climate and are considered a great success story.

Whilst these four countries share a common history of Soviet control, since independence their RE markets have developed in some distinctive ways. The growth of independent economies has had a significant affect on the RE markets of each respective country. Additionally, the diverse environmental characteristics of each State has led to the varied RE-related priorities and initiatives implemented within each case study country. As these countries have, over time, exposed themselves to globalisation through accession to the European Union and the relaxation of restrictions on foreign investment, their respective RE markets and RE values have been greatly influenced. Moreover, the ongoing changes to restrictions on investment and the continued recovery of the global economy will continue to shape and influence these markets.

References

Berdár B., Fenyő G., Márkus B., Nyiri J. (2005) Costs of Real Estate Transactions in Hungary. COST meeting – WG 3 Economy, Grange-over-Sands, England.

Berdár B., Mizseiné Nyiri, J. (2000) Az ingatlanértékelés gyakorlata Magyarországon. LIME projekt keretében készített oktatási jegyzet.

Bunda, I., and Ca'Zorzi, M. (2010) Signals From Housing and Lending Booms. *Emerging Markets Review*, 11, pp. 1–20.

Cirman, A. (2010) Interconnection of Real Estate Market and the Banks: Why So Fatal? Povezanost med nepremičninskim trgom in bankami: zakaj tako usodna? Paper presented at the 6th Slovenian Real Estate Conference: Portorose, 23 and 24 September. Ljubljana: Planet GV, pp. 4–6.

Court of Auditors of the Republic of Slovenia (2013) Summary of the Audit Report on the Efficiency of Stimulation of Foreign Direct Investments / Povzetek revizijskega poročila Učinkovitost spodbujanja tujih neposrednih investicij.

Damijan, J. P. (2013) Why Is Slovenia the Least Attractive for Foreign Investment? Zakaj je Slovenija najmanj privlačna za tuje investicije? [online]. Available at: http://damijan.org/2013/06/27/zakaj-je-slovenija-najmanj-privlacna-za-tuje-nalozbe/ [Accessed 8 September 2014].

Dixon-Gough R., Hernik J., Gawroński K., Taszakowski J. (2013) Peri-Urban Problems around Krakow, in: Hepperle, E., Dixon-Gough, R., Maliene, V., Reinfried Mansberger, R., Paulsson, J., a Pödör, A. (Eds), *Land Management: Potential, Problems and Stumbling Blocks*. Wyd. vdf Hochschulverlag AG an der ETH, Zürich, pp. 133–140.

European Council for Real Estate Professional: Polska [online]. Available at: http://www.cepi.eu/index.php?page=polska&hl=en [Accessed 6 March 2014].

EUROSTAT (2014) [online]. Available at: http://epp.eurostat.ec.europa.eu/statistics_explained/index.php/Construction_production_(volume)_index_overview) [Accessed 8 September 2014].

Fórizs, Z. (2000) (Ed.) *Gyakorlati ingatlan-tanácsadó* (*Practical Real Estate Adviser*). Kézikönyv. Verlag Dashöfer Szakkiadó Kft. Budapest.

GARS (2014) Report on Slovenian Property Market in 2013 / Poročilo o slovenskem trgu nepremičnin za leto 2013, Geodetic Administration of the Republic of Slovenia.

Global Property Guide [online]. Available at: http://www.globalpropertyguide.com/Europe/Hungary/Price-History [Accessed 1 August 2014].

Główny Urząd Statystyczny (in English: Central Statistical Office) [online]. Available at: http://stat.gov.pl/ [Accessed 1 August 2014].

Hajnal, I. (1995, 1999): Az ingatlan értékelés Magyarországon (*Real Estate Valuation in Hungary*). BME Mérnöktovábbképző Intézet, Budapest.

Hartvigsen, M. (2014) Land Reform and Land Fragmentation in Central and Eastern Europe. *Land Use Policy*, 36, pp. 330–341.

Hernik J., Gawroński K., Dixon-Gough R. (2013) Social and Economic Conflicts between Cultural Landscapes and Rural Communities in the English and Polish Systems. *Land Use Policy*, 30, pp. 800–813.

Hungarian Chamber of Commerce and Industry. (No date.) Benefits of EU Membership [online]. Available at: http://www.mkik.hu/en/magyar-kereskedelmi-es-iparkamara/benefits-of-eu-membership-2630 [Accessed 18 March 2015].

Ingatlannet (2014) [online]. Available at: http://www.ingatlannet.hu/statisztika/Magyarorsz%C3%A1g?#table [Accessed 1 August 2014].

Instytut Rozwoju Miast (2013) 23 lata polskiego rynku nieruchomości – Monitoring za lata 1990–2012 (23 Years of the Polish Real Estate Market 1990–2012). Instytut Rozwoju Miast (Institute of Urban Development), Kraków.

Knight Frank (2011) Broszura – Rynek nieruchomości w Polsce.

Kosmulska S. (2009) Warsztat rzeczoznawcy majątkowego - metody wyceny [online]. Available at: http://pirm.pl/wypowiedzi-ekspertow-pirm/135-warsztat-rzeczoznawcy-majtkowego-metody-wyceny.html [Accessed 5 August 2014].

Kotsis, E. (1927) Épületek értékelése. Budapesti Műszaki Egyetem, Budapest.

Kotsis, E. (1942) Épületek értékelése. BME Mérnöki Továbbképző Intézet, Budapest.

Kotus, J. (2006) Changes in the Spatial Structure of a Large Polish city – The Case of Poznań. *Cities*, 23, pp. 364–381.

Kucharska-Stasiak, E. (2006) Nieruchomość w gospodarce rynkowej. Wydawnictwo Naukowe PWN, Warszawa.

Lavrac, V. and Majcen, B. (2006) Economic Issues of Slovenia's Accession to the EU. Ljubljana: Institute for Economic Research [online]. Available at: http://www.esi-web.org/pdf/slovenia_LavracMajcen_econ%20issues%20of%20SLO%20EU%20accession-2006.pdf [Accessed 18 March 2015].

Lietuvos Statistikos Departamentas (2011) m. visuotinis gyventojų ir būstų surašymas [online]. Available at: http://osp.stat.gov.lt/web/guest [Accessed 10 July 2014].

Márkus, B. and Nyiri, J. (2005) Land Property Transactions and the Hungarian Land Consolidation Strategy, Thessaloniki, COST project.

Márkus, B. and Nyiri, J. (2003) Organisational Structures and Their Role in the Processes of the Real Property Transactions, Hungarian National Report, COST project.

Michniak, L. (2011) Polski rynek nieruchomości komercyjnych, III kwartał 2011 [online]. Available at: http://www.morizon.pl/blog/raporty-i-analizy-wgn/polski-rynek-nieruchomosci-komercyjnych-iii-kwartal-2011 [Accessed 7 March 2014].

Ministerstwo Infrastruktury i Rozwoju (2014) Rejestr Rzeczoznawców majątkowych [online]. Available at: http://www.mir.gov.pl/strony/zadania/budownictwo/rejestr-rzeczoznawcow-majatkowych/ [Accessed 1 August 2014].

Ministry of Foreign Affairs of the Republic of Lithuania (2014) The 10th Anniversary of Lithuania's Accession to the European Union [online]. Available at: http://www.urm.lt/default/en/the-10th-anniversary-of-lithuanias-accession-to-the-european-union [Accessed 18 March 2015].

Mizseiné Nyiri J. (2008) A földegyenérték, mint birtokrendezési elem. PhD értekezés, Nyugat-magyarországi Egyetem, Erdőmérnöki Kar, Roth Gyula Erdészeti és Vadgazdálkodási Tudományok Doktori iskola, Erdővagyon-gazdálkodás Program, Sopron.

Office of the Committee for European Integration (2008) Four Years of Poland's Membership in the UE. Office of the Committee for European Integration, Warsaw [online]. Available at: http://www.msz.gov.pl/resource/11ce765c-027d-45bc-8947-e01e2811d120:JCR [Accessed 18 March 2015].

Organisation for Economic Co-Operation and Development (OECD) (2014) OECD Economic Surveys: Hungary 2014 [online]. Available at: http://dx.doi.org/10.1787/eco_surveys-hun-2014-en ISSN 1999-0529 [Accessed 1 August 2014].

Palacin, J. and Shelburne, R. C. (2005) The Private Housing Market in Eastern Europe and the CIS. United Nations Economic Commission for Europe, discussion paper series no 2005.5, Geneva, Switzerland.

Pödör, A and Nyiri. J. (2010) GIS Application in Real Estate Investment. Paper presented at the 51th Riga Technical University Scientific Conference on Economics and Entrepreneurship (SCEE), Riga, 15 October.

Polski Związek Firm Deweloperskich (Polish Association of Developers) (2012) Sytuacja mieszkaniowa w Polsce 2012 (The Housing Situation in Poland in 2012).

Rakvács, J. (1999) Az ingatlanforgalmazók kézikönyve (Handbook of Real Estate Sellers) (Jogi Ismeretek), HVG-ORAC Lap- és Könyvkiadó, Budapest.

Reinhardt, A. (2014) New Hungarian Civil Code [online]. Available at: http://www.warwicklegal.com/news/73/ [Accessed 1 August 2014].

Rose, A. K. and Spiegel. M. M. (2012) Cross-country Causes and Consequences of the 2008 Crisis: Early Warning. *Japan and the World Economy*, 24, pp. 1–16.

Sipos A. Szücs I. (1992) A mezőgazdasági termőföld komplex értékelése (Valuation of Agricultural Land), Közgazdasági Szemle.

Soós, J. (2002) (Ed.) Ingatlangazdaságtan, KJK-KERSZÖV Jogi és Üzleti Kiadó Kft, Budapest.

SORS (2014) Statistical Office of the Republic of Slovenia [online]. Available at: http://www.stat.si [Accessed 8 September 2014].

TEGoVA (2012). European Valuation Standards. TEGoVA, Brussels.

Valstybės Įmonė Registrų Centras, Nekilnojamojo turto kadastras ir registras [online]. Available at: http://www.registrucentras.lt/ntr/. [Accessed 15 July 2014].

Verdick, S. and Islam, I. (2010) The Great Recession of 2008–2009: Causes, Consequences and Policy Responses. The Institute for the Study of Labor (IZA), Bonn. [online]. Available at: http://ftp.iza.org/dp4934.pdf [Accessed 18 March 2015].
Wikipedia (2013) List of Regions of Hungary [online]. Available at: http://en.wikipedia.org/wiki/List_of_regions_of_Hungary [Accessed 20 Febraury 2015].
Wootton, Jones Lang, Estate Gazette Limited and South Bank Polytechnic (1989) *The Glossary of Property Terms*. Estate Gazette Limited, London.
World Bank (2010) Global Economic Prospects, Crisis, Finance, and Growth. The World Bank, Washington DC.
Zwoelf, L. (2009) Apartment Rent Analysis in the Republic of Slovenia – Case Study of the Municipality of Ljubljana / Analiza najemnin v Republiki Sloveniji – primer mestne občine Ljubljana. University of Ljubljana, Faculty of Civil and Geodetic Engineering, Ljubljana.

Websites

http://browarlubicz.pl/pl/ [Accessed 5 August 2014].
http://www.mir.gov.pl/Budownictwo/Gospodarka_nieruchomosciami/Uprawnienia_i_licencje_zawodowe/Rejestr_Rzeczoznawcow_Majatkowych/Strony/ [Accessed 5 August 2014].
http://www.morizon.pl [Accessed 7 March 2014].

Legislation and Regulations

Hungary

Act No. CXXII of 2013 concerning agricultural and forestry land trade in Hungary.
Act No. VII. of 1875 on the land tax regulation in Hungary.
Hungarian Civil Code (2013).

Lithuania

Civil Code of the Republic of Lithuania of July 18, 2000, Law No. VIII-1864 (Last amended on April 12, 2011, No XI-1312).
Lietuvos Respublikos civilinis kodeksas, Žin., 2000, Nr. 74-2262.
Lietuvos Respublikos Konstitucija *(priimta Lietuvos Respublikos piliečių 1992 m. spalio 25 d. referendume)* [online]. Available at: http://www3.lrs.lt/home/Konstitucija/Konstitucija.htm. [Accessed 17 July 2014].
Lietuvos Respublikos nekilnojamojo turto įstatymas Nr. X-233, 2005-06-07. Žin., 2005, Nr. 76-2741.
Lietuvos Respublikos nekilnojamojo turto ir verslo vertinimo pagrindų įstatymas Nr. VIII-1202, 1999-05-25. Žin., 1999, Nr. 52-1672.
Lietuvos Respublikos piliečių nuosavybės teisių į išlikusį nekilnojamąjį turtą atkūrimo įstatymas Nr. VIII-359, 1997-07-01. Žin., 1997, Nr. 65-1558.
Lietuvos Respublikos žemės įstatymas Nr. IX-1983, 2004-01-27. Žin., 2004, Nr. 28-868.
Lietuvos Respublikos Aplinkos ministro įsakymas Dėl statybos techninio reglamento STR 1.01.09:2003, Statinių klasifikavimas pagal jų naudojimo paskirtį" patvirtinimo" Nr. 289, 2003-06-11. Žin., 2003, Nr. 58-2611.

Poland

Baza cen nieruchomości mieszkaniowych (III kw. 2006 - II kw. 2014) http://www.nbp.pl/.
Rozporządzenie Rady Ministrów z dnia 21 września 2004 r. w sprawie wyceny nieruchomości i sporządzania operatu szacunkowego, Dz.U. 2004 nr 207 poz. 2109. (in English: Regulation of the Council of the Ministers from 21 September 2004 on the valuation of the property and the preparation of the appraisal).
Rozporządzenie Rady Ministrów z dnia 14 lipca 2011 r. zmieniające rozporządzenie w sprawie wyceny nieruchomości i sporządzania operatu szacunkowego, Dz.U. 2011 nr 165 poz. 985 in English: Regulation of the Council of the Ministers from 14 July 2004 which amends the regulation on the valuation of the property and the preparation of the appraisal).
Ustawa z dnia 21 sierpnia 1997 r. o gospodarce nieruchomościami. Dz.U. 1997 nr 115 poz. 741, tekst jednolity (in English: Act on Real Estate Management from 21 August 1997).

Slovenia

Act regarding the siting of spatial arrangements of national significance in physical space 2010: UL RS 80/10, 106/10 and 57/12.
Agricultural Land Act 1996: UL RS 59/96, 71/11 and 58/12.
Construction Act 2002: UL RS 110/02, 1032/04, 14/05. 92/05 - ZJC-B, 93/05 - ZVMS, 111/05 - odl. US, 126/07, 108/09, 61/10 - ZRud-1, 20/11 - odl. US, 57/12, 101/13 - ZDavNepr and 110/13.
Courts Act 1994: UL RS 19/94, 94/07, 45/08, 96/09, 86/10 - ZJNepS, 33/11, 75/12 - ZSPDSLS-A and 63/13.
Fiscal Balance Act 2012: UL RS 40/12,96/12 - ZPIZ-2, 104/12 - ZIPRS1314, 105/12, 25/13 - odl. US, 46/13 - ZIPRS1314-A, 56/13 - ZŠtip-1, 63/13 - ZOsn-I, 63/13 - ZJAKRS-A, 99/13 - ZUPJS-C, 99/13 - ZSVarPre-C, 101/13 - ZIPRS1415, 101/13 - ZDavNepr, 107/13 - odl. US and 85/14.
Inheritance and Gift Taxation Act 2006: UL RS 117/06.
Personal Income Tax 2006: UL RS 117/06, 10/08, 78/08, 125/08, 20/09, 10/10, 13/10, 28/10, 43/10, 51/10, 106/10, 13/11, 24/12, 75/12 and 94/12.
Personal Income Tax 2012: UL RS 24/12, 75/12 and 94/12.
Promotion of Foreign Direct Investment and Internationalisation of Enterprises Act 2006: UL RS 107/06, 11/11 and 57/12.
Real Estate Agencies Act 2003: UL RS 42/03, 72/06 and 49/11.
Real Estate Mass Valuation Act 2006: UL RS 50/06, 87/11, 40/12 - ZUJF and 22/14 - odl. US.
Real Estate Tax Act 2013: UL RS 101/2013 and 22/14 - odl. US.
Real Estate Turnover Tax Act 2006: UL RS 117/06.
Spatial Planning Act 2002: UL RS 110/02, 8/03 - popr., 58/03 - ZZK-1, 33/07 - ZPNačrt, 108/09 - ZGO-1C and 80/10 – ZUPUDPP.
Spatial Planning Act 2007: UL RS 33/07, 70/08 - ZVO-1B, 108/09, 80/10 - ZUPUDPP, 43/11 - ZKZ-C, 57/12, 57/12 - ZUPUDPP-A, (109/12) and 76/14 - odl. US.
Value Added Tax Act 2006: UL RS 117/2006, 13/11, 18/11, 78/11, 38/12, 83/12 and 86/14.
Slovenian Business and Financial Standards 2001: UL RS 56/01.
Slovenian Business and Financial Standards 2013: UL RS 106/13.

16 Real estate, construction and economic development in emerging market economies

Past, present and future

Raymond T. Abdulai, Franklin Obeng-Odoom,
Edward Ochieng and Vida Maliene

1. Introduction

This book has provided a critique of separatist[1] themes in built environment research, explained why the separatism leads to unsatisfactory and disempowering explanations and, in turn, offered a new approach cast more in a holistic and integrating methodology on the basis of which the various chapters have investigated the nexus between real estate (RE) and construction, on the one hand, and economic development, on the other. In applying this *methodological reorientation*, the book's emphasis has been on the experiences of the materially poor and middle-income countries collectively called "emerging economies". This delimitation is important because, unlike many other countries, emerging economies have important distinguishing features. While becoming the home of many a new economic interest, this "newness" is often overtaken by rapid technological changes, generating unique processes of tensions between forces of change and inertia, new modes of governance and old forms of "governmentality".

These complexities and peculiarities notwithstanding, as the experience in Nigeria has shown, corporate RE management, hitherto assumed non-existent in Africa, has become a major focus of attention for various RE actors in some emerging economies. Indeed, the wave of interest in emerging economies – within which is the so-called "Africa on the rise" debate (Obeng-Odoom, 2014) – together with the contemporary challenges mounted against GDP (Fioramonti, 2013) as the primary measure of economic development in mainstream circles, and the need to consider the social dimensions of this "progress" (UNRISD, 2010; Hujo, 2012) makes this book an imperative. The aim of this last chapter is to highlight the issues arising from the book's chosen approach in order to tease out the implications for learning and teaching, research and policy, and to contemplate what challenges lie ahead in the transition from now to the future.

2. Highlights

As shown in the opening chapter, it can be shown how the construction industry generates economic activities for investment and development of RE. These articulated investment markets of RE types, in turn, generate socio-economic activities that interact with the general economic processes in various emerging economies. While one starts from this rather simple framework of analysing the built environment in market terms, cast as the interactions between the demand and supply of rational actors, prices and the establishment of equilibrium in the built environment, the detailed empirical studies in the book offer more nuanced perspectives related to the nexus between RE, construction and economic development, the nature of the institutions undergirding these relationships, and what such relationships mean for the society, economy and environment.

The book highlights that the interconnections between RE, construction, and economic development do not lend themselves to linear analysis: multi-linear, interlocking and interdependent analyses reveal much more. While the mainstream construction research in "construction management and economics" tends to start its analysis from a construction perspective, RE is a much bigger category and, while it can be a function of construction, not all RE activities directly involve construction activities, even when such non-construction specific RE processes make important contributions to economic development. For instance, the linkages between RE agents and economic development are not operationalised through construction channels but rather through social networks. This important highlight poses a difficult riddle for analysts who claim that there is such a thing as an autonomous RE market. RE markets work very differently in the various countries considered in this book. In an economy such as Turkey (Chapter 3), where there is not a huge public sector investment in construction, it is rather the boom in the national economy that drives the construction sector. It does not follow, however, that there is a "trickle down" or that mere economic growth translates into economic development. The case of Slovenia (Chapter 4) strongly articulates this view.

Thus, the book questions established norms and views about the workings of RE markets. Such markets do not simply respond to the dictates of demand and supply, neither will they generate propitious ends if they did. On the contrary, the book has shown that socially embedded markets in Ghana (Chapter 11) work quite well in the sense that they contain accountable measures, checks and balances on the actors in the built environment. The idea of "community", then, is not to be confused with backwardness: it merely connotes the nature of social relations, which evolve over time. When this evolution is slow relative to changes in the built environment, questions arise as to whether the pace of social change is in step with the speed of technological advancement in the built environment in emerging economies and what are the resulting implications. Whether this community check is generalisable to other countries depends on how cohesive and dynamic are the communities under discussion.

Regardless, this book shows that RE and construction markets do not merely spring into existence in the economic development process: they are created and managed. While often thought about as natural, the experiences of China analysed in Chapter 5 of the book show how state bureaucracy can create and shape the workings of the RE market either for good or for ill. It is evident, though, that creating the conditions for the market without tightly embedding it within institutions and social relations can give room for speculation and a phantom economic growth with no strong links with the productive and other sectors of the economy. Yet, it is such processes that are triggered in the question of how to treat the "dearth" of information in emerging economies. Insisting on formalisation as a way to reduce transaction costs, which are deemed to be too high in the "traditional" sectors, can itself be costly both directly and indirectly. Directly, the cost entailed in creating huge registers only to be dogged by credibility crises and the difficulty of implementing them can amount to waste. Similarly, reliance on formal agents can be costly. Indirectly, trying to reduce these transactions costs can, and often does, lead to high social costs: dispossession, scapegoating, and marginalisation without any real improvement in the material and social conditions in the countries where they have been implemented, as many of the chapters (for example, Chapters 10 and 11) show.

Even worse, such social disturbances create dynamic cycles in which inequality leads to further inequality. In this scenario, housing for the poor only gets worse as speculation arising from the lowering of transaction costs elevates house prices to heights that can only be afforded by the rich. The prize for the rich, however, is the burden of the poor who struggle to "keep just a single foot" in the proverbial door, which only opens ajar for them anyway. The case of Latin America discussed in Chapter 13 best illustrates the problem. Thus, analyses of the standard studies on the economic development nexus leave much out in their discussions of correlations and causations relative to GDP, even when they take into account questions of employment. The book shows that mainstream stories should be significantly extended to include the conditions of labour and the effects of the processes in which they are involved on the environment because growth will not necessarily improve well-being for people, nor the environment.

3. Implications

The findings and highlights developed in the book have important implications for learning and teaching, research and policy. Learning and teaching in the built environment has increasingly emphasised either a technical, business or professional ethos paralleling the field's vocational, pragmatic, and hands-on orientation on which is built the "specialist" fervour. In turn, cross-disciplinary learning and teaching is residual when in fact it should be front and centre. Also front and centre should be a holistic approach which emphasises critical investigation and discussion of ideas and practices across the entire field. Doing so will help the field to overcome disciplinary rivalries over whose turf is A, B, or C, for A, B, and C must be holistically studied and understood. Pulling down the boundaries is likely to be challenging, among others because of differences in language and terminologies. However, a conscious effort at minimising the use of jargon will help, as will stressing the importance of the interconnections between RE, construction and economic development.

Joint or interdisciplinary research can be useful in this process. The mixing of disciplines within the built environment and beyond on common problems – as this book has shown – will help to untangle much complexity. None of this is going to be easy of course. Institutional categorisation and "wars" about how disciplinary publications should be categorised is a step backwards. Similarly retrogressive is the multiplicity of journals that are specialist and inattentive to holistic and pluralist analyses. The emphasis on epistemological positivism in recent times and the elevation of quantitative research over and above all others is yet another hindrance to greater and deeper insights in the field. Not all is lost, however. Certain journals – such as the Hong Kong-based *Surveying and Built Environment,* Australian-based *Property Management*, and Finland-based *Nordic Journal of Surveying and Real Estate Research* – publish a range of papers that are pluralist. Of all these, however, it is the *Journal of International Real Estate and Construction Studies* (JIRECS), which takes the boldest step in pursuing this line of thinking. In setting its mandate, JIRECS notes that "Issues in real estate and construction are interconnected, though they are often treated as if they are not. . . . [I]t is important that research be conducted in an integrated fashion."

Apriori, if teaching and research are conducted in this "integrated fashion", policy prescriptions would be expected to follow in that same mode of thinking. However, there is no reason to assume that *the pace* of these spheres of influence will change synchronously. Thus, the policy space will also need to be specifically targeted for change. In this sense, the "RE" dimensions of "construction" and the "social" dimensions of "RE" will need to be systematically interwoven in policy in/about the built environment. The institutional separatism can be addressed through continuing professional development of practitioners and revisions of the work of professional bodies, which tend to be structurally organised to enforce this separatism even in very similar fields (for example, RE management and land economy; see Obeng-Odoom and Ameyaw, 2011). At a much bigger level, however, policies geared towards using the RE and construction sectors to reinvigorate local and national economies will need to be designed in such a way that they are sensitive to the social, economic, environmental and physical aspects of this process simultaneously.

4. Prospects

This journey through learning and teaching, research and policy is unlikely to be easy and the road ahead is likely to be quite bumpy. Signs of the days to come are quite clear. Even in this volume there has been no shortage of tension. Attempts at sustainability are useful, but they are not necessarily connected very carefully to the management of housing, although they have been connected to construction. It is not clear to what extent sustainability is of interest to using student housing to stimulate local economies (Chapter 12). There is clearly the need to do more empirical work on how or why social housing cannot/can attain a similar goal of driving local economic development with more redistributive twist. Correlations and causations about RE and economic development or economic development and RE are fascinating, but are more public, community-based and managed RE greater or lesser drivers, or are they driven by economic development – and how does this relationship change over time?

There are other pressing questions. While the focus of the volume has been on the emerging economies, it is important to ask whether these economies are being treated merely as geographical and economic "others" or as dialectically related to the materially rich countries. Establishing links are important but to what end? Will the links be strong enough to enable the emerging middle-income countries escape the so-called "middle-income trap"? In this volume, the chapters that look at the world system and its complicity in shaping particular relationships in emerging economies suggest that much critical, historical and political economic work in the built environment is required in future studies. Such future work can usefully start by looking at alternative measures of "economic development" or broader ideas of progress such as "happiness" and well-being and the built environment, alternative energy in the built environment, and critical perspectives on building-specific climate change-mitigating strategies. As editors, it is hoped this interest, this methodological holism, and this audacity in *Real Estate, Construction and Economic Development in Emerging Market Economies* has been sparked and readers will, hopefully, find the volume intellectually stimulating and insightful.

Note

1 Much like what Charles Gore did in his *Regions in Question* (1984) classic, although the interest has been in the built environment and not necessarily on cities and regions.

References

Fioramonti, L. (2013) *Gross Domestic Problem: The Politics Behind the World's Most Powerful Number*, London: Zed Books.
Gore, C. (1984) *Regions in Question: Space, Development Theory and Regional Policy*, London: Methuen.
Hujo, K. (2012) (Ed.) *Mineral Rents and the Financing of Social Policy: Opportunities and Challenges*, New York: Palgrave.
Obeng-Odoom, F. (2014) Africa: On the rise but to where? *Forum for Social Economics*.
Obeng-Odoom, F. and Ameyaw, S. (2011) The state of surveying in Africa: A Ghanaian Perspective, *Property Management*, 29(3), pp. 262–284.
UNRISD (2010) *Combating Poverty and Inequality: Structural Change, Social Policy and Politics*, Geneva: UNRISD.

Index